Models in Geomorphology

Models in Geomorphology

Edited by
Michael J. Woldenberg
State University of New York at Buffalo

Boston
ALLEN & UNWIN
London   Sydney

Allen & Unwin Inc.,
Fifty Cross Street, Winchester, Mass. 01890, USA

George Allen & Unwin (Publishers) Ltd,
40 Museum Street, London WC1A 1LU, UK

George Allen & Unwin (Publishers) Ltd,
Park Lane, Hemel Hempstead, Herts HP2 4TE, UK

George Allen & Unwin Australia Pty Ltd,
8 Napier Street, North Sydney, NSW 2060, Australia

First published in 1985

Library of Congress Cataloging in Publication Data

Main entry under title:
    Models in geomorphology.
Includes bibliographies and index.
1. Geomorphology – Models – Congresses.   2. Geomorphology –
Congresses.   I. Woldenberg, Michael J.
GB21.M56   1985   551.4′0724   85-1257
ISBN 0-04-551075-X (alk. paper)

British Library Cataloguing in Publication Data

    Models in geomorphology. – (The "Binghamton"
symposia. International series, ISSN 0261-3174; 14)
1. Geomorphology
I. Woldenberg, M.   II. Series
551.4   GB401.5
ISBN 0-04-551075-X

Set in 10 on 12 point Times by Computape (Pickering) Ltd, N. Yorkshire
and printed in Great Britain by Mackays of Chatham

Preface

We approach geomorphology from many directions and we often begin by learning about structure, process and stage. The 'Binghamton' Symposia in Geomorphology have focused on various approaches, some traditional and some new and different. We have essentially abandoned stage as a research topic and we have de-emphasized structure. Although the fifteenth symposium will feature tectonic landforms, no other symposium in this series has been devoted exclusively to 'structure.' The first fourteen symposia have been organized around process. There have been traditional symposia on coastal, fluvial (two), glacial and karst processes. Climatic geomorphology was examined by the symposia on arid lands and glacial geomorphology. Instead of examining the processes relating to an agency of erosion, processes may be studied comparatively. One symposium compared the variation of process in time and space; another explored the concept of thresholds for different processes. Processes may also be studied in an applied context. Thus there have been symposia on environmental engineering and applied geomorphology.

While the symposia have dealt with process, they have simultaneously used a wide variety of models, theories and techniques or approaches which may be important in themselves, or as a record of the progress of the discipline. Methodology has therefore been the subject of three symposia; one on quantitative methods, another on theories of landform development and this volume on models in geomorphology.

The authors were invited to present papers that illustrated the use of models in their research areas. The distinction between models and theories did not trouble them. Indeed, we can avoid arguments about definitions if we simply say that geomorphic models and theories are hypotheses about system form, and/or process and/or behavior.

The scientific endeavor is to build knowledge objectively by asking questions and by constructing and testing hypotheses to answer them. Hypotheses simplify the real world, examining critical factors and ignoring others (Clark, Ch. 5, this volume). This often leads to a trade-off between the generalizing power of a model or theory and its application to specific cases. We try to achieve simplicity, while maintaining or augmenting explanatory power. This is the esthetic of science. When an hypothesis is overly complex in relation to what it is trying to explain it is probably wrong.

Hypotheses must be tested, and this means that they must be expressed in ways which permit them to be falsified. If we have confidence that an hypothesis is false, it can be rejected. This is not a bad thing, for it can lead to the repair or improvement of the model or it can redirect our inquiry in other directions. If the hypothesis cannot be rejected, it remains, for the time being, as a possible explanation of some phenomenon(a).

During the process of hypothesis testing, the investigator should always be aware of the possibility that the falsification can itself be false (Baker, Ch. 13, this volume). While such an error could occur because of inadequate or incorrectly interpreted data, often it is due to an uncritical acceptance of authority or belief. When the catastrophists were overthrown by the uniformitarians, one dogma was substituted for another. J. Harland Bretz' explanation of the channeled scablands (Bretz 1923, 1969, Baker 1973) was rejected because he attributed these landforms to the effects of a catastrophic flood; later, the field data, not dogma, proved him right.

The authors in this volume present and test models (hypotheses) of varying scope and generality. In some papers the emphasis is on model building and in others on model testing. Some models are simulations by computers while others are physical analogs. Some models are mathematical and others are verbal. Some papers include field work to test or create models or to define rates and boundary conditions; others are based on remotely sensed data. Some papers offer an analysis of the physics involved in the short-term processes being investigated; others concentrate on the evolution of form over longer periods of time.

Rather than grouping the papers according to the varieties of models or hypotheses discussed, I have chosen a more traditional format. There are four groups of four papers. The groups cover aspects of glacial, coastal, and fluvial geomorphology, followed by papers on the geomorphology of Mars. Although I will now review the papers as members of groups related by process or environment, I will conclude this preface with a classification of the papers according to model type.

The reader will see that the papers cluster around areas of scientific concern. The glacial papers deal with the form and dynamics of present and past ice sheets. Two coastal papers investigate the interrelations of glacial and tectonic history on sea level; the other two present models of deltaic and beach sedimentation. The fluvial papers are divided into two studies on hillslope erosion and form, one on the topology of deltaic networks and the other on branching angles of river networks. The papers on Mars discuss the causes and consequences of climatic change on that planet. One study deals with the morphology of landforms related to ground ice, including curvilinear landforms. There is a paper on channels caused by catastrophic flooding and another on sapping networks. Finally there is a major review of wind abrasion on the Earth and Mars.

A summary of the papers

The first four papers deal with the Laurentide and Antarctic Ice Sheets. They all focus on the form and flow of the ice in the present (Antarctica) and the past (the Laurentide Ice Sheet). David Drewry, Neil McIntyre and Paul Cooper discuss the use and potential of new satellite-based radar which will measure, with great accuracy, the surface elevations of the ice in Antarctica. This will allow the calculation of changes in ice volume and could make possible monitoring of climatic change in real time. Knowing the slopes on the ice surface will make it possible to infer the dynamics of ice flow and the pattern of katabatic winds. Mathematical models describe the surface topography, incorporating large-, medium- and small-scale features. These equations applied to satellite data can be used to generate a series of block diagrams which are visual models for various features of the ice sheet.

The Antarctic Ice Sheet is used as an analog model for the paleo-ice sheets of the northern hemisphere in the article by Terry Hughes, George Denton and James Fastook. Their approach is to identify features in Antarctica and then to suggest analogous features in the Arctic during the Quaternary. They locate terrestrial and marine components of the ice sheets, divides with their domes and saddles, and the ice streams and their relation to changing sea level. They propose a late Wisconsin ice divide over western and southern Hudson Bay and a major ice stream through Hudson Strait. Finally, they suggest that while the Antarctic Ice Sheet's mass balance is affected by precipitation and by calving caused by changing sea level, the decline of the northern ice sheet was mainly caused by summer melting on the margins and to a lesser degree by marine instability mechanisms and precipitation.

Because of the great size, long history and dynamic nature of the ice sheets of the northern hemisphere, it is difficult to put forward such a sweeping model without stimulating vigorous scientific debate. William Shilts develops a model of the Laurentide Ice Sheet based on field observations. Using evidence from erratics he concludes there were at least two domes or centers on both sides of Hudson Bay. Dated marine fossils indicate that during Wisconsin time the ice sheet melted enough to allow at least one episode of marine deposition near its center between readvances of the glacier. Shilts suggests that the distribution of eskers on the west side of Hudson Bay can only mean that the final deterioration of that part of the ice sheet was by the retreat of thin stagnant ice. The eskers contradict the proposed major ice stream flowing eastward through Hudson Strait during the late Wisconsin. Marine instability mechanisms could have caused surges which reduced the volume of the ice, leaving this layer of stagnant ice. This sequence agrees with the hypothesis of Hughes *et al*. that the major mode of disintegration of Laurentian ice was by marginal melting coupled with marine instability mechanisms.

If the Laurentide Ice Sheet had a late Wisconsin ice ridge in Hudson Bay, and if collapse occurred, then the discharge of ice through Hudson Strait, Cumberland Sound and Frobisher Bay would have been enormous. These areas were examined by John Andrews, Jay Stravers and Gifford Miller for evidence of the predicted great ice streams. Andrews *et al*. point out that all three troughs may be descended from pre-Cambrian structures, although they experienced later glacial modification. Cumberland Sound probably underwent most of its glacial erosion about 3 million years ago and Frobisher Bay was probably covered by a floating ice shelf throughout the Quaternary; ice streams were not a factor for either one in the late Wisconsin. The areas adjacent to Hudson Strait suggest considerable glacial scour in the Quaternary. However, the patterns of striations and erratics are discordant with the direction of flow of the hypothesized ice stream and suggest source areas different from those proposed by Hughes *et al.* Andrews *et al.* offer two patterns, either of which could reflect ice flow near and within Hudson Strait at different times. They call for more mineralogical studies in this crucial area.

As the ice masses waxed and waned, sea level rose and fell and the land moved isostatically in response to the loading and unloading of the ice. James Clark uses a mathematical model to predict sea level change at any location on the earth at different times in the Pleistocene. The model depends on our current knowledge of the extent of glaciation and earth rheology. Clark also uses known sea level histories to reconstruct the past ice sheets which are represented as a series of discs.

Clark finds that in North America it is easy to predict sea level from the ice sheet, but the ice sheet thickness cannot be defined with any reasonable precision from the sea level data. Sea level data can, however, give the range of allowable thicknesses for different parts of the ice sheet and this range is interpreted as a good estimate of the constraints that sea level data place upon ice sheet reconstruction.

The problem of unravelling the combined effects of sea-level change and tectonic (or isostatic) uplift is attacked by Arthur Bloom and Nobuyuki Yonekura. They investigate dated coral terraces on the Huon Peninsula in Papua New Guinea. Terrace VIIb was formed about 125 000 years ago when sea level was about 6 m higher than at present. By regressing the height of a given terrace, i, against Terrace VIIb in six locations, they are able to generate a predictive equation with an extremely high correlation coefficient. From this they can deduce the height of the sea at the time of formation of terrace i. The difference between this sea level and the current terrace elevation is attributed to uplift. Bloom and Yonekura are then able to generate uplift histories at several locations. They point out that uplift has been sporadic over 1000 year intervals, but when aggregated, uplift has been constant over 20 000 year intervals. The theme of the impact of cold temperatures and paleo-climates on land forms will be taken up again in the four papers on Mars.

We now consider some very different models relating to sedimentation in marine environments. Paul Komar illustrates the use of computer modeling to simulate headland erosion and the movement of sand into an adjacent embayment. The shore line is divided into a series of cells; the computer traces the erosion, deposition and movement of sand through the cells with time. Komar investigates the plan profile of the graded shoreline. Because there is less sand transported as one approaches the bay center, the shore adjusts to become less curved so that breaker angles decrease, thereby decreasing longshore drift capacity. Komar then models the introduction of an additional supply of sand from a river, and he traces the development of a delta for a constant wave climate and a constant rate of delivery of sand.

Modeling sedimentation by computer with a limited number of variables can lead to some elegant and simple solutions and can suggest some general principles, but modeling of actual sedimentation can be much more complex. Charles Adams, Jr., John Wells and James Coleman investigated sediment transport off shore of the developing Atchafalaya Delta of the Mississippi River. Data on currents at various depths were collected from an offshore platform. This information, viewed in the light of physical transport laws, was used as an aid in identifying the significant events where the threshold for movement of sand was exceeded. During a 4–5 month winter period the vector plot of water displacement was to the west and north. The motion of sand-sized particles, which moved only during and after 8 winter storms (these had strong westerly wind components) was to the east and south. Because the morphology of the delta is controlled by the distribution of coarse sediment components, sand transport and distribution should provide a clue to future patterns of delta development. The authors also confirm that common winter storms of moderate strength are more important than the relatively infrequent hurricanes.

The section on fluvial processes in geomorphology contains two papers on slope modeling and two on networks. Larry Band studies the effect of creep and sheet wash on slope evolution under the influence of a fluctuating rainfall regime. Band simplifies his problem by carefully selecting his study area, an abandoned hydraulic gold mine in California. His goal is to simulate slope evolution using initial conditions obtained from old photographs, and from tree-ring dating along the drainage lines. He measured creep and slope-wash rates in the field, and had access to excellent precipitation records from which to generate rainfall magnitude and frequency data. Thus Band is able to use fairly realistic data in simple simulation models. His models duplicate slope forms which currently exist, and go on to predict the slopes which will eventually evolve. He finds that more geomorphic work is done during the average rainfall event than during storms of any other magnitude.

Michael Kirkby creates a slope simulation model of greater complexity. He makes some assumptions in mathematical form about functional relationships between slope form, parent material, and soil, and he then uses the computer to generate possible outcomes. Kirkby separately considers the interaction of lithology with chemical weathering and solution, hillslope hydrology, mass movement, and wash processes. He builds these processes into models of slope formation under conditions of landsliding, creep, solifluction, and wash. Finally he integrates all factors into a general computer slope-simulation model.

Marie Morisawa analyzes the topology of deltaic networks. Three vertex types (forks, joins and outlets) generate six link types. Given the number of vertices of each type, it is possible to calculate the expected number of links of each type, assuming random combinations of vertices. Morisawa reports that the numbers actually found differ significantly from random expectations in 5 of 20 natural networks and in 4 of 13 computer generated networks. The ratio of joins to forks (called alpha) oscillates (a property termed resonance) with increasing distance from the apex for real deltas but not for simulated networks. She suggests that alpha behaves differently in actual and simulated deltas because the simulations fail to take changing power requirements into account. Morisawa speculates that variations in stream slope, discharge, amount and caliber of load could induce topological change to minimize the cost of sediment transport.

André Roy tests the principle that the angular geometry of networks varies to minimize the cost, perhaps in terms of power, of operating and or maintaining the system. The general model is widely used in locational problems in economic geography, and it is based on the parallelogram-of-forces model, where the forces or weights are equivalent to some cost per unit length. Each variety of cost (power, resistance, volume, etc.) when minimized, will usually produce different branching angles for the same two tributaries and main stream. Roy finds that the cost per unit length can be expressed as a function of discharge raised to a constant, $1 + k$, where k is a function of z, the exponent in the equation relating slope to a power of discharge. The data indicate that rivers are probably minimizing power losses, so that $k = z$.

Roy finds that he can predict the average angles of entry, $\bar{\theta}_1$ and $\bar{\theta}_2$, quite well from the average of the ratios of the flows of the two tributaries (symmetry ratio) and the exponent (a substitution for z) taken from the slope versus area relationship. He attempts to predict individual junctions angles from the junction specific values of z (taken from the slopes of the main stream and tributaries at a junction). The inaccuracies of map contouring and stream mapping probably prevents close agreement of observed to expected angles. When he uses field data he obtains a much better agreement.

The study of landforms on Mars depends on photography from space and the planet's surface and from some observations of the atmospheric conditions and the surface materials. Because of the limited possibility for 'field work' by landed space craft, we must be especially ingenious in our use of models. The models are often based on analogies to processes and landforms found on the Earth.

Victor Baker reports on models of fluvial activity on Mars and simultaneously conveys his insights on the scientific endeavor of model making and model testing. New discoveries such as those being made on Mars generate puzzles. These, in turn, stimulate new models that expand understanding beyond the limited range of existing theory.

Improved images from the Viking missions reveal many features, including large-scale sinuous channels, streamlined forms and evidence of ponding. Baker uses Chamberlin's method of multiple working hypotheses to evaluate competing explanatory models. Rejecting other agents of erosion, Baker concludes that these features can only be explained by a model that invokes vast quantities of flowing water.

Ice-rich permafrost is considered to be the source of the water. The mechanism of release cannot be known on the basis of present information, although several plausible explanations have been suggested. Valley networks are morphometrically similar to terrestrial and simulated sapping networks, and thus have a genesis that differs from the large channels (Kochel *et al.*, Ch. 14, this volume). The valley networks imply a hydrologic cycle; this means that the atmosphere in the first several hundred million years of Martian history should have contained more water vapor than exists now. The sun supposedly radiated less energy during this time and Mars would have been too cold to maintain such an atmosphere and thus too cold to permit a hydrologic cycle. On the other hand, volcanic outgassing of CO_2, and/or water vapor could have produced a greenhouse effect. This, coupled with large variations in the tilt of the spin axis, might have allowed the required seasonal warming.

As the ground ice on Mars alternated between the liquid and solid state, networks, curvilinear land forms, chaotic terrain, debris flows and polygonally fractured ground could have formed. Craig Kochel, Alan Howard, and Charles McLane compare the morphometry and morphology of Martian networks to terrestrial stream networks and to groundwater sapping networks produced at a greatly reduced scale in a laboratory. They also make morphological comparisons to sapping networks on the Colorado Plateau. Finally, they present an extended discussion of the physics involved in sapping erosion. They find that Martian networks are similar to experimental and terrestrial sapping networks but all differ from terrestrial stream networks. Sapping networks are elongate and have relatively short first-order streams. Bivariate regressions for several pairs of variables have weaker correlations for sapping networks than for terrestrial river networks. This is probably due to some inhomogeneous property of the subsurface; such a condition is commonly found in terrestrial sapping networks. Branching angles for sapping networks are less than for river networks. The U-shaped channel cross sections and cirque-like valley heads and the similarity of tributary and master-channel cross-sectional areas also serve to differentiate the sapping, and Martian, networks from stream networks.

Lisa Rossbacher investigates curvilinear landforms on Mars. Curvilinear landforms can be classified as ridges and troughs, rimless arcuate depressions and features with no apparent relief. They are found on the northern plains, mainly near the base of the slope from the cratered uplands. These curvilinear landforms may be unique to Mars. Clues to their mode of formation may be found in the associated landforms—debris flows, chaotic terrain, channels, and polygonally fractured ground. These are all features consistent with the alternate freezing and thawing of subsurface water. Rossbacher evaluates models for the stratigraphy underlying the surface and two models for the formation of curvilinear landforms. One implies they are erosional remnants where the surface was lowered by mass wasting, wind, and perhaps fluvial activity. The other model is based solely on mass wasting. If liquid water had once been present, then solifluction could have taken place on low slopes. This appears to be the case. Most of the evidence points to an origin for the curvilinear ground involving the growth and decay of segregated ground ice coupled with solifluction.

Because the presence of liquid water on the surface of Mars was probably very sporadic, eolian abrasion should also have played an important role in modifying the landscape. Ronald Greeley, Steven Williams, Bruce White, James Pollack and John Marshall present an extended review of their work simulating processes of wind abrasion on the Earth and on Mars. Using a wind tunnel, and simulating the atmosphere of Mars, they compare the properties and behavior of target materials and windblown particles on Earth and Mars. They also compare the dynamics of particle transport by wind for both planets and they assess threshold wind frequencies. They conclude that the rate of wind abrasion at the Viking Lander 1 site should have been enormous, and yet geomorphic evidence for high rates of wind abrasion are missing here and elsewhere. They account for the discrepancy in observed and projected rates of abrasion by suggesting that aggregates of fine dust are formed and create a veneer to protect target rocks. Aggregates can only become abrasive at free stream wind speeds in excess of 75 m/s, which are thought to be rare events on Mars. This inhibiting factor, coupled with protective burial and later exhumation of parts of the surface, probably account for the excellent preservation of very ancient small craters and other features.

Having reviewed the papers in terms of their geomorphological content, it is appropriate now to classify them according to the kinds of models used. I will follow, with slight modifications, a classification introduced by Chorley (1967) in his paper on 'Models in Geomorphology.' To save space I will outline Chorley's typology and include citations to the papers in this book. Some authors will appear in more than one class because they use models in more than one way.

Models in geomorpholgoy: a classification (after Chorley 1967)

I   Models used predominantly to analyze systems
  A   Natural analogs
    1   Historical analogs

      (a)   'The present is the key to the past.' Uniformitarian model. Many papers imply this. (Hughes *et al.*, Shilts, Andrews *et al.*, Baker, Kochel *et al.*, Rossbacher, Greeley *et al.*).
      (b)   Old landscapes can predict the future form of present landscapes. Denudation chronology (e.g. Davisian models. These can be used for synthesis. See II.C.2. below).

    2   Spatial analogs. Understanding a geomorphic process or landscape in one locale helps in understanding process and form in another place. (See citations in I.A.1(a) above). Note that Shilts and Andrews *et al.* deny parts of the analogy after examination of the evidence of Hughes *et al.*).

B   Abstract analogs
  1   Hardware models

    (a) Geometrically and dynamically similar model using identical materials to those found in nature. Sometimes a limited portion of a natural system is used (Greeley *et al.*, Adams *et al.*, Band, Roy).
    (b) Dynamically similar but geometrically dissimilar. Dimensionless ratios involving dynamic parameters are constant. Use of natural materials.
    (c) Substitution of analogous materials to simulate form and dynamic behavior (Kochel *et al.*, Greeley *et al.*).

  2   Mathematical models

    (a) Deterministic: often based on knowledge of, or assumptions about physical or chemical process laws (Clark, Adams *et al.*, Komar, Band, Kirkby).
    (b) Probabilistic models

      1   Markov simulation. Based on previous system states (Morisawa).
      2   Monte Carlo simulation. Independent of previous system states.
      3   Expectation based on calculation of probabilities (Morisawa).

    (c) Optimization models—maximization or minimization of some criterion.

      1   Maximally probable state or maximum entropy. See I.B.2(b)3 above.
      2   Maximum efficiency or minimum cost models. (Roy).

  3   Experimental design models

    (a) Regressions between variables suggest strength of relations and directions of causation. (Bloom and Yonekura, Roy, Kochel *et al.*, Greeley *et al.*).
    (b) Relations between variables allow predictions of form. Causation not necessarily implied (Drewry and McIntyre).
    (c) Nonparametric observations, including presence and abscence data create tests for multiple working hypotheses (Chamberlin 1890). These observations eliminate hypotheses and create puzzles demanding new hypotheses (Hughes *et al.*, Shilts, Baker, Kochel *et al.*, Rossbacher, Greeley *et al.*).

II Models that synthesize systems

A   Attempts at specification of complete system with inputs, outputs, feedback relations and evolution. Often based on experimental design work (Hughes *et al.*, Band, Kirkby).

B Partial systems (gray box) approach. Focus on inputs and outputs and feedbacks without examining all the internal mechanisms (Komar).

C Black box approach. Processes poorly understood. Reliance on intuitive leaps.

1 Equilibrium systems dependent on negative feedback. Form attains steady state geometry (Morisawa).

2 Time-bound historical systems involving positive feedback. See I.A.1(b), above.

The membership lists for the several classes indicate the wide range of models employed. The natural analog models have a long tradition and they are well represented. The more modern approaches exemplified by hardware and mathematical models are also common. There is only one paper using optimality criteria. Surprisingly, statistical and experimental design models are not emphasized. The heyday for denudation chronology is past and we find no examples in our list. There are few papers describing negative feedback mechanisms leading to steady state geometries; this may be an anomaly.

Looked at in another way, many of Chorley's categories can be interpreted as illustrating goal-seeking behavior, either as a presumed property of natural systems, or as a property of a class of models to which natural systems are being compared. Davisian models of landscape evolution, equilibrium-seeking models, probabilistic models and optimization models are all examples. My personal taste leads me to encourage others to attempt to explain process and form through the use of optimization models.

Since a purpose of presenting this classification is to identify the range of models which might be useful in our work, then we should also consider models from other disciplines. Although some of our models are indigenous to geomorphology, some are adopted from other fields and perhaps modified. Some of our models have been exported to other disciplines (Holland 1969, Woldenberg *et al.* 1970). For examples from all three categories relating to networks see Jarvis and Woldenberg (1984) and McDonald (1983).

A future symposium on the cross-disciplinary use of models would give us new perspectives that could lead to solutions for outstanding problems and could raise new questions for geomorphology.

Michael J. Woldenberg
Department of Geography
State University of New York at Buffalo

References

Baker, V. R. 1973. *Paleohydrology and sedimentology of Lake Missoula flooding in eastern Washington*. Geol. Soc. Am. Spec. Pap. 144.

Bretz, J. H. 1923. The Channeled Scablands of the Columbia Plateau. *J. Geol.* **31**, 617–49.

Bretz, J. H. 1969. The Lake Missoula floods and the Channeled Scabland. *J. Geol.* **77**, 505–43.

Chamberlin, T. C. 1890. The method of multiple working hypotheses. *Science* **15** (old series), 92–6.

Chorley, R. J. 1967. Models in geomorphology. In *Models in geography*, R. J. Chorley and P. Haggett (eds), 59–96. London: Methuen.

Holland, P. G. 1969. The maintenance of structure and shape in three mallee eucalypts. *New Phytol.* **68**, 411–21.

Jarvis, R. S. and M. J. Woldenberg 1984. *River networks*. Stroudsburg: Hutchinson Ross.

MacDonald, N. 1983. *Trees and networks in biological models*. Chichester: Wiley.

Woldenberg, M. J., G. Cumming, K. Harding, K. Horsfield, K. Prowse, S. Singhal 1970. *Law and order in the human lung*. Harvard Pap. Theor. Geog. Nr. 41, Natl Tech. Inf Serv. Nr AD 709602.

Acknowledgments

Grateful acknowledgement is made of grant support for the Symposium from the 'Conference in the Disciplines' program of the Graduate School of the State University of New York at Buffalo and from the 'Conversations in the Disciplines' program, a state-wide award made by the Faculty Senate Committee on University Programs and Awards administered from Albany. Thanks go to Jim McConnell, Chairman of the Department of Geography of SUNY at Buffalo who furnished seed money and a reservoir of student labor, and to Lorraine Welch and Pam Nuss who provided secretarial assistance.

I would especially like to express my gratitude to the following organizations for personal support during various phases of the project: the Johnson Fund of the American philosophical Society, the Minna-James Heineman Stiftung of Hannover, administered by the Senior Scientists Program of NATO, and the Collaborative Research Grant Program of NATO.

My thanks go to Marie Morisawa and Donald Coates, the founders of this series, for most welcome advice and support. Jack Vitek suggested the theme of the conference. Parker Calkin was very helpful in planning the glacial session. I wish to thank Lorraine Oak of SUNY Buffalo for her extensive assistance with the technical editing. Greg Theisen redrafted the maps in Chapter 2; Tom Kress, Dean Temlitz and Greg redrafted other maps and many figures. They did an excellent job. I also thank Gordon Cumming, Norman Harris, Christine McKechnie, Karen Wadey, Heather Inman Beard, and John Griffiths of The Midhurst Medical Research Institute for their kind assistance. I wish to thank the reviewers: Athol Abrahams, James Allen, Ray Arvidson, Parker Calkin, Michael Carr, Michael Carson, Richard A. Davis, Jr., William T. Fox, Dale Gillette, Richard Jarvis, David Mark, Ernest Muller, William Nickling, Norbert Psuty, Lisa Rossbacher, Ronald Shreve and Samuel Smart.

Finally, I am indebted to the authors who have, after all, written this book.

Michael J. Woldenberg

Contents

Tables

Contributors

Charles E. Adams, Jr.
Coastal Studies Institute, Louisiana State University, Baton Rouge, LA 70803

John T. Andrews
Institute for Alpine and Arctic Research, and Department of Geological Sciences, University of Colorado, Boulder, CO 80309

Victor R. Baker
Department of Geosciences, University of Arizona, Tucson, AR 85721

Lawrence E. Band
Department of Geology and Geography, Hunter College, City University of New York, New York, NY 10021

Arthur L. Bloom
Department of Geological Sciences, Cornell University, Ithaca, NY 14853

James A. Clark
Department of Geology, Geography, and Environmental Studies, Calvin College, Grand Rapids, MI 49506

James M. Coleman
Coastal Studies Institute, Louisiana State University, Baton Rouge, LA 70803

A. P. R. Cooper
Scott Polar Institute, Cambridge, CB2 1ER, England

George H. Denton
Department of Geological Sciences and Institute for Quaternary Studies, University of Maine at Orono, Orono, ME 04469

David J. Drewry
Scott Polar Institute, Cambridge, CB2 1ER, England

James L. Fastook
Department of Geological Sciences and Institute for Quaternary Studies, University of Maine at Orono, Orono, ME 04469

Ronald Greeley
Department of Geology, Arizona State University, Tempe, AR 85287

Alan D. Howard
Department of Environmental Sciences, University of Virginia, Charlottesville, VA 22903

Terry J. Hughes
Department of Geological Sciences and Institute for Quaternary Studies, University of Maine at Orono, Orono, ME 04469

Michael J. Kirkby
School of Geography, The University, Leeds, LS2 9JT, England

R. Craig Kochel
Department of Geology, Southern Illinois University, Carbondale, IL 62901

Paul D. Komar
School of Oceanography, Oregon State University, Corvallis, OR 97331

John R. Marshall
Department of Geology, Arizona State University, Tempe, AR 85287

Charles McLane
Department of Environmental Sciences, University of Virginia, Charlottesville, VA 22903

Neil F. McIntyre
Scott Polar Institute, Cambridge, CB2 1ER, England

Gifford H. Miller
Institute for Alpine and Arctic Research, and Department of Geological Sciences, University of Colorado, Boulder, CO 80309

Marie Morisawa
Department of Geological Sciences and Environmental Sciences, State University of New York, Binghamton, NY 13901

James B. Pollack
Space Sciences Division, NASA Ames Research Center, Moffett Field, CA 94035

Lisa A. Rossbacher
Earth Science Department, California State Polytechnic University, Pamona, CA 91768

André G. Roy
Department of Geography, University of Montreal, Case Postale 6128, Succursale A, Montreal, P.Q. H3C 357

William W. Shilts
Department of Energy, Mines and Resources, 601 Booth St., Ottawa, K1A 0E8

Jay A. Stravers
Institute for Alpine and Arctic Research and Department of Geological Sciences, University of Colorado, Boulder, CO 80309

John T. Wells
Coastal Studies Institute, Louisiana State University, Baton Rouge, LA 70803

Bruce R. White
Department of Mechanical Engineering, University of California at Davis, Davis, CA 95616

Steven H. Williams
Department of Physics, University of Santa Clara, CA 95053

Nobuyuki Yonekura
Department of Geography, University of Tokyo, Hongo, Tokyo 113, Japan

1

The Antarctic Ice Sheet: a surface model for satellite altimeter studies

David J. Drewry, Neil F. McIntyre and Paul Cooper

Introduction

At the beginning of the International Geophysical Year in 1956 little was known of the configuration of the Antarctic Ice Sheet and still less about the dynamic and thermodynamic processes that govern the flow and stability of large ice masses. The continuous presence of a number of nations in Antarctica has enabled major programs of research to be undertaken to understand the largest ice sheet on Earth. Oversnow traverse campaigns, carried out principally by American and Soviet scientists, were able to determine ice thickness at scattered locations by seismic shooting and gravity observations that revealed an ice sheet of significant depth overlying a land surface of diverse topography.

Surface altitudes were measured crudely by barometry and occasionally by leveling. A pattern emerged, similar to that discovered earlier in Greenland, of a central flattened dome up to 4 km in elevation, with steep coastal regions and a mosaic of flanking ice shelves (the latter unique to Antarctica). Early surface measurements also included mean annual temperature and (by a variety of steadily improving techniques) mean annual snow accumulation on the continent. Few ice flow velocities were measured due to the absence of fixed points of reference such as rock outcrops. The lack of velocity data has proved a continuing, major shortfall since it prevents measured and calculated 'balance' velocities from being compared in order to assess the mass budgets of the ice sheet.

By the mid 1960s investigations were also underway on some of the major ice shelves and the first attempts to drill through the ice sheet were being made. Deep drilling to recover cores for a host of mechanical and chemical studies has proved to be one of the most exciting and scientifically rewarding projects in glaciology; however it has also proved to be one of the most costly and frustrating because drilling technology has consistently failed to keep pace with the expanding demands of scientists. Also, in the late 1960s, methods of measuring ice thickness were developed which use electromagnetic rather than acoustic energy sources.

At the commencement of the 1970s considerable detail was emerging on the character of the ice sheet. Drilling had penetrated over 2000 m at Byrd Station in West Antarctica revealing a glacioclimatic history stretching back into the last interglacial. Radio echo sounding (RES) from aircraft by a British–American project became a routine activity for mapping ice thickness and subglacial bedrock over the continent. In addition a number of international programs were being formulated to take glaciological research a significent step forward by focusing upon specific problem areas. The Ross Ice Shelf Project developed as a U.S.-led study of the $0.5 \, M \, km^2$ floating ice shelf in the Ross Sea and involved geophysical studies (seismic, gravity, resistivity measurements), glaciological investigations (velocities using the new satellite doppler technique, accumulation and ice core drilling) and oceanographic observations both beneath and in front of the ice shelf. The International Antarctic Glaciological Project (IAGP) began as a series of coordinated national activities in East Antarctica, undertaking velocity, mass balance, surface glaciochemistry, radio echo sounding ice thickness and drilling work over an area of some $5 \, M \, km^2$. Although there is still no complete reconnaissance of Antarctica for certain parameters (e.g. ice thickness, surface accumulation, velocity), sufficient data at a first order of precision are now available to allow ice dynamics and deep temperatures to be moderately well understood for the paleoenvironmental interpretation of ice cores (Robin 1983) and modeling of the response of some sectors of the Antarctic to climatic and ocean forcing (Budd & Smith 1981, 1982). The next stages are to increase the level of accuracy of glaciological measurements (see Table 1.1) from that presently achievable, in order to obtain full continental coverage of critical parameters, to address key scientific questions raised during the preceding 30 years of research (such as the stability of West Antarctica), to determine significant changes in ice volume and to integrate ice sheet behavior with models of global climate (Warren 1982).

It is clear that remote sensing from space can offer the possibility of gathering much of the necessary critical data, continuously and on a continental basis. Robin et al. (1983) have listed the contributions that satellite techniques can make towards collecting the glaciological parameters listed in Table 1.1. Of particular importance to glaciology has been the development of high-accuracy radar altimetry. With an instrument precision of a few tens of centimeters and a final height measurement accuracy of 1–2 m over smooth ice areas radar altimetry has the promise of yielding, within a few months, more extensive and accurate data on ice surface topography and slopes, grounding line location, ice fronts and even ice shelf thickness and bottom melting rates. The quality and quantity of the data would be superior to any obtained so far and, in contrast with the past, these measurements could be collected in only a few months.

Such data would provide a synoptic view of ice sheets which in turn would allow ice-volume changes to be carefully monitored. Indeed, no other method

Table 1.1 Required observations for study of ice sheet shelves.

Parameter	Required Resolution	Currently Achieved Resolution	Current Antarctic Coverage (%)
(1) Surface morphology			
(i) large-scale	1.0m	5.0m (satellite altimetry, geoceiver)	20
		60m (barometric altimetry)	70
(ii) small-scale	0.5m	2–10m	1
(2) Ice thickness	5.0m	10–30m	40
(3) Temperature	1K	1K	50 (low density)
(4) Accumulation/ablation			
(i) surface	0.1m	0.5m	50 (low density)
(ii)bottom	0.1m	1.0m	a few points
(5) Velocity field			
(i) surface	0.5 m/a	0.5–2.0m/a	a few points
(ii) bottom & internal	0.1m/a	–	–
(6) Dimensional fluctuations			
(i) thickness	0.1–0.5m/a	0.5–10m/a	a few points
(ii) margin position	10m/a	10m/a (survey, aerial photo) 100m/a (Landsat)	a few points & coastal sections
(7) Snowline	10m	10–100m (Landsat)	some blue ice areas

comes near to providing such extensive and accurate data, and the capability of monitoring elevation changes over the required time scales of months to decades. The precision of radar altimetry is such that the principal component of the world's water balance (i.e. ice sheets) can be monitored considerably more accurately than the remaining components. Changes could be detected long before an unambiguous sea level signal from the melting of ice sheets was recognizable (NERC-SERC 1983).

Satellite altimetry of the Antarctic Ice Sheet

Ice sheet surface topography
Ice flow is in the direction of the maximum regional surface slope. The principal driving force area is that required to overcome the basal shear stress, τ_b:

$$\tau_b = \rho_i \, g \, h \sin \alpha \qquad (1.1)$$

where ρ_i is the density of ice, h is the ice thickness, g is the gravitational acceleration and α is the surface slope.

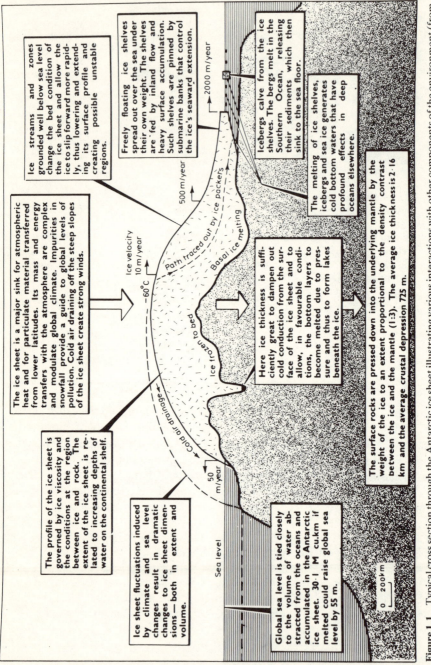

The ice sheet is a major sink for atmospheric heat and for particulate material transferred from lower latitudes. Its mass and energy transfers with the atmosphere are complex and modulate global climate. Impurities in snowfall provide a guide to global levels of pollution. Cold air draining off the steep slopes of the ice sheet create strong winds.

Ice streams and zones grounded well below sea level change the bed condition of the ice sheet and allow the ice to slip forward more rapidly, thus lowering and extending its surface profile and creating possible unstable regions.

Freely floating ice shelves spread out over the sea under their own weight. The shelves are fed by inland flow and heavy surface accumulation. Such shelves are pinned by submarine banks that control the ice's seaward extension.

Icebergs calve from the ice shelves. The bergs melt in the Southern Ocean, releasing their sediments which then sink to the sea floor.

The profile of the ice sheet is governed by ice viscosity and the conditions at the region between ice and rock. The extent of the ice sheet is related to increasing depths of water on the continental shelf.

The melting of ice shelves, icebergs and sea ice generates cold bottom waters that have profound effects in deep oceans elsewhere.

Ice sheet fluctuations induced by climate and sea level changes result in dramatic changes to ice sheet dimensions — both in extent and volume.

Here ice thickness is sufficiently great to dampen out cold conduction from the surface of the ice sheet and to allow, in favourable conditions, the bottom layers to become melted due to pressure and thus to form lakes beneath the ice.

Global sea level is tied closely to the volume of water abstracted from the oceans and accumulated in the Antarctic ice sheet. 30·1 M cu.km if melted could raise global sea level by 55 m.

The surface rocks are pressed down into the underlying mantle by weight of the ice to an extent proportional to the density contrast between the ice and the mantle (1:3). The average ice thickness 2·16 km and the average crustal depression 725 m.

Ice velocity 10 m/year

Path traced out by ice packets

Basal ice melting

−60°C

Cold air drainage

Ice trapped to bed

Sea level

2000 m/year

500 m/year

50 m/year

0   200 km

Figure 1.1 Typical cross section through the Antarctic ice sheet illustrating various interactions with other components of the environment (from Drewry 1982. Copyright © by New Science Publications. Used by permission).

Radar altimetry can yield very accurate measurements of surface height thus providing a key indicator of ice dynamics. On a small scale, ice flow over rough bedrock can be interpreted from surface relief since longitudinal stress gradients in the ice are not averaged out (Robin 1967, Budd 1970, Cooper *et al.* 1982, McIntyre & Cooper 1983). Surface slopes may similarly be used to infer the gross aspects of subglacial topography.

Surface slopes (at both large- and small-scales) are important for the understanding of ice sheet-related atmospheric phenomena. Steep ice margins are responsible for the flow of cold, dry air in the form of katabatic winds from the continental interior to the sea (Parish 1981). Katabatic air flow is important in the energy balance of coastal areas. Unstable lapse rates in cold air advected over warmer water will produce considerable transfers of sensible heat and water vapor from the ocean to the atmosphere. In addition, frictional coupling between strong off-shore air flow and rough sea ice results in opening of shore-leads, ice break-up and coastal ice movement. Strong katabatic winds also transport large quantities of snow precipitated on the ice sheet out to sea and are thus important factors controlling coastal mass budgets.

Transition zones and grounding lines
Figure 1.1 shows a typical cross section through the ice sheet. It is apparent that marked changes in surface slope occur along a flowline and reflect fundamental transitions in ice behavior. One of the most marked changes occurs where grounded ice comes afloat in deep water to form ice shelves. The surface gradient of the ice shelf is extremely small (N.B. water cannot support a shear stress so that, in Eq. 1.1 $\tau_b = 0$) with a slope proportional to the ice thickness gradient. If the ice is fully grounded and frozen to bedrock there will be a

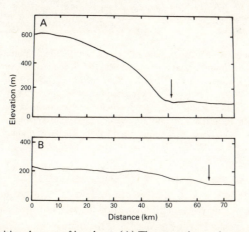

Figure 1.2 Transitional zones of ice sheet. (A) The upper ice-surface profile illustrates a typical transition between grounded ice (no basal sliding) and ice shelf. (B) The lower diagram represents a transition between ice stream (with basal sliding and presence of sub-glacial water) and ice shelf. The figure demonstrates the effect of basal shear stress (τ_b) on surface slopes. Data are taken from airborne radar altimetry conducted in the vicinity of the Shirase Coast and the Ross Ice Shelf.

Figure 1.3 (a) The Lennox-King Glacier, an outlet glacier of the East Antarctic ice sheet which transects the Transantarctic Mountains at approximately 83.5°S. Ice flow is from right to left (i.e. from the East Antarctic ice sheet to the Ross Ice Shelf which is shown in the far field of the photograph). Note the surface flowlines and crevasse systems. (b) Landsat image of heavily crevassed ice stream feeding Pine Island Glacier as it enters the Amundsen Sea in West Antarctica. Cloud obscures some of the rough coastal topography. The ice stream is approximately 25 km wide as it goes afloat (E-1185-13530, band 7). The main part of the glacier is about 12 km wide.

considerable break of slope between ice shelf and ice sheet (Fig. 1.2A). If coupling between ice and bedrock is weakened, by the introduction of water due to basal sliding, the magnitude of the slope change is diminished in proportion to the decrease in τ_b (Fig. 1.2B). This occurs where high velocity ice streams (Fig. 1.3a,b), which often flow into ice shelves, generate considerable volumes of water by strain heating and sliding motion (Drewry 1983).

Subglacial lakes are also known to exist beneath the Antarctica ice sheet (Oswald & Robin 1973; Robin *et al*. 1977). Those lakes with lateral dimensions

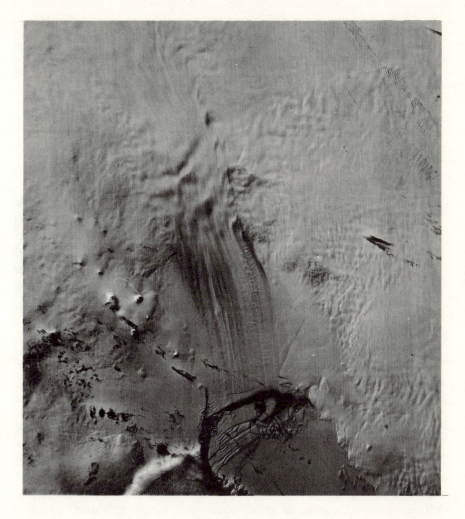

many times that of the average ice thickness will have an affect on surface slopes similar to that of basal melt water due to their influence on τ_b.

Changes in surface gradient induced by either subglacial lakes or ice shelf grounding lines should be easily detected by satellite radar altimetry and their continued monitoring could yield information on long-term ice stability.

Ice sheet margins
The fronts of ice shelves and ice sheet margins are generally marked by an ice cliff several tens of meters in height (Fig. 1.4). It is important to take account of the sudden jump in elevation at the ice edge for radar altimeter tracking. Cliff positions can be accurately mapped within a few hundred meters from radar altimeter data (Thomas *et al*. 1983). Such information is a useful ancillary source to ice edge mapping undertaken from Landsat MSS imagery.

Figure 1.4   Ice cliffs typical of grounded portions of the Antarctic ice sheet (Taylor Glacier, southern Victoria Land). The figures in bottom left give scale.

Ice sheet and ice shelf mass balances
It is not known whether the Antarctic ice sheet is growing or shrinking; at present the differences between snow accumulation and ablation (i.e. mass balance) are not known to better than ±50%. Despite the lack of accurately leveled lines it is clear that changes in surface height are not simple, with some regions showing rapid thinning and others exhibiting small increases. For instance, crude mass-balance calculations for the whole continent, are weakly positive implying an elevation rise of 0.03 m/a (Bentley 1981). In certain drainage basins, however, marked reduction in surface elevations of up to 1.0 m/a has been observed (Mae & Naruse 1979) although interpretation is complex (Morgan & Jacka 1979). The rate of change that will require measurement by satellite systems is in the range of 0.01 to 0.5 m/a. Any long-term estimates of mass balance of the ice sheet from satellite radar altimetry will require the use of data gathered over at least 10 or possibly 20 years; this is more than six times the lifetime of a single satellite. Such studies will necessitate the comparison of carefully evaluated datasets from a variety of space platforms.

Experience of Seasat altimetry over ice

Our earliest radar altimetry data comes from Seasat which was optimized for operation over the oceans (MacArthur 1978), yet successfully acquired a considerable dataset for ice sheets in Greenland and Antarctica (some 0.6 M values) (Zwally *et al.* 1983). Nevertheless, the Seasat altimeter was frequently unable to track the ice surface over areas of rough ice topography as detailed by Rapley *et al.* (1983), Brenner *et al.* (1983) and Martin *et al.* (1983). Rapid changes in range rate induced by surface irregularities, and the presence of several discrete reflecting regions within the gate-limited footprint created a multi-peaked waveform which caused loss of tracking (Fig. 1.5). In addition, steep slopes (i.e. >2.8°) resulted in pronounced degradation of the leading edge again inhibiting tracking (Rapley *et al.* 1983).

To improve the ability of future satellite radar altimeter systems to acquire accurate information from ice sheets (i.e. range, returned pulse strength, pulse shape) it will be necessary to base their design specifications on realistic ice sheet surfaces. This procedure involves first the use of a waveform generator to synthesize a string of radar pulses or waveforms from characteristic ice surfaces; and secondly a tracker simulator that tests the effectiveness of various tracking algorithms with respect to the waveforms. In order to assist in the first of these developments an ice sheet surface model has been constructed for use with the waveform generator based upon airborne measurements over Antarctica and an examination of Landsat imagery. This activity has been undertaken in support of the radar altimeter mission onboard the European Space Agency

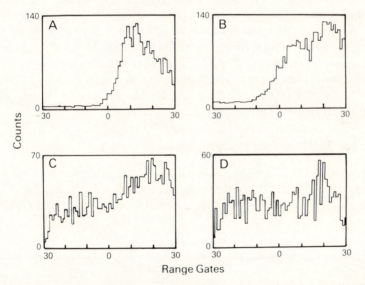

Figure 1.5 Seasat radar altimeter waveforms during one crossing of Antarctica illustrating a sequence in which the altimeter failed to track the surface return. (A) Typical return from ice surface. The altimeter is tracking the steep leading edge which is positioned in the middle gates. (B) Rough surface elements begin to corrupt the returned pulse and the leading edge is degraded: the tracked point has moved over to the right. (C) Complex, rough surface terrain generates a complex pulse with a long tail. The leading edge is now almost out of the tracking gates. The returned power is also significantly lower. (D) Loss-of-tracking is now complete and no leading edge is discernible. Moments later the altimeter went into its acquisition sequence.

(ESA) earth remote sensing satellite, ERS-1 to be launched in 1988. (A resumé of ERS-1 specifications is given in Appendix A).

Measurement of ice shelf surface elevations from satellites to an accuracy of ±1.0 m should make it possible to determine their thickness to ±10 m, a useful accuracy for ice shelf dynamic studies (Robin *et al.* 1983). Improved accuracy to ±0.1 m in elevation, as may be possible over level ice shelves, will yield ice thicknesses to ±1.0 m, an accuracy sufficient to determine significant changes of ice thickness from bottom melting within a few years (Robin *et al.* 1983). Radar altimetry should thus open up a new method of studying mass balances of ice shelves.

Airborne and surface datasets for Antarctica

Drewry (1983a) and Drewry *et al.* (1982) have reviewed available information on the surface configuration of the Antarctic ice sheet. The primary sources and accuracies are shown in Table 1.2.

Two other main sources of Antarctic data are used in the present study of the surface topography of the ice sheet. The principal one is the extensive collection of altimetric profiles recorded during two Antarctic seasons of airborne radio echo sounding by the joint program of the U.S. National Science Foundation and the Scott Polar Research Institute, Cambridge (Drewry *et al.* 1980).

Surface elevations with relative accuracies of 2 m were derived from aircraft pressure altimeter and terrain clearance measurements. The second data set is Landsat imagery (available north of 82°S). Given suitable solar elevations and absence of clouds, these images complement the altimetry data by adding a plan view to vertical profiles. Correlation of the two sources shows good agreement (McIntyre 1983).

Antarctic Ice Sheet surface topography

Key characteristics of the ice sheet surface have been extracted for modeling purposes and form a hierarchy of three distinctive terrain types: large-scale ice sheet geometry (typical scale of 10^2 km), dynamic surface topography (scale of

Table 1.2 Accuracy of ice-sheet surface elevation measurements.

Type of measurement	Resolution (m)
balloon altimetry	±40–60
oversnow barometry	±50
airborne radio echo sounding	±30
geodetic leveling station	±10
satellite doppler	±2–5
satellite radar altimetry	±3–15

10 km) and transient surface features such as sastrugi and snow dunes (scale of $\sim 10^2$ m).

Ice sheet geometry
For an ice sheet in steady state, frozen to a flat bed and with no ablation area, the surface profile is given by:

$$(Z/H)^{2+2/n} + (x/L)^{(1+1/n)} = 1 \tag{1.2}$$

where H = height of ice sheet at center, L = ice sheet half-width (radius if circular), Z = surface elevation at a distance x from center, and n = visco-plastic flow law exponent $(1.5 < n < 4.5)$.

If the ice is assumed to be perfectly plastic, $n \to \infty$ and Equation 1.2 for the half profile becomes:

$$Z = \left[\left(\frac{2\tau_b}{\rho_i g} \right) L - x \right]^{1/2} \tag{1.3}$$

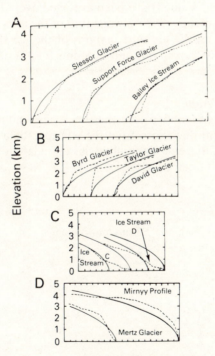

Figure 1.6 Surface profiles (dashed lines) down flowlines of the Antarctic ice sheet compared with theoretical profiles (solid lines) assuming an ice 'yield stress' of 50 kPa, flat bedrock and steady-state. (A) Major ice-sheet outlets showing good correlation between theory and reality. Deviations may be caused by fringing mountains, (B); sliding at bed, (C); and rough subglacial terrain, (D).

where τ_b = basal shear stress given in Eq. 1.1.

Such a quasi-parabolic profile is characteristic of much of the Antarctic ice sheet. In Figure 1.6A profiles inland of Mirnyy, Slessor and Support Force Glaciers and the Bailey Ice Stream are compared with Eq. 1.3. There is good, large-scale correlation between reality and theory. Significant deviations from the convex-upward profile, however, arise from violation of the assumptions inherent in Equation 1.3. These occur where the ice sheet is fringed by mountains that dam ice flow (Fig. 1.6B) or has irregular subglacial topography (Fig. 1.6D., Paterson 1981, Drewry 1983a, b).

Other significant deviations from the quasiparabola tend to occur in association with fast-flowing ice streams and outlet glaciers which develop when ice is funneled toward the periphery of the ice sheet. These often result in embayments in the otherwise steep coastal zones. In West Antarctica, where such features dominate the ice sheet, an overall concavity is produced in the lower parts of the ice sheet profile (Fig. 1.6c).

The distribution of surface slopes is comparatively uniform when considered on the large scale; Figure 1.7 shows the relative proportions of regional gradients in Antarctica. Values have been averaged over 100 km. More than 1/2 of the area has gradients <0.30% while 9/10 of the area has gradients less than 1.5%. It is only in the extreme margins of the ice sheet that average gradients rise above 3.00% and this only accounts for 3% of the grounded ice sheet.

Dynamic surface topography

The principal features identified at a mesoscale are undulations in the ice surface which represent deviations from the quasiparabola due to subglacial topography. The main trend discernible is a gradation from a virtually featureless surface at the center of the ice sheet (i.e. close to the ice divide) to

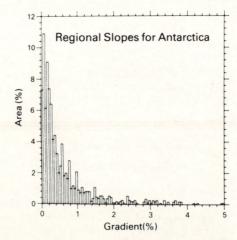

Figure 1.7 Regional ice sheet surface slope distribution in Antarctica.

Figure 1.8 Landsat image showing rough surface terrain as it is channeled toward Dibble Glacier between two local domes of smooth appearance. Note the areas of bare ice (dark grey tones) due to scour by katabatic winds (E-1447-23214, band 7).

an uneven, irregular morphology at the peripheries (Fig. 1.8). With distance from the ice divide, wavelengths of the undulations typically decrease while amplitudes increase. It is thus possible to identify three broad categories which represent the continuum of surface roughnesses found on an ice mass (McIntyre 1983).

Type 1 surfaces are illustrated in Figure 1.9A. The bandpass-filtered profile from the polar plateau of East Antarctica is virtually level and shows no deviations from the mean greater than ±5 m; the RMS roughness is 0.9 m. The dominant wavelengths present vary from <5–50 km, but the larger features dominate. The power spectrum, produced by applying a fast Fourier transform to the bandpass-filtered data, show there to be very little energy in the profile. The spectral peaks are at approximately 10 and 20 km. Gradients, calculated by fitting a 1 km-regression to lowpass-filtered data, show a distribution which is

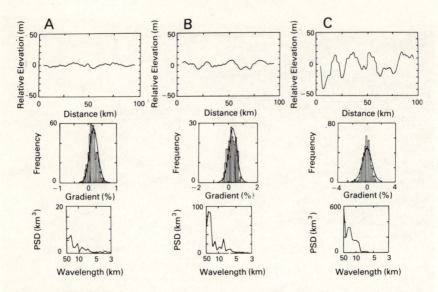

Figure 1.9 Characteristics of ice surfaces in Antarctica from airborne radar altimetry. Each terrain type displays bandpass filtered profile, frequency distribution of surface gradients, and smoothed power spectrum. (A) Type 1 terrain. (B) Type 2 terrain. (C) Type 3 terrain.

only slightly skewed from Gaussian. Superimposed on the low mean gradient of 0.16% is a RMS of only 0.14%. Most local gradients are therefore <0.5%.

Characteristics of Type 2 terrain are shown in Figure 1.9B. They illustrate the topographic developments that occur towards the ice sheet margin. The planar surface is clearly disturbed by a series of undulations with amplitudes up to 16 m; the RMS roughness of this type of terrain is typically 4–5 m. The power spectrum indicates increasingly significant contributions from a variety of wavelengths. There is a dominant peak at 20 km but energy is also present at wavelengths ≤10 km. The distribution of gradients has a mean displaced by 0.25% and its RMS deviation is twice that of Type 1 topography (i.e. 0.27%). The tendency for reduced wavelengths and increased amplitudes gives the terrain a more irregular appearance with a reduced radius of curvature.

Type 3 terrain is demonstrated in Figure 1.9C, the location of which is 150 km from the coast of East Antarctica. Amplitudes of up to 60 m can be seen to dominate the topography. Smaller wavelengths occur but those of 10–15 km are the ones that control the form of the surface. The RMS roughness of the profile is 9.6 m. The spectrum shows there to be a great variety of wavelengths present and clearly the longest undulations no longer totally dominate the terrain. Energy is present in significant proportions down to 5 km and there are high spectral contributions between 8 and 25 km. The gradients of Type 3 terrain are substantially greater then elsewhere. With a RMS deviation of >0.65%, gradients may reach ±3% and are less normally distributed than those in the interior. The steeper gradients, superimposed on shorter wavelengths, produce notably smaller radii of surface curvature. In Antarctica ice sheet terrain does occur which is significantly more irregular than Type 3 topography, although the extent of its area is not very important. Extreme conditions can be found in local areas of thin ice and high velocities. Gradients of up to 10% associated with short wavelengths and high amplitudes have been recorded in the vicinity of outlet glaciers. The limited occurrence of such terrain is, however, insufficient to warrant a separate grouping.

The distribution of the three topographic types shows a clear progression of roughness from the smooth ice divides to the ice-sheet margin which is characterized by the most irregular topography. This general radial pattern is distorted by smooth domes due to localized centers of outflow and also by particularly rough, irregular terrain. This terrain is usually produced by channeling of the flow at the coast or by anomalously thin ice. Thus substantial

Figure 1.10 Photograph of smaller-scale ice surface roughness elements generated by intense crevassing on Robert Scott Glacier, Transantarctic Mountains. The seracs have widths on the order of 100 m.

deviations may occur locally within the topographic classes. If these patterns of surface roughness are extrapolated to the rest of the continental ice sheet, excluding ice shelves, the proportions covered by Type 1, 2 and 3 terrains will be approximately 20%, 52% and 28% respectively. Taking into account the very severe topography which is likely to occur locally around outlet glaciers the coverage of Type 3 terrain becomes 25%, thus leaving 3% for the most extreme roughness grouping.

Small-scale surface features

Information on snow formations discussed in this section was compiled by E. Novotny (Rapley *et al.* 1983). These snow surface features have lengths of <0.5 km. Little accurately measured information is available on such small-scale forms and conclusions provided here are derived partly from measurements of aerial photographs and partly from the published literature. Aerial photographs reveal features with dimensions of tens or hundreds of meters; smaller-scale objects may also be visible, but not measurable (Figure 1.10). Information on objects with lengths of a few meters or less is available only from observations at ground level, usually made during oversnow traverses.

Two principal processes are responsible for creating the small-scale roughnesses considered here. The smallest features are formed primarily by blowing snow; snowfall is redistributed and shaped into irregular waves, ripples, ridges and hummocks which reflect both the prevailing direction and maximum

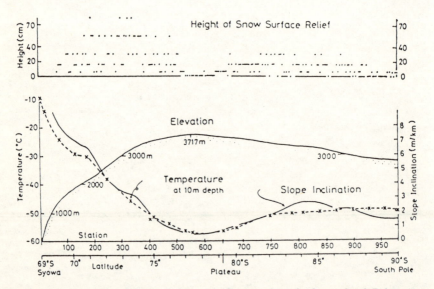

Figure 1.11 Height of small-scale surface snow features along the Syowa-South Pole traverse, East Antarctica. Information on ice sheet altitudes, slope and mean annual surface temperatures are also shown. (From Endo & Fujiwara 1973. Used by permission of the National Institute of Polar Research, Tokyo).

velocity of boundary layer winds, as well as snow properties, particularly crystal size, density and temperature. Such features, at the most general level, may be considered similar to small-scale forms created by blowing sand in arid regions. Dimensions considered typical for these features are from a few centimeters to 1–2 m in height and lengths from a few to several tens of meters.

There is an apparent trend for the height of snow surface features to increase outward from the center of the ice sheet to a maximum, a few hundred kilometers from the coast. Thereafter, they rapidly diminish in height to the ice sheet margin (Fig. 1.11). This pattern accords with the predominant wind strength. Winds are weakest on the low gradients of the ice sheet interior where they are insufficient for the development of pronounced snow forms. Winds are greatest in coastal localities maintaining considerable quantities of snow in motion and inhibiting deposition. High velocity thus allows formation of only the smallest amplitude snowforms.

While recognizing that features at a scale of tens or hundreds of meters will be significant in modulating radar altimeter pulses, it is important to note that there are insufficient measurements available from ice sheets to give representative statistics. Due to the varying degree of 'ice dynamic' control of such features it is likely that the spatial variability of their geometry will be very high.

The ice sheet model

Based upon key satellite altimeter measurements of the Antarctic Ice Sheet outlined in the previous section, it has been possible to construct a model of the ice-sheet surface for evaluating altimeter tracker performance. The model does not attempt to exactly reproduce Antarctic Ice Sheet profiles, rather it incorporates typical and realistic characteristics.

The model is divided into three parts corresponding to those discussed in the previous section, and having horizontal scales in the order of 10^2, 10 and 10^{-1} km respectively. The first and second scales of feature are regional or local ice dynamics, whereas the third is generated by depositional mechanisms in the snow cover. There is, of course, a continuum of features between the very largest and the very smallest.

Large-scale features
Four principal elements are incorporated into the model:

(1)   'normal', quasi-parabolic profile of the form given in Eq. 1.3 and Eq. B1 in Appendix B.
(2)   a cross-flow (y-direction) modifier representing ice streams or outlet glaciers, and given by a sinusoidal function whose amplitude varies down-flow, given by Equations B2, B3 and B4 in Appendix B.
(3)   a fringing ice shelf whose width is 5% of the maximum length of the ice

sheet, and whose height varies from 100 m at the inland edge to 50 m at the seaward margin.

(4)   a vertical cliff at the ice sheet and ice-shelf margin of 50 m height.

The overall equation for the surface is given by Equation 1.4.

$$Z = H(1-P_x^{1.3})^{1/2.7}-[A_x(\cos{(2\pi P_y)}-1]/2 \qquad (1.4)$$

where Z = height of the ice surface, x = x-coordinate of required point (down flow), y = y-coordinate of required point (across flow), L = ice sheet half-width, H = maximum height of ice sheet, A_{max} = amplitude of across-flow sinusoid (Eqs B3 and B4), A_x = amplitude of across-flow sinusoid at x, $P_x = x/L$, $P_y = y/W_y$, and W_y = wavelength of across-flow sinusoid.(See Appendix B).

An isometric view of a typical surface generated by these elements given in Equation 1.4 is shown in Figure 1.12.

Medium scale features (undulations)
Ice surface features within the horizontal range 1–100 km, are modeled as bell-shaped undulations superimposed upon the regional surface. The centers of the undulations are distributed with random offsets from the points on a lattice. The amount of undulation is:

$$dZ = \exp{[(-8x_{off}^2+y_{off}^2)/G]} \qquad (1.5)$$

where dZ = change in surface elevation, x_{off} = x-offset from center of undulation, y_{off} = offset from center of undulation, and G = grid spacing.

Four sets of undulations are used in the model. For each set the amplitude

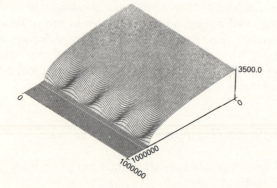

Figure 1.12 Isometric view of an ice sheet of half-width 800 km and maximum elevation of 3500 km as generated by the surface model. 'Ice streams' are shown with a surface amplitude of 1000 m and a spacing of 250 km. Annotations are in meters. No medium or small scale features are shown. Note that there is a fringing ice shelf and associated cliff of 50 m elevation.

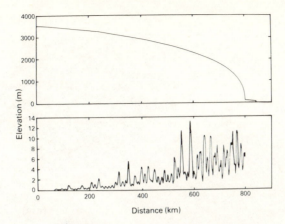

Figure 1.13 Profile of quasiparabolic regional surface profile of the ice sheet (Eq B1) with fringing ice shelf. Also shown is a transect through the combined effects of four types of surface undulation which display a progressive increase in amplitude and decrease in wavelength toward the ice-sheet margin.

increases from the center of the ice sheet to the margin according to some power of the distance from the center (Figure 1.13). If the surface is depressed below the curve given in Equation 1.14 this distance is recalculated using an 'ice stream' component of the model. Isometric views of surfaces created using all four sets of undulations at points along a flowline are shown in Figures 1.14, 1.15 and 1.16. On the fringing ice shelf only one set is used, with a spacing of 10 km and an amplitude of 3 m.

Small scale features
Features such as sastrugi and snow dunes are represented in the model by 'tent-shaped' mounds, randomly scattered over the surface at specified densities. Two sizes of these mounds (6 m and 30 m) have been chosen as representative. Their vertical heights vary linearly with distance from the center of the ice sheet. An isometric view is provided in Figure 1.17.

Acknowledgments

Aspects of this study were carried out under UK Natural Environment Research Council research grant (GR3/4462) and a European Space Agency research contract (5182/82/F/CG(SC)). A full report is available from the authors. We thank C. G. Rapley for his effective coordination of the ESA project. Airborne radio echo sounding data from Antarctica were collected during collaborative missions between the SPRI, U.S. National Science Foundation and the Technical University of Denmark. We thank E. Novotny, M. R. Gorman, and V. A. Squire for their contributions to aspects of this work.

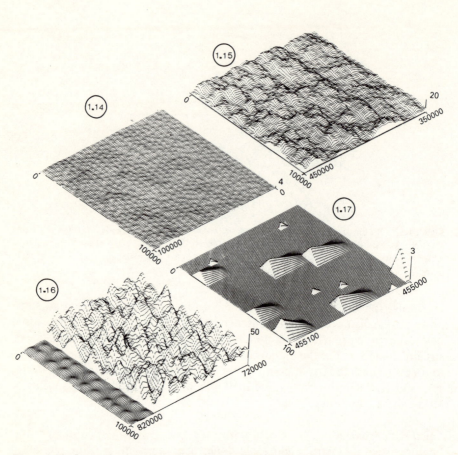

Figure 1.14 Isometric view of medium-scale undulations close to the center of an ice sheet. Annotations are in meters.

Figure 1.15 Isometric view of medium-scale undulations half way between ice sheet center and margin. Annotations are in meters.

Figure 1.16 Isometric view of medium-scale undulations at the margins of an ice sheet. A fringing ice shelf is shown with smaller-scale irregularities.

Figure 1.17 Isometric view of small-scale surface irregularities. Annotations are in meters.

Appendix A

Characteristics of ERS-1 Satellite Radar Altimeter

Orbit	Mean altitude	777 km
	Inclination of orbit	098 deg
	Period	100.46 min
Waveform	Type	Linear FM (Chirp)
	Center frequency	13.7 GHz
	Bandwidth	300 MHz
	Uncompressed pulse width	c.5.0 s
	Compressed pulse width	3.0 ns
	Pulse repetition frequency	c.1000 Hz
Antenna	Type	1 m
	Beamwidth	1.2 degrees
Footprint	Pulse limited diameter	1.6 km
	Beam limited diameter	21.0 km

Appendix B

Equations used in generating the ice surface model, Equation 1.4. Variables are defined in Equations 1.2 and 1.4.
Equation for normal, quasiparabolic regional ice-surface profile:

$$(X/L)^{1.3} + (Z/H)^{2.7} = 1 \tag{B1}$$

If $P_x < 0.75$

$$A_x = 0 \tag{B2}$$

If $P_x \geq 0.75$ and $P_x < 0.95$

$$A_x = A_{max} \, 1-\cos(5 \, (P_x-0.75))/2 \tag{B3}$$

If $P_x \geq 0.95$

$$A_x = A_{max} \, (1-400(P_x - 0.95)^2) \tag{B4}$$

References

Bentley, C. R. 1981. *Response of the West Antarctic Ice Sheet to CO_2-induced climatic warming: a research plan*. Report of Workshop sponsored by American Association for the Advancement of Science and U.S. Dept. of Energy, Orono, Maine 1980.

Brenner, A. C., R. A. Bindschadler, R. H. Thomas and H. J. Zwally 1983. Slope-induced errors in radar altimetry over continental ice sheets. *J. Geophys. Res.* 88, 1617–23.

Budd, W. F. 1968. The longitudinal velocity profile of large ice masses. *Intl. Assoc. Sci. Hyd. Pub.* 79, 58–75.

Budd, W. F. and I. N. G. Smith 1982. Large-scale numerical modelling of the Antarctic ice sheet. *Ann. Glaciol.* 3, 42–49.

Cooper A. P. R., N. F. McIntyre and G. de Q. Robin 1982. Driving stresses in the Antarctic ice sheet. *Ann. Glaciol* 3, 59–64.

Drewry, D. J. 1982. Antarctica unveiled. *New Scientist* 95 (1315), 246–51.

Drewry, D. J. 1983a. Antarctic ice sheet: aspects of current configuration and flow. In *Megageomorphology*, V. Gardiner and H. Skoging (eds), 18–38, Oxford: Oxford University Press.

Drewry, D. J. 1983b. Surface configuration of the Antarctic ice sheet. In *Antarctica: Glaciological and Geophysical Folio*, D. J. Drewry (ed.), Sheet 2. Cambridge: Scott Polar Research Institute.

Drewry, D. J., D. T. Meldrum and E. Jankowski 1980. Radio echo and magnetic sounding of the Antarctic ice sheet, 1978–79. *Polar Record* 20(124), 43–51.

Drewry, D. J., S. R. Jordon and E. J. Jankowski 1982. Measured properties of the Antarctic ice sheet: surface configuration, ice thickness, volume and bedrock characteristics. *Ann. Glaciol.* 3, 83–91.

Endo, Y. and K. Fujiwara 1973. Characteristics of the snow cover in East Antarctica along the route of the JARE South Pole Traverse and factors controlling such characteristics. *Jap. Ant. Res. Exped. Sci. Repts. Ser. C*, No. 7, 4–31, App. 11p.

MacArthur, J. L. 1978. Seasat-A radar altimeter design description, Applied Physics Laboratory SDO-5232. Baltimore: Johns Hopkins University.

McIntyre, N. F. 1983. Topography and flow of the Antarctic ice sheet. Ph.D. Thesis, Cambridge University.

McIntyre, N. F. and A. P. R. Cooper 1983. Driving stresses within the Antarctic ice sheet. In *Antarctica: Glaciological and Geophysical Folio*, D. J. Drewry (ed.), Sheet 5. Cambridge: Scott Polar Research Institute.

Mae, S. and R. Naruse 1978. Possible cause of ice sheet thinning in the Mizuho Plateau. *Nature* 273, 291–2.

Martin, T. V., H. J. Zwally, A. C. Brenner and R. A. Bindschadler 1983. Analysis and retracking of continental ice sheet radar altimeter waveforms. *J. Geophys. Res.* 88, 1608–16.

Morgan, V. I. and T. H. Jacka 1981. Mass balance studies in East Antarctica. *Intl. Assoc. Sci. Hyd. Publ.* 131, 253–60.

NERC-SERC 1983. *A study of ocean and ice topography from space: a joint NERC-SERC proposal*. Chilton: Rutherford Appleton Laboratory.

Oswald, G. K. A. and G. de Q. Robin 1973. Lakes beneath the Antarctic ice sheet. *Nature*, 245, 251–4.

Parish, T. R. 1981. The katabatic winds of Cape Denison and Port Martin. *Polar Rec.* 20, 525–33.

Paterson, W. S. B. 1981. *The physics of glaciers*, 2nd edn. Oxford: Pergamon.

Rapley, C. G., H. D. Griffiths, V. A. Squire, M. Lefebvre, A. R. Birks, A. C. Brenner, C. Brossier, L. D. Clifford, A. P. R. Cooper, A. M. Cowan, D. J. Drewry, M. R. Gorman, H. E. Huckle, P. A. Lamb, T. V. Martin, N. McIntyre, K. Milner, E. Novotny, G. E. Peckham, C. Schgounn, R. F. Scott, R. H. Thomas and J. F. Vesecky 1983. *A study of satellite radar altimeter operation over ice-covered surfaces*. European Space Agency Report 5182/82/F/CG(SC).

Robin, G. de Q. 1967. Surface topography of ice sheets. *Nature*, 215, 1029–32.

Robin, G. de Q. (ed.) 1983. *The climatic record in polar ice sheets*. Cambridge: Cambridge University Press.

Robin, G. de Q., D. J. Drewry and D. T. Meldrum 1977. International studies of ice sheet and bedrock. *Phil. Trans. R. Soc. London Ser. B* 279, 185–96.

Robin, G. de Q., D. J. Drewry and V. A. Squire 1983. Satellite observations of polar ice fields. *Phil. Trans. R. Soc. London, Ser. A* 309, 447–61.

Swithinbank, C. W. M. 1983. Towards an inventory of the great ice sheets. *Geografiska Ann.* **65A,** 289–94.

Thomas, R. H., T. V. Martin and H. J. Zwally 1983. Mapping ice-sheet margins from radar altimetry data. *Ann. Glaciol.* 4, 283–88.

Warren, S. G. 1982. Ice and climate modelling: an editorial essay. *Climatic Change* 4, 329–40.

Zwally, H. J., R. A. Bindschadler, A. C. Brenner, T. V. Martin and R. H. Thomas 1983. Surface elevation contours of Greenland and Antarctic Ice Sheets. *J. Geophys. Res.* 88, 1589–96.

2

The Antarctic Ice Sheet: an analog for Northern Hemisphere paleo-ice sheets?

T. J. Hughes, G. H. Denton, and J. L. Fastook

Introduction

Features and processes of the Antarctic Ice Sheet have been studied in some detail since International Geophysical Year activities in 1957–58. The size and complexity of the Antarctic Ice Sheet make it a natural laboratory for understanding the behavior of Quaternary paleo-ice sheets in the Northern Hemisphere. Because the Antarctic Ice Sheet has unique geography its configuration is not necessarily an analog to former Northern Hemisphere ice sheets. Rather, we believe that important glaciological features and processes of the Antarctic Ice Sheet are common to ice sheets in general and thus are analogs to paleo-ice sheets of the Northern Hemisphere. In this paper, we draw from our understanding of the Antarctic Ice Sheet to suggest such analogs. We then suggest how glaciological processes can interact to control global ice-sheet fluctuations during a glacial cycle. Figures 2.1, and 2.2 and 2.15 are identification maps for the Arctic and the Antarctic to show where analogies discussed in the following sections can be drawn. Figures 2.12 and 2.13 may also be consulted for place names.

Terrestrial and marine components

For the purposes of this paper, we shall consider an ice sheet to be a continent-size slab of ice thick enough to spread under its own weight. From Antarctic observations, we know that an ice sheet can have grounded marine and terrestrial components as well as floating ice shelves. Marine components are grounded largely below coeval sea level and are drained primarily by seaward flow through marine ice streams; they are generally vulnerable to the marine instability mechanisms modeled by Stuiver *et al.* (1981). Terrestrial components are grounded largely above coeval sea level; they are not vulner-

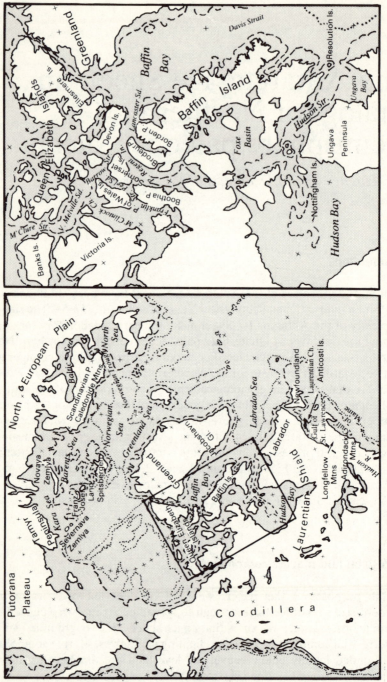

Figure 2.1 Arctic identification map. Dashed lines denote 200 m and 1000 m bathymetric contours.

able to the marine instability mechanisms, even though portions may be grounded below coeval sea level.

Antarctica

The Antarctic Ice Sheet, shown in Figure 2.3 for today (Drewry *et al.* 1982) and in Figure 2.4 for 17000–21000 a BP, rests in deep subglacial basins and over high subglacial plateaus. Interior ice thicknesses currently range from 4300 m above the floor of Bentley Subglacial Trench, 2500 m below sea level, to 1600 m above the summits of the Gamburtsev mountains, 2500 m above sea level. On the basis of differences in basal and surface topography, the Antarctic Ice Sheet can be divided into ice sheets in East and West Antarctica. The Transantarctic Mountains separate these ice sheets.

Although it is grounded over several subglacial basins, the core of the East Antarctic Ice Sheet is largely terrestrial. Most subglacial basins would be above sea level if the East Antarctic Ice Sheet were removed and isostatic rebound of the bed were complete (Bentley 1972, Drewry 1983). Geologic studies imply stability of East Antarctic ice through late Quaternary glacial cycles (Stuiver *et al.* 1981, Denton *et al.* 1971, Mercer 1968a, b). Those portions of the East Antarctic Ice Sheet covering isostatically depressed subglacial basins conceivably could have collapsed in response to marine instability mechanisms triggered when the grounding line attains a critical depth below sea level and lies on a bed that slopes downward into the subglacial basins. The fact that the ice sheet is intact, and apparently remained so during late Quaternary interglaciations, implies either that these conditions of bed depth and bed slope were not satisfied over that interval or that these subglacial basins had seaward sills that were near or above sea level. However, the larger Antarctic Ice Sheet postulated by Denton, Prentice, *et al.* (1984) in Neogene time may have caused the additional isostatic depression necessary to lower some sills enough to give some portions of the East Antarctic Ice Sheet an unstable marine-based character. If so, it follows that the East Antarctic Ice Sheet is terrestrial for present and late Wisconsin ice loads, but that portions can be marine for larger ice loads. For example, added isostatic depression might have allowed the marine instability mechanisms to operate during deglaciation in marine ice streams draining seaward from Wilkes Subglacial Basin. One such ice stream would have entered the embayment now occupied by the Cook Ice shelf, and Ninnis Glacier may have been another. David and Byrd Glacier ice streams could have aided in evacuating the basin. Similarly, Aurora Subglacial Basin possibly could have been deglaciated by marine instability mechanisms acting along Totten and Vanderford Glaciers, and ice draining into the Filchner Ice Shelf through ice streams on either side of Dufek Massif may have triggered marine instability over the Polar Subglacial Basin.

The periphery of the East Antarctic Ice Sheet is now drained by terrestrial ice streams, defined as ice streams on a bed that slopes uphill inland, and is commonly fringed by ice shelves. Only one of these ice streams, the partly marine Lambert Glacier, reaches deeply into the ice sheet (Allison 1979).

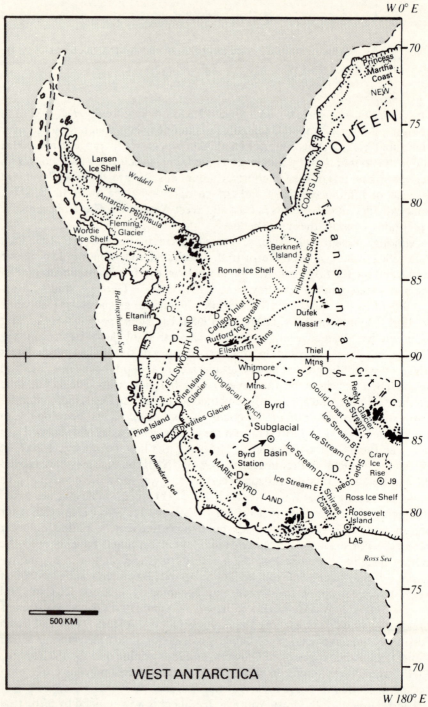

Figure 2.2 Antarctic identification map. Dashed outer line denotes 500 m bathymetric contour, dashed interior lines denote ice divides with domes D and saddles S, solid lines denote tidewater ice sheet margins, hachured lines denote ice shelf calving margins, dotted lines denote ice shelf grounding lines, black areas denote mountains above the ice sheet.

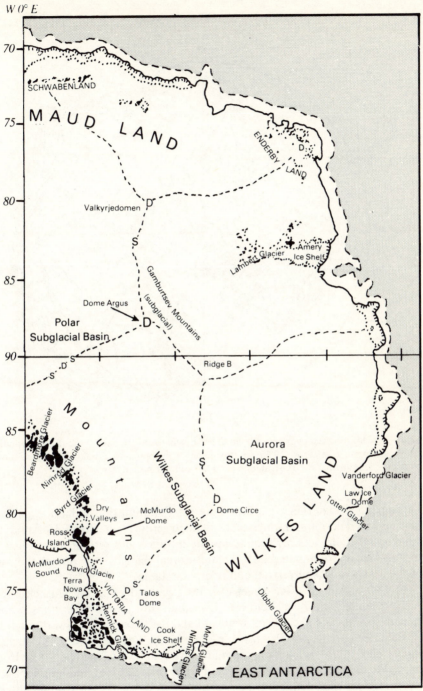

70

SCHWABENLAND

M A U D   L A N D

75

ENDERBY
LAND

D

80

Valkyrjedomen   D

S

Lambert Glacier   Amery
                   Ice Shelf

85

Gamburtsev Mountains (subglacial)

Dome Argus

Polar
Subglacial Basin   D

90   S   D   Ridge B

S   D

85

Beardmore Glacier

M
o
u
n
t
a
i
n
s

Aurora
Subglacial Basin

Nimrod Glacier

Wilkes Subglacial Basin   Vanderford Glacier

S   Law Ice
                                   Dome
Byrd Glacier

80   Dry   McMurdo   D   Dome Circe   Totten Glacier
        Valleys   Dome

Ross
Island

W I L K E S   L A N D

McMurdo
Sound   David Glacier
                    S
Terra   D   Talos
75 Nova   Dome
    Bay   VICTORIA
             LAND   Dibble Glacier
    Rennick Glacier   Cook
                           Ice Shelf

Ninnis Glacier
Mertz Glacier

70   EAST ANTARCTICA

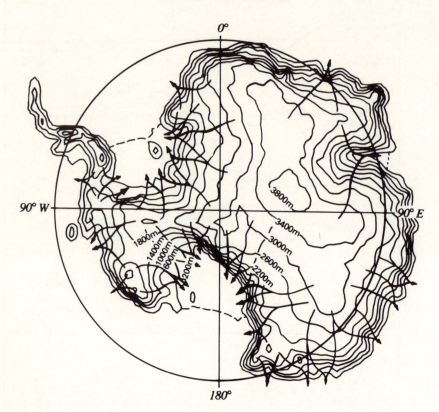

Figure 2.3   Ice surface topography and major ice streams in present-day Antarctica. Elevations in meters are from Drewry *et al.* (1982). Ice streams and their approximate drainage areas are shown by arrows with forked tails. (Orientation error not author's responsibility).

During late Quaternary glaciations, East Antarctic ice fed an expanded marine-based peripheral fringe that extended across the narrow East Antarctic continental shelf. Hence, grounding-line advance and retreat, controlled in turn by eustasy, dominated East Antarctic areal changes even though the core of the ice sheet was terrestrial and relatively stable.

The West Antarctic Ice Sheet overlies a generally rugged bedrock floor, much of which is now well below sea level and, with several exceptions, would remain so if the ice sheet were removed and the floor were allowed to adjust isostatically (Bentley & Ostenso 1961, p. 892–95, Drewry 1983). The surface topography of the ice sheet is irregular, owing in part to the rugged bedrock floor. Floating ice shelves in protected embayments fringe the ice sheet and marine ice streams, defined as ice streams on a bed that slopes downhill inland, control most seaward ice drainage (Hughes 1977). Mercer (1968a) identified the West Antarctic Ice Sheet as a marine ice sheet, as opposed to the predominantly terrestrial ice sheet in East Antarctica. Both geologic and glaciologic investigations indicate that this marine ice sheet can be unstable and that its fluctuations are controlled largely by sea level (Hollin 1962, Denton *et al*. 1971, Weertman 1974, Thomas & Bentley 1978, Stuiver *et al*. 1981). Today, thick marine-based ice is still grounded over deep interior subglacial basins. Island groups enclose much of this core of grounded marine ice and help to keep it from collapsing. The large pinned ice shelves floating in the Ross Sea and Weddell Sea embayments aid in preserving the West Antarctic Ice Sheet, for they buttress seaward-flowing marine ice streams that drain unenclosed marine portions. In fact, Mercer (1968a, 1978) postulated that recession of these ice shelves during particularly warm interglaciations would lead to rapid collapse of grounded marine ice in West Antarctica.

Northern Hemisphere
Figure 2.5 is a sketch of Northern Hemisphere ice sheets at the maximum of late Wisconsin glaciation (Hughes *et al*. 1981). The extent of these ice sheets is controversial (Denton & Hughes 1981a). Particularly at issue is the existence of ice sheets in the Arctic (Blake 1970, England 1976, Grosswald 1980, Boulton *et al*. 1982). As shown in Figure 2.5 we follow Blake (1970), Schytt *et al*. (1968), and Grosswald (1980) in placing extensive ice sheets in Arctic Canada, as well as on the continental shelves now occupied by the Barents and Kara Seas of northern Eurasia. Our selection is central to the discussion that follows. There is also lively debate about the placement of ice-sheet margins elsewhere (Denton & Hughes 1981a, Andrews 1982), but their exact position is not central to the suggestions in this paper. Therefore, Figure 2.5 should not be taken to represent these margins exactly.

We think that Northern Hemisphere ice sheets depicted in Figure 2.5 had terrestrial and marine components. Grounded marine components existed in the Queen Elizabeth Islands (Innuitian Ice Sheet) of Arctic Canada and in the Barents and Kara Seas of Arctic Eurasia. The Laurentide Ice Sheet had southern and western terrestrial margins, but had extensive marine-based components in its interior and Arctic sectors. The Scandinavian Ice Sheet was largely terrestrial, except along its Atlantic fringe but had a subglacial basin in the Baltic Sea much like those in present-day East Antarctica. The Greenland Ice Sheet was largely terrestrial, although it had marine-based fringes. The Cordilleran Ice Sheet (or intermontane glacier complex) was largely terrestrial, as were smaller glaciers. Overall, the Northern Hemisphere ice-sheet complex was largely terrestrial along its southerly margins and largely marine in the Arctic.

There were major differences and similarities in the geographic arrangement of marine and terrestrial components in Antarctica and the Northern Hemisphere at the maximum of late Wisconsin glaciation. Northern Hemisphere ice sheets could reach far enough south to attain extensive melting margins, which are absent on the Antarctic Ice Sheet. Mass balance was controlled by surface melting along the southern margin of the former, and is controlled by iceberg carving along all margins of the latter. The importance of this distinction will be made clear later in the section on bimodal versus trimodal responses of these

(a)

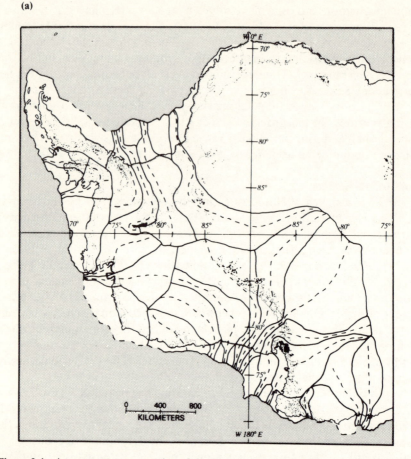

Figure 2.4 A computer reconstruction of the West Antarctic Ice Sheet at 17 000–21 000 a BP. Ice-surface elevations were computed along flowlines at 100 km intervals shown as broken lines in (a) for ice-stream drainage systems enclosed by heavy solid lines, and these elevations are shown at 500 m contour intervals referred to present-day sea level in (b). Snow accumulation data for this reconstruction was taken from Bull (1971). The reconstruction is compatible with known glacial geologic field data.

ice sheets. Further, the Northern Hemisphere ice sheets were arrayed with a central core of floating ice shelf or sea ice, fringed largely by marine and then terrestrial components; just the reverse of the Antarctic Ice Sheet. Thus the core of the Northern Hemisphere ice-sheet complex was marine whereas the core of the Antarctic Ice Sheet was and is terrestrial.

The best geographical comparison to the predominantly marine West Antarctic Ice Sheet is the Eurasian Arctic continental shelf from the Taymyr Peninsula to Scandinavia (Hughes 1982b). Novaya Zemlya is a massif that divides the Eurasian continental shelf into the Kara Sea embayment to the east and the larger Barents Sea embayment to the west, just as the Ellsworth

(b)

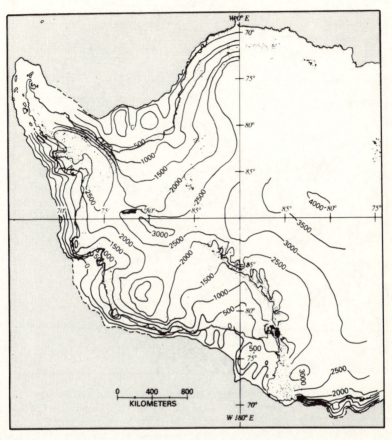

Mountains and subglacial highlands divide the Antarctic continental shelf into the Weddell Sea embayment to the east and the larger Ross Sea embayment to the west. Byrd Subglacial Basin is that part of the Ross Sea embayment over which the West Antarctic ice sheet is still grounded. The 300 m deep East Novaya Zemlya Trough lies alongside Novaya Zemlya, just as Bentley Subglacial Trench lies alongside the Ellsworth Mountains. Island groups border the northern ends of both continental shelves. From east to west, there are Severnaya Zemlya, Franz Josef Land, and Spitzbergen in Eurasia; and the mountain and island groups of the Antarctic Peninsula, Ellsworth Land, and Marie Byrd Land in Antarctica.

If a marine ice sheet covered the Eurasian Arctic continental shelf between the Taymyr Peninsula and Scandinavia during the late Weichselian glaciation, it would be most vulnerable to the marine disintegration mechanisms in the wide gaps between Severnaya Zemlya and Franz Josef Land in the Kara Sea and between Spitzbergen and Scandinavia in the Barents Sea. We postulate that the West Antarctic Ice Sheet during this glaciation was vulnerable to the

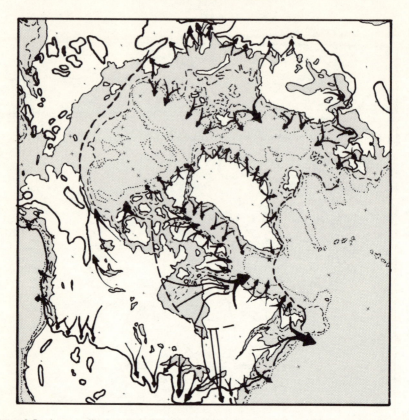

Figure 2.5 A generalized outline of Northern Hemisphere ice sheets at 17 000–21 000 a BP. Solid lines show approximate limits of terrestrial ice sheets and mountain glaciers. Dashed lines show approximate limits of floating ice shelves and perennial sea ice. Arrows with forked tails suggest all potential major ice streams and their ice drainage areas. Note ice divide traversing Hudson Bay. Compare to Figure 2.1.

marine disintegration mechanisms in the wide gaps between Coats Land and the Antarctic Peninsula in the Weddell Sea and between Marie Byrd Land and Victoria Land in the Ross Sea (Stuiver *et al.* 1981). The Filchner/Ronne and Ross Ice Shelves today are the collapsed and floating remnants of the marine West Antarctic Ice Sheet that had been grounded on the Antarctic continental shelf in the Weddell and Ross Seas as late as 17 000 a BP (Stuiver *et al.* 1981).

The closest Northern Hemisphere comparison to the geographical setting of the predominantly terrestrial East Antarctic Ice Sheet is Europe, from Scandinavia across the Baltic Sea to the North European Plain, areas covered by the largely terrestrial Scandinavian Ice Sheet. The Caledonide Mountains were the seaward barrier to this ice sheet, although large ice streams passed through the mountains and onto the Norwegian continental shelf (Anderson 1981). In analogous situations today, the Transantarctic Mountains and the coastal mountains of Queen Maude Land are seaward barriers to the East Antarctic

Ice Sheet, and are traversed by outlet glaciers which coalesce to form a marine fringe of floating or grounded ice. The Baltic Sea lies behind the Caledonide Mountains and is a rebounding basin that had been isostatically depressed below sea level by the late Weichselian Scandinavian Ice Sheet. An analogous situation today is Wilkes Subglacial Basin, which lies behind the Transantarctic Mountains and is isostatically depressed well below sea level by the East Antarctic Ice Sheet.

The interior Laurentide Ice Sheet had late Wisconsin subglacial basins in Hudson Bay and Foxe Basin that were vulnerable to the marine instability mechanisms acting on a marine ice stream in Hudson Strait. During deglaciation, the grounding line was able to migrate over the bedrock sill at the Labrador Sea entrance to Hudson Strait, and this triggered marine instability. The East Antarctic Ice Sheet is intact today because the ice-stream sills stabilizing its subglacial basins were high enough to block grounding-line retreat.

Ice streams and downdraw

Ice streams are fast currents of ice, typically up to a few hundred km long and under 50 km wide, that develop within an ice sheet near its margin. They begin where ice flow is strongly convergent; they commonly develop striking zones of shear along their sides, and the fastest ones are heavily crevassed. Both terrestrial and marine ice streams exist, the difference being that marine ice streams are grounded farther below sea level on a bed that slopes downhill into marine subglacial basins. Perhaps the most important new perspective that Antarctica provides on the behavior of ice sheets concerns ice streams, since they are the vehicles through which marine instability mechanisms operate to collapse marine ice sheets.

Antarctica

Radar sounding and balloon altimetry traverses over the Antarctic Ice Sheet allow detailed contour maps of ice elevation (Drewry *et al.* 1982, Drewry 1983). Ice flowlines can be drawn perpendicular to surface contour lines, since the force of gravity is greatest in the maximum downslope direction. Most of these flowlines dovetail into numerous ice streams that develop near grounded margins of the ice sheet; thus interior sheet flow typically becomes stream flow toward the margin (Fig. 2.3). These ice streams are the most dynamic component of the ice sheet, they drain most of the ice, and the marine ice streams are vulnerable to the marine instability mechanisms (Hughes 1973, 1975, 1977, Rose 1979, Robin *et al.* 1970, Stuiver *et al.* 1981). Stream flow is usually much faster than sheet-flow. Antarctic ice streams become heavily crevassed when the surface velocity exceeds about 500 m/a as seen in Figure 2.6 (Swithinbank 1964, 1977). This figure compares the rough surface of ice stream B, moving over 500 m/a (Thomas 1976) with the smooth surface of Rutford Ice Stream, moving 450 m/a (Stephenson & Doake 1982).

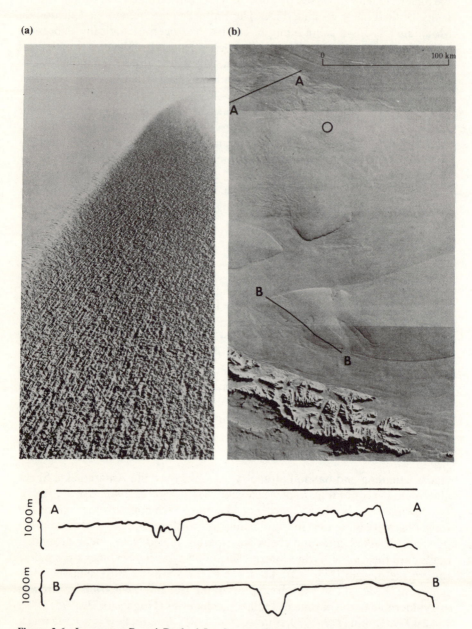

Figure 2.6 Ice stream B and Rutford Ice Stream. Heavy crevassing in ice stream B (left) compared to Rutford Ice Stream (right) is a result of its higher velocity (500 vs 450 m/a) and reflects the sharp thermal boundary betwen ice sliding over the bed in the ice stream and lateral ice frozen to the bed. Ice thickness profiles are shown along transects A–A and B–B in (b) to show that the ice streams flank plateaus that are dissected by fjord-like features. No such subglacial topography exists in (a). (From Swithinbank 1977. Used by permission of the Royal Society, London).

Figure 2.7 Landsat image of Byrd Glacier as it enters the Ross Ice Shelf. Diverging lateral rifts show the buttressing capacity of the ice shelf. Lateral shear zones do not coincide with the rock sidewalls in Byrd Glacier fjord. Darwin Glacier (center) and Mulock Glacier (lower left) are also shown. ERTS E-1542-18435-701, 16 January 1974.

Shallow elongated troughs—probably largely structural in origin—extend from the grounding lines of many present-day Antarctic ice streams to the edge of the Antarctic continental shelf, where the troughs commonly have sills. These troughs are particularly well mapped in the Ross Sea and beneath the Ross Ice Shelf (Bentley & Jezek 1981). We infer that the ice streams occupied these troughs when the grounding lines were perched on the sills at the edge of the continental shelf, and that the grounding lines have subsequently retreated

Figure 2.8   The lateral shear zone of Byrd Glacier. The abrupt truncation of local Merrick Glacier ice drained from the Transantarctic Mountains and the smooth ice surface between the fjord sidewall and the shear zone shows that basal sliding and erosion by Byrd Glacier begins well out into the fjord channel.

into the troughs behind the sills. However, radar sounding across present-day ice streams does not always detect a trough, and the bed under the ice stream can be rough (Rose 1979). For example, Rutford Ice Stream apparently occupies a tectonic trough, but no trough lies beneath ice stream B. Other ice streams, particularly those that drain the East Antarctic Ice Sheet, occupy smooth troughs that are commonly fjords with structural control. Byrd Glacier ice stream, shown in Figure 2.7, occupies one such fjord through the Transantarctic Mountains. Stream-flow within Byrd Glacier begins well out from the fjord wall, as seen in Figures 2.7 and 2.8. Glacial sliding along the fjord wall, and hence striations as down-fjord flow indicators, occur on rocks below sea level, at the base of the lateral shear zones, which are some distance from the fjord wall. The sharp side boundaries of ice streams that are evident in Figures 2.6 through 2.8 must lie above boundaries between thawed and frozen beds, since ice-stream velocities imply basal sliding and there are no basal topographic features along these side boundaries that would channel ice flow.

We infer that marine ice streams can migrate if they lie on a bed that slopes downward into a subglacial basin and if they are not confined by steep fjord

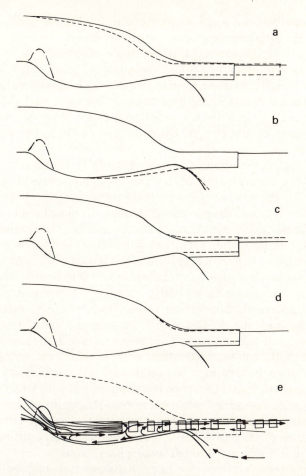

Figure 2.9 Marine instability mechanisms. For explanation, see text. (From Denton and Hughes 1981. Copyright by John Wiley. Used by permission.)

sidewalls. Figure 2.9 illustrates our concept of the marine instability mechanisms for the general case of a marine ice stream whose grounding line is just beyond a sill at the end of a trough. For further background on the marine instability mechanisms, see Weertman (1974), Thomas and Bentley (1978), Thomas (1979) and Denton and Hughes (1981a). The grounding line will move to the top of the sill as (Fig 2.9a) longitudinal creep of the floating ice stream tongue causes vertical ice thinning, (b) on-going isostatic sinking of the trough beneath the ice stream lowers the sill to the depth of the grounding line, (c) rising sea level raises the grounding line to the height of the sill, and (d) ablation at the surface and base of the floating ice tongue causes vertical ice thinning. Responses (a) through (d) are reversible so long as the grounding line moves upslope to the top of the sill. However, any further response moves the grounding line to the downslope side of the sill. The grounding line will then retreat irreversibly until it again becomes anchored to the uphill slope at the rear end of the foredeepened trough, as shown in (e).

The surface of the ice sheet is lowered substantially as an ice-stream grounding line retreats along a trough, as shown in Figure 2.9 (e). This reduces the gravitational potential energy of the ice sheet, which reduces the discharge velocity of the ice stream. Iceberg calving will then carve back the floating ice stream tongue faster than ice is supplied to the calving margin, and a calving bay will migrate up the trough. Estuarine circulation in the trough ferries icebergs seaward.

In the absence of responses (a) through (d) in Figure 2.9, interior ice transported to a marine margin of an ice sheet can thicken·ice at the grounding line. This forces the grounding line to move into deeper water beyond the sill at the end of the channel. Responses (a) through (d) must therefore overcome this ice thickening due to advective ice transport before the grounding line will retreat. When the grounding line lies atop the sill, and one of the responses (a) through (d) triggers an irreversible retreat, this retreat will be greatly slowed as the interior surface lowering shown in (e) causes ice to be advected across the grounding line at a rate in excess of the rate needed to maintain mass-balance equilibrium. Therefore, during initial surface lowering the increased ice advection rate may push the floating ice stream tongue forward faster than the iceberg calving rate carves it back. The ice sheet, taken as the sum of its ground and floating parts, may then actually increase in area at the very time that its interior elevation is lowering. We think that this combination of interior lowering and possible areal advance is accomplished by ice streams. We refer to this process as 'downdraw' when interior ice-sheet collapse is a result of ice being downdrawn into ice streams faster than it is replaced by surface accumulation (Denton & Hughes 1981b). Downdraw can be particularly rapid if interior ice drains seaward through surging ice streams.

From these considerations, we infer that downdraw may be a dominant process during the first part of a deglaciation hemi-cycle, and may take place with an expansion of floating marine margins of an ice sheet accompanied by little if any retreat of terrestrial margins, which are not affected by the marine instability mechanisms until downdraw shifts interior ice divides, thereby upsetting the terrestrial mass balance. In the second part of a deglaciation hemi-cycle, downdraw has lowered the ice sheet surface so much that the gravitational potential energy cannot maintain ice-stream discharge velocities that match or exceed iceberg calving rates. The floating ice shelves that were created by the collapse of marine ice downdrawn into ice streams are then carved away until the iceberg calving rate once again matches the ice discharge velocity. Our field observations and modeling studies suggest to us that these processes have produced the present-day Ross Ice Shelf in Antarctica, including the relatively stable position of its calving front (Stuiver et al. 1981). We suspect that the same conclusion holds for other Antarctic ice shelves, but field work has yet to be undertaken to test this suggestion.

An ice stream loses its marine character when its grounding line lies on an uphill slope in water much shallower than the depth needed to float the ice. Retreat of these terrestrial ice streams is by iceberg calving until the calving front retreats beyond the tidewater zone (Iken 1977, Fastook & Schmidt 1982, Sikona 1982). Further retreat is by surface ablation. Antarctic examples of terrestrial ice streams occur along tidewater calving margins, particularly in East Antarctica. They play a minor role in the dynamics of the Antarctic Ice Sheet.

A surge is a sudden increase in ice velocity that transforms an ice surface into a jumbled mass of crevasses and seracs (Paterson 1981). Many, but not all, Antarctic ice streams have this appearance (Swithinbank 1977). From Antarctic field observations, we suggest that the surge condition of an ice sheet is restricted to its ice streams (Hughes 1973, 1975, 1977). All ice streams need not surge continuously. Surging ice streams imbedded in a pinned and confined ice shelf inject thick tongues of floating ice into the ice shelf (Hughes 1975, Bentley et al. 1979). One such tongue exists in the Ross Ice Shelf beyond ice stream C. Although this ice stream is now nearly stagnant, it must have surged at some time in the past (Hughes 1975). A surging ice stream that is unconfined produces a floating tongue that may break up in shallow water beyond the ice-stream grounding line. The huge Dibble Iceberg Tongue lies beyond Dibble Glacier, which is now small and relatively inactive, and suggests that this ice stream surges. We therefore downplay theories of Northern Hemisphere deglaciation that depend on a massive surge of an entire ice sheet, but we assign an important role to surges of major ice streams. Such surges can be particularly effective in downdrawing marine-based interior ice (Denton & Hughes 1981b, 1983).

Northern Hemisphere

Present-day Northern Hemisphere terrestrial ice streams occur in Greenland. Unlike their Antarctic counterparts, the Greenland terrestrial ice streams play an important role in ice-sheet dynamics. All of them terminate below the surface equilibrium line, most in shallow water where the tidewater calving mechanism operates. Consequently, ablation of terrestrial ice streams that drain much of the Greenland Ice Sheet is typically by both surface melting and iceberg calving. Ablation over the remainder of the ice sheet occurs along margins where sheet-flow terminates on land.

The only present-day Northern Hemisphere ice stream that might be partly marine is Jakobshavns Glacier, which drains about five percent of the Greenland Ice Sheet and has a floating terminus in Jakobshavns Isfjord (Lingle et al. 1981). At 8000 m/a, Jakobshavns Glacier moves three times faster than the fastest-known Antarctic ice stream (Thwaites Glacier) and, like terrestrial ice streams in Greenland, experiences massive surface melting during the summer. The surface is heavily crevassed, as shown in Figure 2.10 where sheet-flow becomes stream-flow. Surface meltwater passes through these crevasses and may lubricate the bed in the manner proposed by Weertman (1973). This phenomenon does not occur in Antarctic ice streams at present, at

Figure 2.10   The beginning of Jacobshavns Glacier ice stream. Ice is pouring over the headwall of Jakobshavns Isfjord, with strongly converging flow, extensive crevassing, and substantial melting of surface and basal ice to produce the world's fastest ice stream (8000 m/a). Photograph by Richard L. Cameron.

least not on the scale observed on Jakobshavns Glacier. Moreover, the grounding line of Jakobshavns Glacier cannot retreat beyond the headwall of Jakobshavns Isfjord, where the ice stream would lose any marine character it now has.

We have seen that numerous ice streams form important dynamic components of the Antarctic Ice Sheet, particularly of marine-based West Antarctic ice (Hughes 1977, Stuiver *et al.* 1981). Yet their significance in Northern Hemisphere paleo-ice sheets have been virtually ignored. Recently we postulated an important role for ice streams in the dynamics of Northern Hemisphere ice sheets (Denton & Hughes 1981a,b). Part of our problem was recognizing the location of former ice streams in the Northern Hemisphere from geologic evidence, particularly if they can migrate as suggested by Antarctic studies. As a first approximation we postulated the locations of paleo-ice streams largely by trough topography, (Denton & Hughes 1981b), despite the fact that many Antarctic ice streams do not occupy troughs. We did this because thin ice will stream into topographic troughs and, since ice is thin near the margin of an ice sheet, ice streams will develop in troughs that point toward the margin. This is common in both Greenland and Antarctica today. Therefore, it is reasonable to suggest that paleo-ice streams occupied marginal troughs of Northern Hemisphere ice sheets. However, it should be clear from Antarctic evidence that all paleo-ice

streams need not have followed troughs. Moreover, troughs oriented trans-
verse to ice-flow directions were not occupied by ice streams.

Paleo-ice streams are difficult to reconstruct from geologic evidence. Ice
distribution is a common problem. For example, there are two schools of
thought concerning the presence of marine ice sheets grounded on Arctic
continental shelves during the last glacial maximum, let alone the existence of
ice streams within these marine components (Denton & Hughes, 1981a,
Andrews 1982). One of our most important Antarctic analogs which is also
supported in Greenland, is our view that Northern Hemisphere ice sheets were
drained largely by ice streams wherever troughs now exist across high-latitude
continental shelves. As seen in Figure 2.5, these troughs are numerous.

Another problem concerns interpretation of striations in terms of ice
movement and possible ice streams. As an example, take Hudson Strait, which
we suggested was the funnel for a huge ice stream that drained interior
Laurentide ice (Denton & Hughes 1981b). Blake (1966) suggested that an
outlet glacier of the Laurentide Ice Sheet occupied Hudson Strait during the
last glaciation. Moreover, Sugden (1977) showed major ice drainage through
Hudson Strait. In sharp contrast, Osterman et al. (1982) postulated that late
Wisconsin ice from Labrador flowed diagonally across the mouth of Hudson
Strait, which would preclude an ice stream draining interior Laurentide ice
(Andrews & Miller 1983). From this evidence Andrews and Miller (1983, p. 10)
denied any major downdraw of ice through Hudson Strait in late Wisconsin
time. Also, striations perpendicular to the north shore of Hudson Strait (Blake
1966) on first glance seem to preclude an ice stream.

We think that care should be exercised in denying a late Wisconsin ice stream
in Hudson Strait. Striations nearly perpendicular to the strait can be expected
on lateral walls, as shown by Antarctic ice streams. In our previous example,
Byrd Glacier has lateral shear zones that are well out from the fjord walls but
local ice in Merrick Glacier moves perpendicular to the fjord wall, as seen in
Figure 2.8. Ice from Merrick Glacier crosses the fjord wall and undergoes a 90°
change in flow direction at the Byrd Glacier lateral shear zone. Likewise, an ice
stream in Hudson Strait would probably have similar shear zones well within
the strait. Hence, the pertinent places to study striation trends to reconstruct
ice streams are on islands within the strait. Moreover, the northeastward
trending striations on southern Baffin Island, associated with carbonate-rich
drift and taken by Andrews and Miller (1983) to preclude an ice stream, could
reflect diverging flow of the ice stream at the end of the strait due to a blocking
ice shelf in the Labrador Sea. Similar diverging flow occurs where Byrd Glacier
meets the Ross Ice Shelf (Fig. 2.7).

The chronology of ice movement is important in reconstructing a former ice
stream. The age of any postulated blocking ice at the mouth of Hudson Strait
must be known accurately before an ice stream in late Wisconsin time can be
denied. For example, interior Laurentide ice could have poured seaward
through Hudson Strait early in late Wisconsin deglaciation. Later in deglaciation, when prolonged downdraw had collapsed marine ice over Hudson Bay,

flow from the residual ice cap in northern Quebec could have streamed into Ungava Bay and moved northeastward across the mouth of Hudson Strait.

One additional comment pertains to potential Laurentide ice streams. Dyke *et al.* (1982) defined the important M'Clintock ice divide of the northern Laurentide Ice sheet. Eastward ice flow from this divide passed diagonally across Franklin Strait, Peel Sound, Somerset Island, and Boothia Peninsula. These flow indicators challenge our postulated flowlines in this area (Denton & Hughes 1981a), which dovetailed into ice streams that occupied Franklin Strait and Peel Sound. Dyke *et al.* (1982) concluded that the Laurentide Ice Sheet ended at the northern margin of their reconstructed ice divide. However, based on Antarctic glaciological observations, we suggest an alternate ice configuration. We argue that the documented eastward ice flow from the M'Clintock ice divide across deep straits would not have occurred without thick blocking ice in Viscount Melville Sound and Barrow Strait to the north. Otherwise, we suggest that ice would have flowed northward through Franklin Strait and Peel Sound. If we are correct, a thick, eastward-flowing ice stream probably occupied Lancaster Sound, while a westward-flowing ice stream filled M'Clure Strait. Northern Somerset Island, Brodeur Peninsula, and east Devon Island would have supported local ice domes, much as local ice domes now occur between Antarctic ice streams. Just as in Antarctica (Rose 1979), the base of the ice streams were at the melting point, whereas slow-moving ice of the local domes was frozen to underlying terrain (Fig. 2.6).

Figure 2.5 shows our depiction of Northern Hemisphere marine and terrestrial ice streams of late Wisconsin time, compared with those of today's Antarctic Ice Sheet in Figure 2.3. We have located these ice streams according to our Antarctic experience, and suggest that marine ice streams were the drain pipes of Northern Hemisphere marine components. Obviously, the final configuration of Northern Hemisphere ice streams must await field testing of this hypothesis. However, the important point, which should not be buried in detailed arguments, is that if marine ice streams existed in any frequency, the marine components of Northern Hemisphere ice sheets were susceptible to the marine instability mechanisms that operate in Antarctica. In particular, the existence of marine ice streams sets the background for extensive downdraw of Northern Hemisphere ice sheets during late Wisconsin deglaciation, a process that we suggested as being especially important (Denton & Hughes, 1981b, 1983). We believe that collapsed sectors of the West Antarctic Ice Sheet, as well as the Northern Hemisphere ice sheets of 17 000–21 000 a BP that are now gone, were vulnerable to the marine instability mechanisms (Stuiver *et al.* 1981, Denton & Hughes 1981a,b) whereas the Greenland and East Antarctic Ice Sheets were largely immune to these mechanisms, and therefore remain intact.

Ice shelves and ice rises

Ice shelves are the floating components of ice sheets. Antarctic ice shelves are important because they constitute about eleven percent of the area of the

Antarctic Ice Sheet and discharge most of its ice (Drewry *et al*. 1982). Antarctic ice shelves average about 500 m in thickness, compared to an average thickness of 2000 m for the grounded part of the Antarctic Ice Sheet. We know from Antarctic studies that an ice shelf confined in an embayment and pinned by numerous islands and shoals (shoaling is revealed by an ice rise on the ice shelf) can have a strong stabilizing effect on the adjacent grounded parts of the ice sheet (Thomas 1979, Stuiver *et al*. 1981, Hughes 1982a, 1983, Thomas & MacAyeal 1983). On the other hand, an ice shelf that is unconfined and unpinned is itself unstable, and cannot stabilize adjacent grounded portions of the ice sheet, whether marine or terrestrial (Weertman 1957, 1974). Finally, Antarctic studies suggest to us that many present-day Antarctic ice shelves are collapsed and floating remnants of former grounded marine components of the Antarctic Ice Sheet, and ice rises above shoals were formerly local ice domes between ice streams (Anderson *et al*. 1980, Kellogg, Osterman *et al*. 1979, Kellogg, Truesdale *et al*. 1979, Stuiver *et al*. 1981, Elverhøi 1981). We may therefore expect ice shelves and ice rises to be prominent features during deglaciation of the Northern Hemisphere in marine sectors where the ice sheet on continental shelves and adjacent continental interiors had isostatically depressed the bed well below sea level, so that these areas were vulnerable to the marine instability mechanisms.

Antarctica

Giovinetto (1964) showed that nearly half of the Antarctic Ice Sheet flows into the Ross Sea and Weddell Sea embayments, which are now occupied by the Ross and the Filchner-Ronne Ice Shelves. It is as though the ice shelves are plugs that keep the ice sheet from draining into these two seas. Whether or not this is true is perhaps the greatest glaciological question in Antarctica. But there is no question that the ice shelves buttress ice streams that supply them and that drain much of the ice sheet. The laterally diverging rifts alongside Byrd Glacier, shown in Figure 2.7, strongly suggest such buttressing. The rifts would not exist, nor would they be bent laterally, if the Ross Ice shelf were not a powerful check on ice discharged by Byrd Glacier. Buttressing reduces the discharge velocity (Hughes & Fastook 1981, Brecher 1982). Thomas and MacAyeal (1983) have computed the back-stresses whereby the Ross Ice shelf buttresses the ice streams that feed it.

By itself, an ice shelf will spread continuously under its own weight (Weertman 1957). What gives the Ross Ice Shelf its buttressing capability are pinning points that anchor the ice shelf to the floor of the Ross Sea, creating ice rises on the surface of the ice shelf. Figure 2.11 shows the pinning-power of Buffer Ice Rise on Wordie Ice Shelf, which buttresses Fleming Ice Stream on the Antarctic Peninsula. Proof that ice rises are effective pinning points is also seen at Crary Ice Rise on the Ross Ice Shelf (Barrett 1975). The fact that the whole ice-shelf thickness, over 500 m, is split clear through in the lateral rifts of Byrd Glacier and in the horst and graben structures in the lee of Crary Ice Rise demonstrates that most of the back-stress buttressing ice streams are supplied

Figure 2.11 Wordie Ice Shelf by Fleming Glacier and pinned by Buffer Ice Rise. Converging flow, surface crevassing, and lateral shear are evident in the ice stream. Upstream ice buckling and downstream ice fracturing are evident at the ice rise. U.S. Navy photograph for the U.S. Geological Survey (cover photo for Annals of Glaciology, vol. 3, 1982).

at ice rises. The remainder is supplied by the confining sides of the embayment occupied by the ice shelf. Jezek and Bentley (1983) have mapped eight ice rises or ice rumples on the Ross Ice Shelf, including six that were located by bottom crevasses opened in the lee of these pinning points. Ice rumples form when the ice shelf scrapes over a pinning point, instead of flowing around it.

If ice streams link an ice shelf to the grounded ice sheet and ice rises link the ice shelf to the sea floor, crevasses alongside ice streams and in the lee of ice rises show that these links are weak. Should the crevasses grow, allowing the ice stream to punch through the ice shelf or the ice shelf to break from its pinning points, the buttressing capacity of the ice shelf would be wholly or largely lost (Hughes 1982a, 1983).

All present-day ice shelves in Antarctica are probably the floating remnants of collapsed marine sectors of the grounded Antarctic Ice Sheet that existed 21 000–17 000 a BP (compare Figs. 2.3 and 2.4). This is a new insight into Antarctic ice shelves, because they had previously been viewed as permanent features of the Antarctic ice sheet (Anderson *et al.* 1980, Kellogg, Osterman *et al.* 1979, Kellogg, Truesdale 1979, Kellogg & Kellogg 1981, 1983; Stuiver *et al.* 1981, Elverhøi 1981). However, the new view originated with Voronov (1960) and found early support from Hollin (1962), Denton and Armstrong (1968), Mercer (1968a,b), Denton *et al.* (1971), Anderson (1972), Hughes (1972), and

Weertman (1974). The turning-point in reaching the new understanding of Antarctic ice shelves was the concept of marine ice sheets and their inherent instability that Mercer (1968a) applied to the present-day ice sheet in West Antarctica.

Modeling studies show that present-day Antarctic ice shelves are inherently unstable if their grounding lines are on a downhill slope into interior subglacial basins, they are not confined in embayments, and they are not pinned to islands and shoals (Thomas & Bentley 1978, Stuiver *et al*. 1981). They are stable if the opposite conditions hold. If some conditions apply, but not others, their stability is uncertain. However, an ice shelf whose grounding line is on a downhill slope into a large subglacial basin is stabilized only by pinning points on its underside (lateral confinement has meaning only if lateral grounding is on uphill slopes). The surface expression of basal pinning is an ice rise. If pinning points are removed by rising sea level, melting of the ice shelf, creep thinning of the ice shelf, or reduced ice advection onto the ice shelf, then grounding lines along the ice-shelf perimeter will retreat irreversibly downslope to the bottom of the subglacial basin and up the other side until thinning and thickening tendencies are again in balance. Before this balance is attained, a calving bay may carve away the ice shelf if its spreading rate is unable to offset the iceberg calving rate (Hughes 1982a, 1983).

The small ice shelves fringing East Antarctica probably resulted from grounding lines retreating upslope as sea level rose after 17 000 a BP. Except for the Amery Ice Shelf, which has a calving front narrower than its length and which is supplied by the huge Lambert Glacier (Allison 1979), East Antarctic ice shelves have a wide calving front compared to their length and they must depend upon pinning points to remain intact (Swithinback 1955). The pinning points and lateral support arise from the uphill slope behind these grounding lines. A downhill slope behind grounding lines destabilizes the ice shelf at both pinning points and lateral boundaries.

In West Antarctica, huge ice shelves occupy vast embayments in the Ross Sea and the Weddell Sea, and small ice shelves fringe the Amundsen Sea and Bellingshausen Sea coasts. In general, the small ice shelves are of the East Antarctic type. Their grounding lines are on uphill slopes inland. Two critically important exceptions, however, are Thwaites Glacier and Pine Island Glacier, shown in Figure 2.12. The floating tongues of these ice streams are either unconfined (Thwaites Glacier) or weakly confined by thermally softened and strain-softened ice in lateral shear zones (Pine Island Glacier). Their rear grounding lines lie on a downhill slope on the side of Bentley Subglacial Trench (Drewry 1979a, Crabtree & Doake 1981). From what we now understand about the marine instability mechanisms, these grounding lines may retreat irreversibly and a third huge ice-shelf embayment may form over Bentley Subglacial Trench (Hughes *et al*. 1980, Stuiver *et al*. 1981). This ice shelf would complete the collapse of the marine West Antarctic Ice Sheet that began 17 000 years ago when the Ross Ice Shelf and the Filchner-Ronne Ice Shelf began to form (Hughes 1981b).

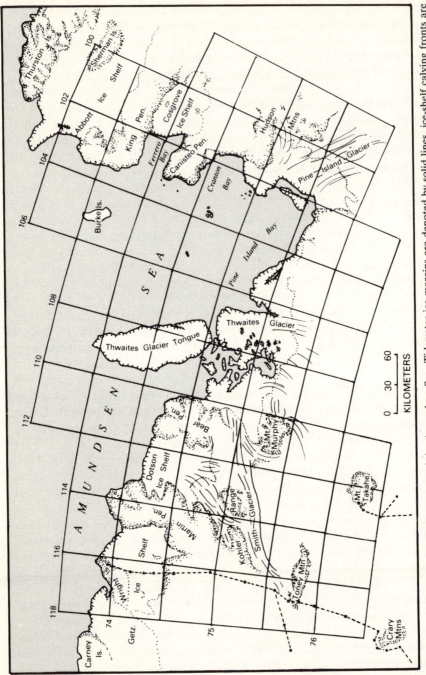

Figure 2.12 The Pine Island Bay sector of the Amundsen Sea. Tidewater ice margins are denoted by solid lines, ice-shelf calving fronts are denoted by hachured lines, ice-shelf grounding lines are denoted by dotted lines, and nunataks are denoted by dotted areas. Thwaites Glacier has punched through a narrow ice shelf and Pine Island Glacier is weakly buttressed by the ice shelf.

The Ross Ice Shelf occupies the Ross Sea embayment and the Filchner-Ronne Ice Shelf occupies the Weddell Sea embayment of the West Antarctic Ice Sheet. Pinning points beneath these ice shelves lie on the uphill slopes of shoals, and the ice shelves are grounded on uphill slopes alongside the Transantarctic Mountains, the Ellsworth Mountains, and the Antarctic Peninsula. The only grounding lines on downhill slopes lie along the Sharase Coast, Siple Coast, and Gould Coast of the Ross Ice Shelf and the Ellsworth Land coast of the Ronne Ice Shelf. All of these grounding lines are at least 500 m below sea level and lie on downhill slopes toward Bentley Subglacial Trench. It is this mixture of stable upslope grounding lines and unstable downslope grounding lines which makes the present-day stability of the Ross and Ronne Ice Shelves so difficult to assess. All grounding lines of the Filchner Ice Shelf are probably upslope, so that ice shelf should be stable.

A third kind of Antarctic ice shelf occupies the southern part of McMurdo Sound in the Ross Sea (Swithinbank 1970). It is fed by Koettlitz Glacier flowing out of the Transantarctic Mountains and by floating ice moving out of the Ross Ice Shelf between Minna Bluff and Ross Island. Katabatic winds ablate the top surface of this McMurdo Ice Shelf, while ice freezes onto the bottom surface. Material from the floor of McMurdo Sound appears on the ice-shelf surface through the combination of bottom freezing and top melting. Bottom freezing ultimately accounts for the entire ice thickness at the calving front because all the glacial ice has, by then, melted or sublimated. We think that the McMurdo Ice Shelf provides a present-day example of the marine origin of an ice shelf and, by rapid thickening should surface accumulation be added to basal freezing, would be an analog for the marine origin of Northern Hemisphere paleo-ice sheets (Denton & Hughes 1981b).

The collapse of a grounded marine ice sheet to produce a floating ice shelf has been studied in the Ross Sea sector of the West Antarctic Ice sheet (Denton et al. 1971, 1975, Thomas & Bentley 1978, Kellogg, Truesdale et al. 1979, Kellogg, Osterman et al. 1979, Kellogg & Kellogg 1981, 1983, Stuiver et al. 1981). At 17 000 a BP, the grounding line lay near the edge of the Antarctic continental shelf and ice streams occupied the shallow linear troughs on the floor of the northern Ross Sea (Fig. 2.4). As Northern Hemisphere ice sheets began shrinking, rising sea level caused grounding lines of ice streams to back off sills at the seaward ends of these troughs, to retreat along the troughs and to merge behind the submarine banks between the troughs. Local ice domes on these banks then became ice rises, and individual buttressing of ice streams by lateral shear alongside narrow ice shelves in linear troughs was replaced by group buttressing of all ice streams by a single ice shelf pinned at the submarine banks. This was the first appearance of the Ross Ice Shelf. It developed into its present configuration as the grounding line and the calving front both retreated southward (Stuiver et al. 1981). That local ice domes between ice streams become ice rises on an ice shelf is implicit in the eastern Ross Ice Shelf. Present-day ice streams D and E formerly occupied the troughs on either side of Roosevelt Island, which is an ice rise where the Ross Ice Shelf is locally grounded on a high part of the submarine bank between the troughs (Bentley & Jezek, 1981). This would have been the site of a local ice dome when the troughs were occupied by ice streams D and E, before the Ross Ice Shelf formed. The Siple Coast local ice dome between ice streams C and D is an analogous situation today, and any continued retreat of the Ross Ice Shelf grounding line will convert this ice dome into an ice rise. In this way, an ice shelf creates its own buttressing capability as its grounding line retreats; first it develops lateral traction alongside former ice-stream channels, and then it develops basal traction at ice rises between these channels. Its buttressing capability, whether by lateral or basal traction, depends on grounding lines that lie on uphill slopes at these sites. The only effective ways to eliminate the buttressing capability of an ice shelf are to float the ice shelf off the uphill slopes where lateral and basal grounding occur, or to carve away the ice shelf with an accelerated rate of iceberg calving (Hughes 1982a, 1983). Rising sea level from 17 000 to 6000 a BP promoted ungrounding beneath the present-day Ross Ice Shelf, and rapid iceberg calving carved the ice shelf back from its pinning points over the submarine banks in the northern Ross Sea.

Northern Hemisphere

A reliable reconstruction of paleo-ice shelves in the Northern Hemisphere is not yet feasible because direct geologic evidence is difficult to interpret except in a few isolated places where nearly horizontal moraines lie alongside fjords. Given this lack of constraints, we can make only several suggestions based on the geography and history of Antarctic ice shelves.

Hughes *et al*. (1977) and Denton and Hughes (1981b) suggested an extreme case for ice-shelf extent during the maximum of late Wisconsin glaciation. They endorsed the concept of an ice shelf on the Arctic Ocean (Thomson 1888, Mercer 1970, Broecker 1975) and included extensions of this ice shelf across the Greenland, Norwegian, and Labrador Seas into the North Atlantic Ocean. These extensive ice shelves were postulated largely to buttress ice streams draining the marine portions of adjacent grounded ice sheets, but also to explain the pattern of Northern Hemisphere deglaciation. However, at least two alternative scenarios are possible. One is that an extensive ice shelf was confined to the Arctic Ocean and the Greenland Sea, buttressing ice streams from the Innuitian, Barents, and Kara Ice Sheets on Arctic continental shelves. It is also possible that extensive ice shelves did not exist during maximum glaciation, but that smaller ice shelves fringed portions of Arctic ice sheets and were pinned by islands on arctic continental shelves, particularly in the Barents Sea and among the Queen Elizabeth Islands.

The mechanisms of formation of Antarctic ice shelves during late Wisconsin deglaciation of grounded marine components may well apply in the Arctic. Present-day Antarctic ice shelves and their associated ice rises are collapsed remnants of the late Wisconsin Antarctic Ice Sheet. Pinned and confined ice shelves may also have formed in channels during collapse of the Innuitian Ice Sheet and of northern and interior Laurentide ice. Further, Eurasian ice shelves could have formed in embayments in collapsing Barents and Kara Sea ice. During their existence, these ice shelves would have slowed grounding-line recession and even may have temporarily stabilized adjacent grounded marine components. Thus such ice shelves would have served as a regulator between sea level and grounded marine components.

Divides, domes, and saddles

Ice divides separate ice moving down opposite flanks of an ice sheet and thereby partition the ice sheet into several ice drainage basins (Giovinetto, 1964). Interior domes and saddles identify high and low places along ice divides. Ice flowing down the flanks of the ice divide diverges from domes and converges toward ice streams; these flow from saddles (Warntz 1975).

Antarctica
The main ice divide of the Antarctic ice sheet is Y-shaped, with a high central dome at the center of the Y and branching ice divides at each extremity of the Y as shown in Figure 2.3 (Drewry 1983). The high central dome is called Dome Argus and is over 4000 m in elevation. It covers the Gamburtsev Subglacial Mountains of East Antarctica to a depth of 1600 m or more. The longest leg of the Y extends into West Antarctica. Several minor domes exist along this leg, which passes close by the South Pole, and near the end of the leg is a dome 2500 m high between the Whitmore and Ellsworth Mountains. This is the major dome in West Antarctica. From it, the ice divide branches northeast-ward toward a dome over a subglacial basin in Ellsworth Land and northwest-ward toward a dome over the Executive Committee Range in Marie Byrd Land. The ice divide branches again at these two domes, with ridges of ice extending to the coast and separating the drainage basins of important ice streams. These ridges may end at local ice domes along the coast (Fig. 2.13).

The two short legs of the Y-shaped ice divide extend into Wilkes Land and Queen Maud Land, respectively, of East Antarctica. The Wilkes Land leg contains Ridge B and Dome Circe, which have no underlying topographic features that would anchor them to these sites. Ridge B lies on the flank of the Gamburtsev Subglacial Mountains. Dome Circe lies on the flank of the deep Aurora Subglacial Basin, which reaches 1500 m below sea level. No distinct ice ridges extend to the coast from either Ridge B or Dome Circe but a notable local ice dome exists on the Wilkes Land coast nonetheless. This Law Ice Dome lies between Totten Glacier and Vanderford Glacier, two ice streams that drain most of the Aurora Subglacial Basin (Figs. 2.2 and 2.14). Ice ridges that terminate in local ice domes occur on the landward flank of the mountains of Victoria Land (Drewry 1980, 1983, Drewry *et al.* 1982). The local McMurdo Dome is situated inland from the Dry Valleys, and the Talos Dome is adjacent to the mountains of northern Victoria Land (Fig. 2.13).

The short leg of the Y-shaped ice divide that extends into Queen Maud Land

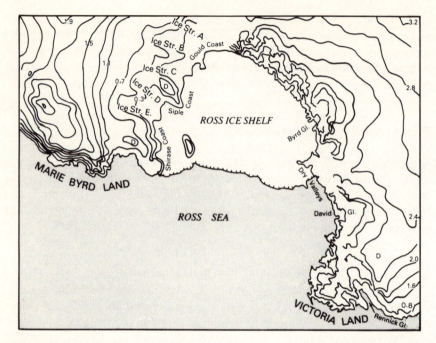

Figure 2.13   Local ice domes bordering the Ross Sea. In East Antarctica (right) local ice domes D lie inland from the Dry Valleys (McMurdo Dome) and between David Glacier and Rennick Glacier (Talos Dome). In West Antarctica (left) local ice domes D lie inland from the Siple Coast between ice streams C and D, and inland from the Shirase coast between ice stream E and Marie Byrd Land. Scale is 1:17 000 000. Contours are in km. with varying intervals. The detailed crenulations can not be mapped at this scale; thus some contours are not continuous. See also Fig. 2.2. (After Drewry, 1983.)

ends at the Valkyrjedomen, which is over 3700 m high. Two prominent ice ridges extend beyond Valkyrjedomen, one northeast toward the mountains of Enderby Land and one northwest toward the mountains of New Schwaben-land. Nunataks in Enderby Land lie close to the sea, where deep fjords are occupied by ice streams that calve without forming extensive ice shelves (Morgan *et al.* 1982). The entire coast of New Schwabenland is fringed by ice shelves, and nunataks appear well inland from the grounding line (Swithinbank 1957). Ice streams which develop between ranges of nunataks merge to form a number of broad ice fans long before the ice becomes afloat. These ice fans constitute Princess Martha Coast, Princess Astrid Coast, and Princes Ragnhild Coast. The subglacial topography of Queen Maud Land is largely unmapped, and we do not know if Valkyrjedomen covers a subglacial highland. Where subglacial mapping has been done, notably in Enderby Land, the ice sheet covers a rugged landscape several hundred meters above sea level (Johnson *et al.* 1980, Drewry 1983).

There appears to be a close relation between surface topography of the Antarctic Ice Sheet and the positions of major ice streams, suggesting a causal relationship. Nearly all saddles along Antarctic ice divides are upslope from major ice streams, and domes that do not cover highlands may be merely places on the ice divide that have suffered less downdraw into ice streams (Hughes, 1981a). Figures 2.2 and 2.3 show these domes and saddles in relation to ice streams. In East Antarctica, the high saddle between Dome Argus and Valkyrjedomen would result from ice draining into Lambert Glacier (Allison 1979). Ice downdrawn into Nimrod Glacier, Beardmore Glacier, and Reedy Glacier, three prominent outlet glaciers through the Transantarctic Mountains, may have created the first three saddles along the ice divide from Dome Argus to West Antarctica. The fourth saddle along this ice divide seems to be

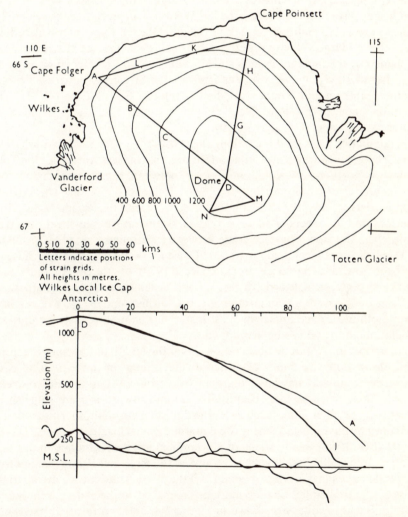

Figure 2.14   The Law Ice Dome in Wilkes Land. This dome is also called the Wilkes Ice Cap. Ice elevations are contoured in 200 m intervals (top). Surface and bed profiles are shown along Transects DA and DJ (bottom). From McLauren (1968), discussed by Budd (1970). (Copyright by the Internat. Assoc. Scientific Hydrology. Used by permission).

linked to ice downdrawn into ice stream B, the fastest West Antarctic ice stream supplying the Ross Ice Shelf. The Byrd saddle between the dome near the Whitmore Mountains and the dome over the Executive Committee Range is linked to ice stream D on the Ross Sea flank and to Thwaites Glacier on the Amundsen Sea flank.

The low saddle between the dome near the Whitmore Mountains and the dome in Ellsworth Land lies over Bentley Subglacial Trench and separates ice flowing into Pine Island Glacier on the Amundsen Sea flank and Rutford Ice Stream on the Weddell Sea flank.

This low saddle lies much closer to Rutford Ice Stream than to Pine Island Glacier (Fig. 2.2), which moves five times faster. This suggests a shift in the ice divide may have occurred during the last Wisconsin deglaciation, provided that more ice was downdrawn into Pine Island Glacier than into Rutford Ice Stream. The large velocity difference between these two ice streams may be explained by the fact that Pine Island Glacier calves into the open water of the Pine Island Bay polynya, whereas Rutford Ice Stream is buttressed by the confined and pinned Ronne Ice Shelf (compare Figs. 2.6 and 2.12). This situation suggests that Pine Island Bay is 'the weak underbelly of the West Antarctic Ice Sheet' (Hughes 1981b).

A number of local ice domes along the coast are apparently the result of ice having been downdrawn into adjacent ice streams. In East Antarctica the Talos Dome in northern Victoria Land exists between saddles associated with Rennick Glacier to the north and David Glacier to the south. The Law Ice Dome exists for two reasons: it lies between Totten Glacier and Vanderford Glacier, and it covers a plateau that is 400 m above sea level. In West Antarctica, the Siple Coast ice dome between ice streams C and D is grounded 1000 m below sea level, and the Shirase Coast ice dome may occur because ice has been downdrawn into ice stream E (Rose 1979). On the other hand, a local ice dome between Rutford Ice Stream and a nearby stagnant ice stream in Carlson Inlet covers a submarine plateau (Swithinbank 1977). The origin of several other local ice domes in West Antarctica cannot be assessed until the subglacial landscape has been mapped.

An important Antarctic observation is that the length of flowlines is comparable down opposite flanks of an ice divide, except in those unusual cases where the ice divide may have migrated away from a very dynamic ice stream on one flank (such as Pine Island Glacier) and toward a more sluggish ice stream on the other flank (such as Rutford Ice Stream). Another exception to this rule may have existed during the Antarctic glacial maximum from 21 000 to 17 000 a BP. If full steady-state equilibrium had not been attained in this time interval, after falling sea level had grounded the Ross and Ronne Ice Shelves, longer flowlines may have extended from the West Antarctic ice divide to the Ross Sea and Weddell Sea ice margins compared to the Amundsen Sea ice margin. Such flowlines are shown in Figure 2.4, and contrast with the quasi-equilibrium flowlines employed by Hughes et al. (1981) for the glacial maximum.

Northern Hemisphere

In the Northern Hemisphere, surface topography and flowlines at any particular time during a glaciation must be reconstructed from geologic features that reflect basal ice flow. These ice-flow features did not necessarily form synchronously over large areas. Thus the problem becomes one of sorting out which ice-flow features are appropriate to use in a particular reconstruction that represents one time slice.

The surface form of the Antarctic Ice Sheet, as well as the configuration and form of its flowlines, can guide Northern Hemisphere reconstructions based on geologic features. For example, it should be evident that a single-dome Laurentide Ice Sheet is not necessarily an Antarctic analog, contrary to the assertion by Dyke *et al.* (1982). Rather, a high saddle may well have existed over Hudson Bay at maximum glaciation because of the influence of an ice stream in Hudson Strait. Likewise, the position of other interior domes and saddles may well have been controlled by ice streams and bedrock topography. Further, numerous peripheral ice ridges and local domes are likely. During deglaciation, saddles at the head of ice streams would have deepened rapidly because of downdraw, and a multidomed configuration would have been dramatically accentuated.

The roughly equal lengths of Antarctic flowlines on opposite flanks of an ice divide casts suspicion on the reconstruction of steady-state Laurentide flowlines by Shilts (1980) (see also Dyke *et al.* 1982). His reconstruction shows flowlines of very unequal length emanating from the Labrador ice divide. Moreover, the flowlines imply an extensive shear zone in the center of the Laurentide Ice Sheet, a highly unlikely situation. We think that the geologic evidence is best explained either by postulating a Hudson Bay dome (Dyke *et al.* 1982) or by assuming that the ice-flow indicators are not synchronous but formed as the interior ice-sheet surface changed configuration during a glacial cycle (Mayewski *et al.* 1981, p. 144–155).

A number of local ice domes, probably all frozen-based, should have existed between ice streams that drained former Northern Hemisphere ice sheets. In the case of the Innuitian, Barents Sea, and Kara Sea Ice Sheets, local ice domes probably occurred over islands on the continental shelf, separating ice streams in inter-island channels. Along the northern Laurentide Ice Sheet, local ice domes could have formed over Victoria Island, Prince of Wales Island, Somerset Island, Brodeur Peninsula, and Borden Peninsula. Along the eastern Laurentide Ice Sheet, local ice domes could have flanked ice streams, and Newfoundland could have been a local ice dome of the kind represented by Law Ice Dome in Antarctica today. That is, ice would have covered a high plateau flanked by two ice-stream troughs. During Laurentide ice retreat, Anticosti Island in the Gulf of St. Lawrence could have been a submarine plateau between two branches of an ice stream from the St. Lawrence Valley, somewhat like the submarine plateau between Rutford Ice Stream and Carlson Inlet in Antarctica (Fig. 2.6). Along the southern Laurentide margin, particularly during retreat, local ice domes could also have formed between terrestrial ice streams that occupied topographic troughs such as the Great Lakes.

Bimodal response of Antarctic Ice Sheet

We now consider the question of how the Antarctic Ice Sheet, with its distribution of glaciological components, reacted during late Quaternary ice ages. We focus on late Wisconsin glaciation, because it is best known.

The Antarctic Ice Sheet is nearly centered on the South Pole, is thermally isolated by the Circumantarctic Current, is surrounded by the Southern Ocean, and occupies an extreme climatological environment. As a result the ice sheet lacks extensive peripheral melting zones, unlike Northern Hemisphere ice sheets which could reach far southward into North America and Eurasia. Therefore, the Antarctic Ice Sheet is relatively immune to summer temperature changes, which so sensitively controlled melting margins of Northern Hemisphere paleo-ice sheets. The exception to this statement is that significant warming (5°–10°C) over today's values during extreme interglaciations might cause recession of fringing ice shelves in the Ross Sea and Weddell Sea embayments, leading to collapse of the marine-based West Antarctic Ice Sheet (Mercer 1978, Hughes 1982a, 1983) through surging ice streams. Such warming could also thin ice streams and cause grounding-line recession in the Amundsen Sea sector of the West Antarctic Ice Sheet (Stuiver *et al.* 1981). It is even possible that such extensive warming could cause recession of the Cook Ice Shelf near northern Victoria Land, perhaps leading to the development of inland marine ice streams that would downdraw and collapse East Antarctic ice over the northern Wilkes Subglacial Basin through the marine instability mechanisms.

In the absence of temperature control of melting margins, three main mechanisms have been proposed to explain ice-age fluctuations of the Antarctic Ice Sheet. All three could have interacted continuously, although any one might have been dominant at a given time or locality. At the turn of the century Scott (1905) proposed that the Antarctic Ice Sheet fluctuated out of phase with Northern Hemisphere ice sheets during ice ages. This hypothesis is based on the premise that worldwide interglacial conditions would allow relatively warm and moist air to penetrate Antarctica, greatly increasing accumulation and causing expansion of the ice sheet. Conversely, intervals of worldwide glaciation would be accompanied by drastically reduced accumulation and consequent ice recession. A second hypothesis proposes that sea level is the dominant factor in determining the area and volume of the Antarctic Ice Sheet, which is basically a huge accumulation area, with most ablation occurring not at melting margins but as icebergs calving from the periphery of the continent. Hollin (1962) suggested that the ice sheet has the capacity to expand but is restricted from doing so by the present position of relative sea level which controls grounding-line positions on Antarctic continental shelves. However, eustatic sea-level oscillations would have allowed expansion and contraction of Antarctic ice during the last Quaternary glaciation. This sea-level hypothesis requires that the Antarctic Ice Sheet fluctuate in phase with major late Quaternary glaciations in the Northern Hemisphere, not because of climatic events but because of mechanical adjustments to fluctuation of sea level. A third hypothesis is that the Antarctic Ice Sheet surges periodically. Wilson (1964) envisioned surges of the entire ice sheet, whereas Hollin (1969) postulated surges of individual drainage basins. We know that individual ice streams can move at surge velocities; beyond this, Antarctic surges are inferred only from modeling studies (Budd & McInnes 1979) and ancient sea-level data (Hollin 1969).

Geological field studies over the past two decades have served to test these hypotheses. Using a glaciological model constrained by geological data available in 1978, Stuiver et al. (1981) reconstructed the late Wisconsin configuration of the Antarctic Ice Sheet. Figure 2.4 shows a more recent reconstruction that is constrained by additional geologic control points obtained within the last five years. The major difference between the two reconstructions is in central West Antarctica, where the new reconstruction shows less thickening and several domes and saddles, as a result of inferences drawn from the Ellsworth Mountains (Rutford et al. 1980; Denton, Backheim et al. 1984). The new reconstruction also shows a higher Talos Dome in northern Victoria Land (Denton & Wilson 1983) and an enlarged peripheral dome over the Executive Committee Range. Both reconstructions are consistent with earlier published views of Antarctic ice-sheet behavior (Denton & Armstrong 1968, Denton et al. 1971, 1975), particularly with respect to widespread grounding in West Antarctic embayments.

As shown in Figure 7–26 of Stuiver et al. (1981), we infer late Wisconsin grounding on the narrow continental shelf fringing the East Antarctic Ice Sheet. Interior East Antarctic ice may have remained nearly constant in elevation, or may even have decreased slightly. However, geologic evidence indicates that the East Antarctic inland ice surface remained nearly unchanged where it abutted the interior flank of the Transantarctic Mountains between the Dry Valleys and Reedy Glacier (Denton 1979, Mercer 1968b, 1972). Two peripheral East Antarctic domes adjacent to the Transantarctic Mountains showed opposite behavior; the small McMurdo Dome directly inland of the Dry Valleys contracted slightly, whereas the Talos Dome in northern Victoria Land thickened by 600 m (Denton & Wilson, 1983).

Drewry (1980) showed that an East Antarctic ice ridge terminates in the local McMurdo Dome inland of the Dry Valleys; Taylor and Wright Upper Glaciers drain this dome. Radiocarbon dates from strategically positioned deltas of Glacial Lake Washburn show that Taylor and local alpine glaciers are now at their maximum extents in the last 17 000–22 000 years, with the exception of a minor advance represented by ice-cored lateral moraines dated to about 3000 a BP (Stuiver et al. 1981, Denton & Wilson 1983). From this, and from the geometric relation of strandlines and moraines, we infer late

Wisconsin contraction of the McMurdo Dome, followed by Holocene expansion.

Denton *et al.* (1970, 1971) inferred that reduced precipitation in the Dry Valleys, due to covering of the local Ross Sea source by late Wisconsin grounded ice, caused recession of Taylor, Wright Upper, and local glaciers. Likewise, renewed precipitation as rising sea level cleared grounded ice from the Ross Sea caused the Holocene expansion. Contrary to assertions by Drewry (1980), Denton *et al.* (1971, p. 283, 289, 292, 299) did not directly apply this local Dry Valleys record to the whole East Antarctic Ice Sheet, even though they were not yet aware of the McMurdo dome. Rather, Denton *et al.* (1971, p. 289), referring to fluctuations of Taylor Glacier, stated 'that the recognized changes in the surface altitude of the ice sheet in East Antarctica *in the McMurdo Sound region* were not synchronous with major northern hemisphere glaciation despite, (4) that the ice sheet *in this area* was smaller than now during Wisconsin (Würm) time, and (5) that the ice sheet *in the McMurdo Sound area* now occupies its maximum surface altitude since some time before the Wisconsin (Würm) glaciation of the northern hemisphere' (emphasis added here).

After considering the three mechanisms listed above, Denton *et al.* (1971, p. 292) concluded that the East Antarctic Ice Sheet most likely showed a 'dual' response to environmental change when they stated that 'the sea-level mechanism exerts the dominant control on the lateral extent of the ice sheet, whereas an alternative mechanism, such as changes in rate of accumulation, dominates thickness variations in the interior.' Drewry (1980) called this a 'bimodal' response (see also Denton *et al.* 1970, Dort 1970, and Robin 1981). In their conclusions, Denton *et al.* (1971, p. 299) implied that inferences drawn from the McMurdo Sound area applied to interior surface-elevation changes over the Antarctic Ice Sheet. Whether or not this is true is still unknown. However, in their conclusions, Denton *et al.* (1971, p. 299) also indicated that any such inferences did not pertain to the areal dimensions of the East Antarctic Ice Sheet when they stated that 'The field evidence that bears on these hypotheses suggests that sea level controlled fluctuations along the periphery of the ice sheet, especially in shelf areas, but that an alternative mechanism, such as periodic surging or changes in accumulation rate, controlled surface level oscillation in interior East Antarctica.'

Figure 2.4 is a revision of Figure 7–26 of Stuiver *et al.* (1981) for West Antarctica. Confirming evidence for late Wisconsin grounded ice in the Weddell Sea comes from sea-floor sediments (Elverhøi, 1981, Anderson *et al.* 1980), and from glacial geologic features in the Ellsworth Mountains (Rutford *et al.* 1980, Denton, Bockheim *et al.* 1984) and at the base of the Antarctic Peninsula (Carrara 1979, 1981). Revisions in location and elevation of interior domes and saddles are from Denton, Bockheim *et al.* (1984).

In Figure 2.4 we reaffirm our original opinion that late Wisconsin ice grounded in the Ross Sea embayments (Stuiver *et al.* 1981). Drewry (1979) suggested an alternative reconstruction with an extensive ice shelf in the Ross

Sea embayment, several local ice domes, and minor advance of the West Antarctic grounding line. Most of our glacial geologic arguments invalidating his view are given in Stuiver *et al.* (1981) and are not repeated here. In addition, we now point out that geologic evidence strongly suggests grounded ice along the front of the Transantarctic Mountains in areas where Drewry (1979) postulated an ice shelf. Recent fieldwork indicates that the lower reaches of Byrd Glacier, as well as adjacent Darwin Glacier, both shown on Figure 2.7 were dammed to considerable thickness (surface elevations 1200 m at glacier mouths) by ice in the Ross Sea in late Wisconsin time (Denton, 1979). This required grounded rather than floating ice over the present western Ross Ice Shelf. Seaward flow would carry such grounded ice northward past Ross Island, and allow the curving flow into the McMurdo oasis required by Figure 7–16 of Stuiver *et al.* (1981). Such an ice-flow pattern affords the explanation for erratics from the Transantarctic Mountains in late Wisconsin drift in McMurdo Sound. A similar argument applied in the Terra Nova Bay region of northern Victoria Land, where thickened ice in the lower reaches of Reeves and Priestley Glaciers was deflected northward against the mountains (Stuiver *et al.* 1981). To us, this implies extensive grounded ice in the western Ross Sea. Farther south in the central Transantarctic Mountains, Mercer (1972) concluded from the configuration of moraines alongside Beardmore Glacier that extensive grounding of the Ross Ice Shelf occurred repeatedly. The last grounding is probably late Wisconsin in age, for it is represented by virtually unweathered moraines with striated surface clasts and internal ice cores (Mercer 1972, personal communication 1983).

What is the evidence from the Ross Sea itself for grounded ice in late Wisconsin time? Drewry (1979) suggested that Unit II sediments on the Ross Sea floor are glaciomarine whereas our hypothesis requires that they be till. Kellogg and Truesdale (1979), Kellogg, Osterman *et al.* (1979), Kellogg, Truesdale *et al.* (1979), and Anderson *et al.* (1980) presented faunal and sedimentological arguments that Unit II, a wide-spread diamicton on the Ross Sea floor, is a basal till deposited by grounded ice in the Ross Sea. Myers (1982) showed that former flowlines of this grounded ice body could be reconstructed from petrographic provinces for Ross Sea basal tills. These flowlines correspond closely with those of the late Wisconsin ice-sheet reconstruction of Stuiver *et al.* (1981). These studies, as well as more recent core collections, show that basal till deposited by grounded ice is widespread over the Ross Sea floor to near the continental shelf edge. The till is Brunhes in age on the basis of enclosed microfossils (Kellogg, Osterman *et al.* 1979, Kellogg, Truesdale *et al.* 1979). Further, Kellogg and Kellogg (1981, 1983) and Anderson *et al.* (1980) argued that the diamicton at drilling site J–9 in the center of the Ross Ice Shelf (see Fig. 2.2) is also basal till of Brunhes age.

This new interpretation of sea-floor sediments points to extensive grounded ice in the Ross Sea embayment, consistent with our dated results from ice-free areas on adjacent land. However, the case for late Wisconsin grounded ice cannot be considered complete until numerous ^{14}C dates are available from sea-floor cores; those that exist favor late Wisconsin and Holocene recession of ground ice from the Ross Sea (Kellogg, Osterman *et al*. 1979). Further, gravity anomalies have been recently reinterpreted to accommodate extensive grounded ice in the Ross Sea embayment during late Wisconsin time (Greischar & Bentley 1980). In conclusion, we prefer the reconstruction of extensive grounded ice in the Ross Sea embayment during late Wisconsin time on the basis of available data. This conclusion is important from a theoretical viewpoint, because it suggests that grounded ice in the Ross Sea responded extensively during global late-Wisconsin glaciation.

Denton *et al*. (1971) and Stuiver *et al*. (1981) both followed Hollin (1962) in invoking marine control for widespread grounding in West Antarctic embayments, with maximum areal extent concurrent with lowest eustatic sea level. This is consistent with radiocarbon evidence from McMurdo Sound and the Byrd-Darwin Glacier area, which indicates maximum ice extent 17 000–21 000 a BP. However, Figure 2.4 reflects our current opinion that the duration of the last Wisconsin sea-level minimum was insufficient for interior ice elevations to increase to full equilibrium values in response to the expanded area. With our improved ice-sheet model we are now able to express this opinion quantitatively. Glaciological modeling studies give the mechanisms for control of marine ice sheets by sea-level variations (Stuiver *et al*. 1981, Thomas & Bentley 1978).

Geological field data and glaciological modeling results, then, support a dual or bimodal response of the Antarctic Ice Sheet during the late Wisconsin glacial cycle (as well as during other late Quaternary glacial cycles). Sea-level change was the most important driving force. Precipitation changes were somewhat less important. Falling sea level caused lateral expansion of Antarctic ice onto surrounding continental shelves. This occurred by grounding-line advance over the narrow East Antarctic continental shelf. In the Ross and Weddell Sea embayments, grounding line advance was augmented by formation of ice rises where ice shelves grounded locally, with consequent increase of back force on ice streams draining into the ice shelf. We envision that at the culmination of late Wisconsin glaciation the Antarctic Ice Sheet was grounded in the Ross Sea, Weddell Sea, and Amundsen Sea embayments, as well as over the narrow East Antarctic continental shelf (Fig. 2.4). Marine ice streams in the Ross and Weddell Sea embayments drained seaward between local ice domes near the edge of the continental shelf (Fig. 2.4). We also think it likely that several short marine ice streams traversed the East Antarctic continental shelf, for example, seaward of the Lambert Glacier.

This sea-level driven expansion of the Antarctic Ice Sheet occurred despite the fact that interior accumulation almost surely decreased at maximum late Wisconsin glaciation (Hendy & Wilson 1981, Lorius *et al*. 1979). Only alpine glaciers and terrestrial portions of the ice sheet in areas not affected by marine ice-sheet dynamics fluctuated in an out-of-phase fashion with overall ice-sheet behavior, due to primary precipitation control. For example, discharge from the isolated peripheral McMurdo Dome inland of the Dry Valleys decreased, and alpine glaciers in the Dry Valley receded, due to reduced precipitation. In sharp contrast, the peripheral Talos Dome in northern Victoria Land thickened substantially in response to the sea-level effect of grounding on the East Antarctic continental shelf. Thus two East Antarctic peripheral domes showed opposite behavior at the maximum late Wisconsin glaciation because of the differing importance of sea-level and precipitation effects in the two areas (Denton & Wilson 1983). Whether the precipitation or the sea-level effect dominated in the interior of the East Antarctic Ice Sheet is not yet known.

Recession of Antarctic ice from the full-bodied configuration shown in Figure 2.4 to its present retracted position shown in Figure 2.3 was driven by late Wisconsin and early Holocene sea-level rise. Large embayments opened over the Ross, Amundsen, and Weddell sea continental shelves as the marine-based West Antarctic Ice Sheet underwent major contraction (Stuiver et al. 1981). Areal recession of the terrestrial East Antarctic Ice Sheet was minor, with ice simply clearing the narrow continental shelf by grounding-line retreat. However, downdraw through the retreating Lambert Glacier ice stream probably lowered the upstream ice-sheet surface. Likewise, ice at the heads of outlet glaciers through the Transantarctic Mountains, such as David, Nimrod, Byrd, and Beardmore Glaciers, may have been downdrawn slightly as grounding lines receded from the Ross Sea into fjords in the Transantarctic Mountains. Ice recession driven by rising sea level generally outweighed increased interglacial accumulation. Again, an exception involved the McMurdo Dome. This dome is isolated from the effect of marine ice-sheet dynamics and responded largely to precipitation-caused increased discharge, leading to expansion of Taylor and Wright Upper Glaciers into the western Dry Valleys. Local alpine glaciers in the Dry Valleys also showed similar precipitation-caused expansion. Holocene advance in the Dry Valleys was interrupted about 3000 a BP, most likely by a period of warmth with increased ablation. Renewed expansion has subsequently occurred, but there are signs that the warming of the past three decades may be bringing the latest expansion to a close. This Antarctic example shows that increased interglacial precipitation can cause general glacier expansion at high latitudes where ablation is low, but that expansion can be halted abruptly by warm intervals that increase surface ablation. Thus the peripheral East Antarctic domes near the Dry Valleys and in northern Victoria Land showed opposite behavior during general late Wisconsin and Holocene recession. The McMurdo Dome expanded because of renewed precipitation. In striking contrast, the Talos Dome fell substantially in surface elevation because of rising sea level during the period of climatic warming. These two domes, therefore, illustrate the separate effect of each mechanism, precipitation/ablation and sea level, in the bimodal response.

Trimodal response of northern hemisphere ice sheets

Figure 2.5 shows a sketch of the Northern Hemisphere ice sheets at the maximum of late Wisconsin glaciation. As the basis for this sketch, we assumed that marine-based components occupied the Barents and Kara Sea of Arctic Eurasia, as well as the Queen Elizabeth Islands (Innuitian Ice Sheet) of Arctic Canada. We have previously identified the major components of the Antarctic Ice Sheet and suggested the location of such components in paleo-ice sheets of the Northern Hemisphere. We now suggest how Northern Hemisphere ice sheets responded during late Quaternary glacial cycles. We do so by referring to the behavior of these components in the Antarctic Ice Sheet. Our analogs are with Antarctic features and processes, and not with unique geographical aspects of the Antarctic Ice Sheet. Figure 2.15 compares ice sheets in both polar hemispheres.

We propose that Northern Hemisphere ice sheets had a trimodal response to sea level, precipitation, and summer temperature, whereas the Antarctic Ice Sheet had a bimodal response to sea level and precipitation. This difference occurred because Northern Hemisphere ice sheets had extensive surface ablation zones along their southerly margins, unlike the Antarctic Ice Sheet. Closely allied with this difference is the fact that Northern Hemisphere ice sheets lay much closer to the equator and nearer a subtropical 'heat engine'. Hence, precipitation could play a more dominant role than in Antarctica. All three influences on Northern Hemisphere ice sheets interacted continuously, but we think that any one might have been dominant; the dominant process would depend on the geographical position on the ice sheet, and the time.

We suggest that summer temperature variations dominated the southern portion of these Northern Hemisphere ice sheets, with sea-level oscillations playing an important but less dominant role. Precipitation was important, but less so than the other two factors. Changes in summer temperature affected the duration and intensity of surface ablation along the extensive southern melting margins. Radiocarbon-dated glacial sequences in northern Europe, when compared to nearby records of vegetation, show the close coupling of temperature and ice-marginal fluctuations (Berglund 1979, Lagerlund *et al.* 1983). Summer temperature also controlled surface ablation zones on large southern terrestrial and marine ice streams, thus influencing interior downdraw. Sea-level oscillations were also important along southerly Atlantic margins, because they influenced the few large marine ice streams there.

Sea level and precipitation changes probably dominated the response of the Arctic portions of Northern Hemisphere ice sheets. Summer temperature oscillations had a smaller impact because ablation zones were small and melting was not as intense as along southern margins. Arctic portions were mostly marine and reacted largely to sea-level fluctuations. Isolated terrestrial Arctic components removed from marine influences could have reacted largely to precipitation in the manner of McMurdo Dome in Antarctica, as long as ablation on their lower portions was not too intense. Andrews *et al.* (1976), following Flint (1943), suggested that precipitation changes were the dominant influence on Arctic ice sheets. We disagree. If Figure 2.5 is at least approximately correct, marine components dominated Arctic ice sheets. By analogy with a similar situation in Antarctica, we suggest that sea-level oscillations, not

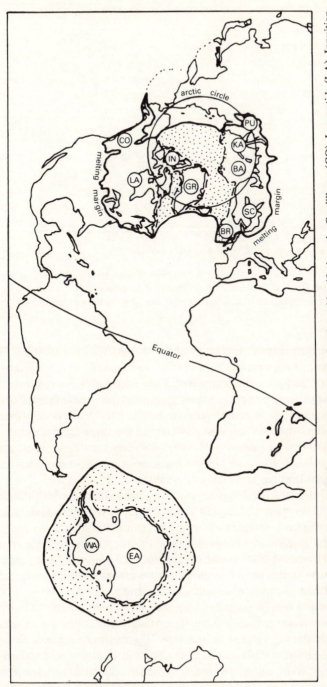

Figure 2.15 Late Wisconsin ice sheets in the Arctic and the Antarctic. Ice sheets identified are the Cordilleran (CO), Laurentide (LA), Innuitian (IN), Greenland (GR), British (BR), Scandinavian (SC), Barents (BA), Kara (KA), Putorana (PU), East Antarctic (EA), and West Antarctic (WA). Dotted areas show the extent of perennial floating ice shelves or sea ice. (From Denton and Hughes 1983 © copyright by Academic Press. Used by permission.)

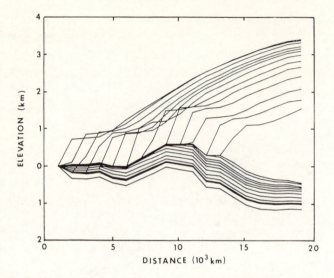

Figure 2.16 Computer-simulated retreat of Laurentide terrestrial ice stream and ice lobe in Hudson Valley. Surface and bed profiles are computed at intervals of 1000 a. Isostatic rebound was computed using a time constant that varied linearly from 3000 a at the ice margin to 12 000 a at the ice divide.

precipitation, dominated Arctic ice-sheet behavior. Precipitation changes were of secondary importance because they influenced interior surface elevations of marine components. Moreover, they may well have controlled some terrestrial margins and peripheral domes removed from the influence of marine ice-sheet dynamics, as we have shown in the Dry Valleys of Antarctica. Summer temperature was the least powerful of the three factors in the Arctic. Variations in extent or intensity of Arctic ablation had the greatest affect on ice-stream surfaces of marine components, because these surfaces had low, concave profiles that could be heavily crevassed.

We suggest that sea level, summer melting, and interior precipitation were all important in the high middle latitudes spanned by Northern Hemisphere ice sheets. Some melting margins extended northward, and marine ice streams were common. Moreover, heavy winter precipitation, under certain conditions (Denton & Hughes 1983), could counterbalance the sea-level influence of downdraw through marine ice streams, or the summer temperature influence on terrestrial and marine ice-stream ablation zones.

As an example of the trimodal response we simulated the behavior of a terrestrial ice stream in the Hudson River valley and a marine ice stream in Hudson Strait during retreat of the late Wisconsin Laurentide Ice Sheet. Retreat of the Hudson Valley ice stream, shown at 1000 year intervals in Figure 2.16, was triggered by climatic warming, which produced an upward shift in the surface equilibrium line. This shift gave the Hudson Valley ice drainage system a negative mass balance, causing the terrestrial ice stream to lower and retreat. Ice melted during each 1000-year interval is the difference between ice thickness profiles integrated over the area of ice draining toward the terrestrial margin. This lost ice, as well as ice lost by other terrestrial ice streams affected by climatic warming, caused sea level to rise. The increase in sea level triggered retreat of marine ice streams draining late Wisconsin marine components. Retreat of the Hudson Strait ice stream triggered by rising sea level is shown in Figure 2.17 at 500-year intervals. Ice lost during each interval is the difference in ice thickness integrated over the marine ice drainage area for all marine ice streams. This combined loss of marine ice, added to the combined loss of terrestrial ice, promotes retreat of the grounding line of the Hudson Strait ice stream, as seen in Figure 2.17.

The ice–ocean interaction of the Hudson Strait ice stream is more vigorous than the ice–atmosphere interaction of the Hudson Valley ice stream, so more ice is downdrawn into Hudson Strait than is melted in Hudson Valley. Figure 2.18 shows the effect of this difference on the elevation of the Laurentide ice divide. More rapid downdraw into the Hudson Strait ice stream causes the ice divide to shift southward. The accumulation zone of southern ice streams is then further reduced, causing a more negative shift in mass balance and a greater retreat rate of terrestrial ice streams, such as that in Hudson Valley. The marine ice stream in Hudson Strait always has a positive mass balance across its grounding line, and the mass balance would become more positive if the grounding line were to retreat more slowly than the ice divide. A slower rate of grounding line retreat would occur if downdrawn ice were to form a confined and pinned ice shelf in Hudson Strait. Conversely, disintegration of this ice shelf would accelerate the rate of grounding-line retreat.

It is worth noting that retreat of the Hudson Strait ice stream begins and ends slowly, but has a rapid partial collapse from 17 500 a BP to 16 000 a BP. This stage of rapid grounding-line retreat is accompanied by rapid downdraw, so that over half of the interior ice is discharged across the grounding line in only 1500 years. It is unlikely that iceberg calving rates could keep up with such a high ice discharge velocity, so an ice shelf would probably form in Hudson Strait beyond the grounding line. Lateral traction along the sides of Hudson Strait would allow this ice shelf partially to buttress the ice stream, and thereby reduce the rate of ice discharge and grounding-line retreat to those depicted in Figure 2.17.

The rapid downdraw stage occurs when the grounding line retreats from the bedrock sill at the entrance to Hudson Strait (represented by Resolution Island) to the bedrock sill at the entrance to Hudson Bay (represented by Nottingham Island and other islands at the end of Hudson Strait). Hudson Strait, therefore, is a trough of the type illustrated in Figure 2.9 and is characteristic of many inter-island channels and straits on the Arctic continental shelf. Therefore, numerous Northern Hemisphere marine ice streams should have the pulse of rapid downdraw modeled in Figure 2.17 for the ice stream in Hudson Strait. Indeed, the record from deep-sea cores in the North Atlantic Ocean seems to record two-step deglaciation, with about half of the ice being lost early in the deglacial hemi-cycle between 16 000 and 13 000 a BP,

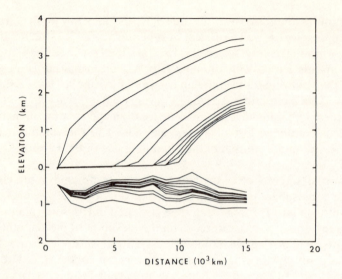

Figure 2.17 Computer-simulated retreat of the Laurentide marine ice stream in Hudson Strait. Surface and bed profiles are computed at intervals of 500 years. Isostatic rebound was computed using the same time constants noted in Figure 2.16. A buttressing ice shelf forms in Hudson Strait as the grounding line retreats. Rapid downdraw occurs in the 2000 years needed for the grounding line to retreat the full length of Hudson Strait. This produces two stages of deglaciation.

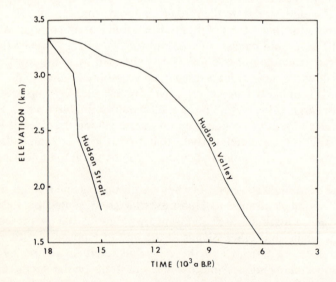

Figure 2.18 Lowering of the interior Laurentide ice divide as a result of recession of a terrestrial ice stream in Hudson Valley and of downdraw through a marine ice stream in Hudson Strait. Downdraw caused by the marine ice stream produces a low saddle in the ice divide over Hudson Bay. Downdraw causes the ice divide to retreat southward in the Hudson Strait drainage basin, thereby reducing the accumulation area of terrestrial ice streams such as the one in Hudson Valley. The two drainage areas and their central flowlines are shown in Figure 2.5.

and the remainder lost after 13 000 a BP (Duplessy *et al.* 1981, Ruddiman & McIntyre 1981).

We think that during ice ages the Antarctic Ice Sheet had a bimodal response to sea level and precipitation, whereas Northern Hemisphere ice sheets had a trimodal response to sea level, precipitation, and summer temperature. This has important implications for the behavior of the global ice-sheet system shown in Figure 2.15. It is sea-level control that is the common denominator for response of these ice sheets. In the absence of sea-level control, there would be no particular reason for these ice sheets to fluctuate in lock step during ice-age cycles. Both precipitation and summer temperature patterns could vary markedly over these ice sheets through time, leading to independent behavior of ice sheets and even of individual drainage basins based on the changing relative influence of each factor. However, the dominance of eustatic sea-level control on the widespread marine components forces an interdependence of these ice sheets, interlocking their behavior on a global scale. By this hypothesis sea level change is a dynamic driving force—rather than a passive recorder—of ice sheet fluctuations. Denton and Hughes (1983) discussed the role of such an interlocked system in ice-age cycles.

Acknowledgments

Our Antarctic field research, as well as our glaciological modeling program, is supported by the Division of Polar Programs of the National Science Foundation. Dr. David Drewry kindly made a surface contour map of the Antarctic Ice Sheet available before its publication in 1983.

References

Allison, I. 1979. The mass budget of the Lambert Glacier drainage basin, Antarctica. *J. Glaciol.* **22**, 223–35.

Andersen, G. G. 1981. Late Weichselian ice sheets in Eurasia and Greenland. In *The last great ice sheets*, G. H. Denton and T. J. Hughes (eds.), 1–65. New York: Wiley Interscience.

Anderson, J. B. 1972. The marine geology of the Weddell Sea. Ph.D. Dissertation, Florida State University.

Anderson, J. B., D. D. Kurtz, E. W. Domack and K. M. Balshaw 1980. Glacial and glacial marine sediments on the Antarctic continental shelf. *J. Geol.* **88**, 399–414.

Anderson, J. B., D. D. Kurtz and F. M. Weaver 1979. Sedimentation on the Antarctic continental slope. *Soc. Econ. Paleont. and Mineral., Special Publication* 27, 265–283.

Andrews J. T. 1982. On the reconstruction of Pleistocene ice sheets: A review. *Quat. Sci. Rev.* **1**, 1–30.

Andrews, J. T., R. W. Feyling-Hanssen, P. E. Hare, G. H. Miller, C. Schluchter, M. Stuiver and B. J. Szabo 1976. Alternate models of early and mid-Wisconsin events, Broughton Island, N.W.T.: Toward a Quaternary Chronology. In *IUGS/UNESCO, International Geological Correlation Program, Project 73/1/24*, L. O. Quom (ed.), 12–61. Bellingham-Prague.

Andrews, J. T. and G. H. Miller 1983. *Quaternary history of the northeast sector of the Laurentide*

Ice Sheet and the mid-Wisconsin sea level problem. Final rept Natl. Sci. Fdtn, Geol., Prog. Grant EAR-79-26061.

Barrett, P. J. 1975. Seawater near the head of the Ross Ice Shelf. *Nature* **256**, 390–92.

Bentley, C. R. 1972. Subglacial rock surface topography, Plate 7, Folio 16. *Antarctic map folio series*. New York: American Geographical Society.

Bentley, C. R., J. W. Clough, K. C. Jezek and S. Shabtaie 1979. Ice-thickness patterns and the dynamics of the Ross Ice Shelf, Antarctica. *J. Glaciol.* **24**, 287–94.

Bentley, C. R. and K. C. Jezek 1981. RISS, RISP and RIGGS: Post-IGY glaciological investigations of the Ross Ice Shelf in the U.S. programs. *J. R. Soc. N. Z.* **11**, 355–72.

Bentley, C. R. and N. A. Ostenso 1961. Glacial and subglacial topography of West Antarctica. *J. Glaciol.* **3**, 882–911.

Berglund, B. E. 1979. The deglaciation of southern Sweden 13 500–10 000 BP. *Boreas* **8**, 89–117.

Blake, W., Jr. 1966. End moraines and deglaciation chronology in northern Canada, with special reference to southern Baffin Island. *Geol. Soc. Canada Paper* 66–29.

Blake, W., Jr. 1970. Studies of glacial history in Arctic Canada, I: pumice, radiocabon dates, and differential postglacial uplift in the eastern Queen Elizabeth Islands, *Can. J. Earth Sci.* **7**, 634–64.

Boulton, G. S., C. T. Baldwin, J. D. Peacock, R. M. McCabe, G. Miller, J. Jarvis, B. Horsefield, P. Worsley, N. Eyles, P. N. Chroston, T. E. Day, P. Gibbard, P. E. Hare and V. von Brunn 1982. A glacio-isostatic facies model and amino acid stratigraphy for late Quaternary events in Spitsbergen and the Arctic. *Nature* **298**, 437–41.

Brecher, H. H. 1982. Photogrammetric determination of surface velocities and elevations on Byrd Glacier, *Ant. J. U.S.*, *1982 Rev.* **17**, 79–81.

Broecker, W. S. 1975. Floating ice cap on the Arctic Ocean. *Science* **188**, 1116–18.

Budd, W. F. 1970. The Wilkes Ice Cap project. *Int. Assoc. Sci. Hydrol. Publn.* 86, 414–29.

Budd, W. F., and B. J. Mcinnes 1979. Periodic surging of the Antarctic ice sheet—an assessment by modeling. *Hydrol. Sci.-Bull. Sci. Hydrologiques.* **24**, 95–104.

Bull, C. 1971. Snow accumulation in Antarctica. In *Research in the Antarctic*, L. O. Quom (ed.), 367–421. Washington: American Association for the Advancement of Science.

Carrara, P. 1979. Former extent of glacial ice in the Orville Coast region, Antarctic Peninsula. *Ant. J. U. S.* **14**, 45–6.

Carrara, P. 1981. Evidence for a former large ice sheet in the Orville Coast—Ronne Ice Shelf area, Antarctica. *J. Glaciol.* **27**, 487–91.

Crabtree, R. D. and C. S. M. Doake 1982. Pine Island Glacier and its drainage basin: Results from radio-echo sounding. *Ann. Glaciol.* **3**, 65–70.

Dagel, M., C. Hendy, G. H. Denton and M. Stuiver 1983. Ur/Th dates of stage 6 glacial events, Marshall Valley, Antarctica. Personal communication, in manuscript.

Denton, G. H. 1979. Glacial history of the Byrd-Darwin Glacier area, Transantarctic Mountains. *Ant. J. U.S.* **14**, 57–8.

Denton, G. H. and R. L. Armstrong 1968. Glacial geology and chronology of the McMurdo Sound region. *Ant. J. U.S.* **3**, 99–101.

Denton, G. H., R. L. Armstrong and M. Stuiver 1970. Late Cenozoic glaciation in Antarctica: the record in the McMurdo Sound region. *Ant. J. U.S.* **5**, 15–21.

Denton, G. H., R. L. Armstrong and M. Stuiver 1971. Late Cenozoic glacial history of Antarctica. p. 267–306. *In Late Cenozoic glacial ages*, K. K. Tinekian (ed.), 267–306. New Haven: Yale University Press.

Denton, G. H., H. W. Borns, Jr., M. G. Grosswald, M. Stuiver and R. L. Nichols 1975. Glacial history of the Ross Sea. *Ant. J. U.S.* **10**, 160–64.

Denton, G. H. and T. J. Hughes (eds.) 1981a. *The last great ice sheets*. New York: Wiley Interscience.

Denton, G. H., and T. J. Hughes 1981b. The Arctic Ice Sheet: an outrageous hypothesis. In *The last great ice sheets*, G. H. Denton and T. J. Hughes (eds), 437–67. New York: Wiley Interscience.

Denton, G. H. and T. J. Hughes 1983. Milankovitch theory of ice ages: Ice sheet link between regional insolation input and global climatic output. *Quat. Res.*, **20**, 125–44.

Denton, G. H. and S. C. Wilson 1983. Late Quaternary geology of the Rennick Glacier area, northern Victoria Land. *Ant. J. U.S.* **17**, 49–51.

Denton, G. H., and J. G. Bockheim, R. H. Rutford and B. G. Andersen 1984. Glacial history of the Ellsworth Mountains, West Antarctica. *Geol. Soc. Am. Memoir*, in press.

Denton, G. H., M. L. Prentice, T. B. Kellogg and D. A. Kellogg 1984. Tertiary history of the Antarctic Ice Sheet: evidence from the Dry Valleys. *Geology*, in press.

Dort, W. 1970. Climatic causes of alpine glacier fluctuation, southern Victoria Land. International Symposium on Antarctic glaciological exploration (ISAGE), Hanover, N.H. *Int. Asoc. Sci. Hydrol. SCAR Publn.* 86, 358–62.

Drewry, D. J. 1979. Late Wisconsin reconstruction for the Ross Sea region, Antarctica. *J. Glaciol.* **24**, 231–44.

Drewry, D. J. 1980. Pleistocene bimodal response of Antarctic ice. *Nature* **287**, 214–16.

Drewry, D. J. 1982. Ice flow, bedrock, and geothermal studies from radio-echo sounding inland of McMurdo Sound, Antarctica. In *Antarctic Geoscience*, C. Craddock (ed.), 977–83. Madison: University of Wisconsin Press.

Drewry, D. J. (ed.), 1983. *Antarctica: glaciological and geophysical folio*. Cambridge: Scott Polar Research Institute, University of Cambridge.

Drewry, D. J., S. R. Jordan and E. Jankowski 1982. Measured properties of the Antarctic Ice Sheet: surface configuration, ice thickness, volume, and bedrock characteristics. *Ann. Glaciol.* **3**, 83–91.

Duplessy, J. C., G. Delibrias, J. L. Turon, C. Piyol and J. Duprat 1981. Deglacial warming of the northeastern Atlantic Ocean: correlation with the paleoclimatic evolution of the European continent. *Palaeogeog. Palaeoclimat. Palaeoecol..* **35**, 121–44.

Dyke, A. S., L. A. Dredge and J. S. Vincent 1982. Configuration and dynamics of the Laurentide Ice Sheet during the last Wisconsin maximum. *Geog. Phys. Quat.* **36**, 5–14.

Elverhøi, A. 1981. Evidence for a late Wisconsin glaciation of the Weddell Sea. *Nature* **293**, 641–2.

England, J. H. 1976. Late Quaternary glaciation of the eastern Queen Elizabeth Islands, N.W.T., Canada: alternative models. *Quat. Res.* **6**, 185–202.

Fastook, J. L. and W. F. Schmidt 1982. Finite-element analysis of calving from ice fronts. *Ann. Glaciol.* **3**, 103–6.

Flint, R. F. 1943. Growth of North American Ice Sheet during the Wisconsin age. *Geol. Soc. Am. Bull.* **54**, 325–62.

Giovinetto, M. B. 1964. The drainage systems of Antarctica: accumulation. In *Antarctic Snow and Ice Studies*, Antarctic Research Series 2, 127–155. Washington: Amer. Geophys. Un.

Greischar, L. L. and C. R. Bentley 1980. Implications for the late Wisconsin/Holocene extent of the West Antarctic Ice Sheet from a regional isostatic gravity map of the Ross Embayment (abs). *Amer. Quat. Assoc., 6th Biennial Mtg. Abstr. and Progr.*, 1980.

Grosswald, M. G. 1980. Late Weichselian ice sheet of northern Eurasia. *Quat. Res.* **13**, 1–32.

Hendy, C. H. and A. T. Wilson 1981. The chemical stratigraphy of polar ice sheets—A method of dating ice cores. *J. Glaciol.* **27**, 3–9.

Hollin, J. T. 1962. On the glacial history of Antarctica. *J. Glaciol.* **4**, 173–95.

Hollin, J. T. 1969. Ice-sheet surges and the geologic record. *Can J. Earth Sci.* **6**, 903–10.

Hughes, T. J. 1972. ISCAP Bulletin Number 1: scientific justification. Ice Streamline Cooperative Antarctic Project. Institute of Polar Studies, Columbus: The Ohio State University.

Hughes, T. J. 1973. Is the West Antarctic Ice Sheet disintegrating? *J. Geophys. Res.* **78**, 7884–910

Hughes, T. J. 1975. The West Antarctic Ice Sheet: Instability, disintegration, and initiation of ice ages. *Rev. Geophys. Space Phys.* **13**, 502–26.

Hughes, T. J. 1977. West Antarctic ice streams. *Rev. Geophys. Space Phys.* **15**, 1–46.

Hughes, T. J. 1981a. Numerical reconstruction of paleo ice sheets. In *The last great ice sheets*, G. H. Denton and T. J. Hughes, (eds), 221–61. New York: Wiley Interscience.

Hughes, T. J. 1981b. The weak underbelly of the West Antarctic ice sheet (letter). *J. Glaciol.* **27**, 518–25.

Hughes, T. J. 1982a. On the disintegration of ice shelves: The role of thinning. *Ann. Glaciol.* **3**, 146–51.

Hughes, T. J. 1982b. Did the West Antarctic Ice Sheet create the East Antarctic Ice Sheet? *Ann. Glaciol.* **27**, 138–45.

Hughes, T. J. 1983. On the disintegration of ice shelves: The role of fracture. *J. Glaciol.* **29**, 98–117.

Hughes, T. J., and J. L. Fastook, 1981. Byrd Glacier: 1978–1979 field results. *Antarc. J. United States, 1981 Rev.* **16**, 86–9.

Hughes, T. J., G. H. Denton and M. G. Grosswald 1977. Was there a late-Würm Arctic Ice Sheet? *Nature* **266**, 596–602.

Hughes, T. J., J. L. Fastook and G. H. Denton 1980. Climatic warming and collapse of the West Antarctic ice sheet. *Annapolis workshop on environmental and societal consequences of a possible CO$_2$–induced climate change. CO$_2$,* **9**, 152–182.

Hughes, T. J., G. H. Denton, G. G. Andersen, D. H. Schilling, J. L. Fastook, and C. S. Ingle 1981. The last great ice sheets: a global view. In *The last great ice sheets*, G. H. Denton and T. J. Hughes (eds), p. 263–317. New York: Wiley-Interscience.

Iken, A. 1977. Movement of a large ice mass before breaking off. *J. Glaciol.* **19**, 595–605.

Jezek, K. C. and C. R. Bentley, 1983. Field studies of bottom crevasses in the Ross Ice Shelf, Antarctica. *J. Glaciol.* **29**, 118–26.

Johnson, G. L., J. R. Vanney, D. J. Drewry and G. de O. Robin 1980. *General bathymetric chart of the oceans (GEBCO)*, sheet 5.18, polar stereogaphic projection 1:6 000 000. Ottawa: Canadian Hydrographic Service.

Kellogg, T. B. and D. E. Kellogg 1981. Pleistocene sediments beneath the Ross Ice Shelf. *Nature* **293**, 130–33.

Kellogg, T. B. and D. E. Kellogg 1983. Reply to H. T. Brady, Interpretation of sediment cores from the Ross Ice Shelf Site J9, Antarctica (Nature 303, 510–11). *Nature* **303**, 511–13.

Kellogg, T. B. and R. S. Truesdale 1979. Late Quaternary paleoecology and paleoclimatology of the Ross Sea: the diatom record. *Marine Micropaleont.* **4**, 137–58.

Kellogg, T. B., L. E. Osterman and M. Stuiver 1979. Late Quaternary sedimentology and benthic foraminiferal paleoecology of the Ross Sea, Antarctica. *J. Foram. Res.* **9**, 322–35.

Kellogg, T. B., R. S. Truesdale and L. E. Osterman 1979. Late Quaternary extent of the West Antarctic ice sheet: new evidence from Ross Sea cores. *Geology* **7**, 249–53.

Lagerlund, E., G. Knutson, M. Åmark, M. Hebrand, L. Jönsson, B. Karlgren, J. Kristiansson, P. Möller, J. Robison, P. Sandgren, T. Terne and D. Waldemarsson 1983. *The deglaciation pattern and dynamics in southern Sweden.* Lund: Lund Univ. Dept of Quaternary Geology, LUNDQUA Rept 24.

Lingle, C. S., T. J. Hughes and R. C. Kollmeyer 1981. Tidal flexure of Jakobshavns Glacier, West Greenland. *J. Geophys. Res.* **86**, 3960–68.

Lorius, C., L. Merlivat, J. Jougel and M. Pourchet 1979. A 30,000-yr isotope climatic record from Antarctic ice. *Nature* **280**, 644–48.

McLauren, W. A. 1968. A study of the local ice cap near Wilkes, Antarctica. *ANARE Scientific Report* 103. Melbourne, Australia: Dept. of Supply, Antarctic Division.

Mayewski, P. A., G. H. Denton and T. J. Hughes 1981. Late Wisconsin Ice Sheets in North America. In *The last great ice sheets*, G. H. Denton and T. J. Hughes (eds), 67–178. New York: Wiley Interscience.

Mercer, J. H. 1968a. Antarctic ice and Sangamon sea level. *Int. Assoc. Sci. Hydrol. Publn* **79**, 217–25.

Mercer, J. H. 1968b. Glacial geology of the Reedy Glacier area, Antarctica. *Geol. Soc. Am. Bull.* **79**, 471–86.

Mercer, J. H. 1970. A former ice sheet in the Arctic Ocean? *Palaeogeog. Palaeoclimat. Palaeoecol.* **8**, 19–27.

Mercer, J. H. 1972. Some observations on the glacial geology of the Beardmore Glacier area. In *Antarctic geology and geophysics*, R. J. Adie (ed.), 427–33. Oslo: Oslo Universitetsforlaget.

Mercer, J. H. 1978. West Antarctic Ice Sheet and CO_2 greenhouse effect: a threat of disaster. *Nature* **271**, 321–5.

Morgan, V. I., T. H. Jacka, and G. J. Ackerman, 1982. Outlet glacier and mass-balance studies in Enderby, Kemp, and MacRobertson Lands, Antarctica. *Ann. Glaciol.* **3**, 204–10.

Myers, N. C., 1982. Marine geology of the western Ross Sea: implications for Antarctic glacial history: M. S. Thesis, Rice Univ.

Orheim O. and A. Elverhøi 1981. Model for submarine glacial deposition. *Ann. Glaciol.* **2**, 123–28.

Osterman, L. E., G. H. Miller and J. S. Stravers 1982. Late Quaternary history of southeastern Baffin Island, N.W.T., Canada. *Geol. Soc. Am. Abstr. with Programs 1982*, **14**, 581. Boulder.

Robin, G. de. 1981. Glaciology of the Ross Sea sector: contributions from the Scott Polar Research Institute. *J. R. Soc. N. Z.* **11**, 349–53.

Robin, G. de, S. Evans, D. J. Drewry, C. H. Harrison and D. L. Petrie 1970. Radio-echo sounding of the Antarctic Ice Sheet. *Ant. J. U.S.* **5**, 299–332.

Rose, K. E. 1979. Characteristics of ice flow in Marie Byrd Land, Antarctica. *J. Glaciol.* **24**, 63–75.

Ruddiman, W. F. and A. McIntyre 1981. The North Atlantic Ocean during the last deglaciation. *Palaeogeog. Palaeoclimat. Palaeoecol.* **35**, 145–214.

Rutford, R. H., G. H. Denton and B. G. Andersen 1980. Glacial history of the Ellsworth Mountains. *Ant. J. U.S.* **15**, 56–7.

Schytt, V., G. Hoppe, W. Blake, Jr. and M. G. Grosswald 1968. The extent of the Würm glaciation in the European Arctic. IUGG, General Assembly Bern, 1967, Commission of Snow and Ice, *Int. Assoc. Sci. Hydrol. Publn* **79**, 207–16.

Scott, R. F. 1905. *The voyage of the Discovery*, 2 vols. New York: Charles Scribner & Sons.

Shilts, W. W. 1980. Flow of patterns in the central North American Ice Sheet. *Nature* **286**, 213–18.

Silkonia, W. G. 1982. Finite-element glacier dynamics model applied to Columbia Glacier, Alaska. *U.S. Geol. Surv. Prof. Pap.* 1258B.

Stephenson, S. N. and C. S. M. Doake 1982. Dynamic behaviour of Rutford Ice Stream. *Ann. Glaciol.* **3**, 295–99

Stuiver, M., G. H. Denton, T. J. Hughes and J. L. Fastook 1981. History of the marine ice sheet in West Antarctica. In *The last great ice sheets*, G. H. Denton and T. J. Hughes (eds.), 319–436. New York: Wiley Interscience.

Sugden, D. E. 1977. Reconstruction of the morphology, dynamics, and thermal characteristics of the Laurentide Ice Sheet at its maximum, *Arctic Alpine Res.* **9**, 21–47.

Swithinbank, C. W. M. 1955. Ice shelves. *Geogr. J.* **121**, 62–76.

Swithinbank, C. W. M. 1957. Glaciology I. The morphology of the ice shelves of western Dronning Maud Land. *Norwegian-British-Swedish Antarctic Expedition, 1949–1952. Scientific Results*, vol. 3, B.

Swithinbank, C. W. M. 1964. To the valley glaciers that feed the Ross Ice Shelf. *Geogr. J.* **130**, 30–48.

Swithinbank, C. W. M. 1970. Ice movement in the McMurdo Sound area of Antarctica. *Int. Assoc. Sci. Hydrol. Pubn.* 86, 472–87.

Swithinbank, C. W. M. 1977. Glaciological research in the Antarctic Peninsula, *Phil. Trans. R. Soc. Lond. Ser. B*, **279**, 161–83.

Thomas, R. H. 1976. Thickening of the Ross Ice Shelf and equilibrium state of the West Antarctic Ice Sheet. *Nature* **259**, 180–83.

Thomas, R. H. 1979. The dynamics of marine ice sheets. *J. Glaciol.* **24**, 167–77.

Thomas, R. H. and C. R. Bentley 1978. A model for Holocene retreat of the West Antarctic Ice Sheet. *Quat. Res.* **10**, 150–70.

Thomas, R. H. and D. R. MacAyeal 1983. Derived characteristics of the Ross Ice Shelf. *J. Glaciol.* **28**, 397–412.

Thomson, W. 1888. Polar ice-caps and their influence on changing sea levels. *Trans. Geol. Soc. Glasgow* **8**, 322–40.

Voronov, P. S. 1960. Opyt restavratsii lednikovogo shehita Antarktidy epokhi maksimalnogo oledeneniya Zemli (Attempt to reconstruct the ice sheet of Antarctica at the time of the maximum glaciation on earth). *Informatsionnyy Byulleten' Sovetskoy Antarkticheskoy Ekspeditsii (Information Bulletin of the Soviet Antarctic Expedition)* **23**, 15–19.

Warntz, W. 1975. Stream ordering and contour mapping. *J. Hydrol. 25*, 209–27.

Weertman, J., 1957. Deformation of floating ice shelves. *J. Glaciol.* **3**, 38–42.

Weertman, J., 1973. Can a water-filled crevasse reach the bottom surface of a glacier? *Int. Assoc. Sci. Hydrol. Publn* 95, 139–45.

Weertman, J. 1974. Stability of the junction of an ice sheet and ice shelf. *J. Glaciol.* **13**, 3–11.

Wilson, A. T. 1964. Origin of ice ages: an ice shelf theory for Pleistocene glaciation. *Nature* **201**, 147–49.

3

Geological models for the configuration, history and style of disintegration of the Laurentide Ice Sheet

W. W. Shilts

Introduction

Since the great size of the Laurentide Ice Sheet was first perceived in the late 19th century, several models relating to its history and configuration have been proposed. These models, until recently, have suffered from a dearth of information about the central three quarters of the ice sheet and have been strongly biased by observations made around its unstable fringes. This bias has been particularly reflected by the stratigraphic models constructed to explain the observed glacial events around the margins in terms of the overall history of the ice sheet. One wonders what form the classic Nebraskan–Kansan–Illinoian–Wisconsin stratigraphic framework would have taken had the early glacial stratigraphers carried out their investigations among the hundreds of kilometers of exposures of the Hudson Bay Lowlands, near the ice-sheet's heart, rather than in south-central Canada and the American midwest, near its periphery.

Two distinct models for the configuration of the Laurentide Ice Sheet[1] have evolved over the past century. One model, which portrays the ice sheet as being composed of several independent centers of ice accumulation and outflow, has been proposed and championed for the most part by geologists who have carried out extensive fieldwork around Hudson Bay (i.e. Tyrrell 1898). The other model, which suggests a simple dome of ice centered more or less on Hudson Bay, has been championed largely by glaciologists and climatologists, although the model was popularized by a glacial geologist (Flint 1943) who used climatological reasoning to support his conclusions.

[1] Laurentide Ice Sheet is used in this paper both (1) in the sense of Prest (1970, p. 705) to describe the Wisconsin Laurentide Ice Sheet(s) and (2) to describe earlier ice sheets that covered the same general area.

J. B. Tyrrell, the first geologist to carry out systematic geological surveys in the arctic 'Barren Lands' west of Hudson Bay, proposed that several independent centers of ice flow coalesced or expanded and contracted independently to produce the glacial deposits of the Hudson Bay region (Tyrrell 1898). His model appears to have been supported by most glacial geologists until about 1943 when Richard Foster Flint published a major paper suggesting that, notwithstanding the geomorphological evidence cited by Tyrrell and others, the Laurentide Ice Sheet consisted, ultimately, of a single dome or center of outflow located roughly over the Hudson Bay depression (Flint 1943). Flint's argument was based largely on climatological theory, starting a trend that has persisted to today—glaciological and climatological theory (and associated computer modeling) are often most easily accommodated within the single dome concept, whereas the multiple-dome concept most easily explains the geological evidence left by the ice-sheet's passage.

In the past six years, advances in mathematical modeling techniques have allowed speculations about the last Laurentide Ice Sheet to be simulated by computers (Hughes *et al.* 1977, Denton and Hughes 1981). As geologic input, these models used Flint's model of the Laurentide Ice Sheet—i.e. one model has been based on a theory predicted by another model.

Coincident with the flourishing of mathematical and glaciological models, a considerable amount of geological and geomorphological data have been obtained from the Canadian arctic and subarctic—regions where little field information bearing on glacial geology had been obtained from the time of Tyrrell's expeditions of the 1890s until the large-scale reconnaissance operations of the Geological Survey of Canada in the 1950s. In the late 1960s and 1970s the author and many colleagues participated in a series of geological projects associated with plans to open the Canadian north to economic development. These projects produced a virtual explosion of geological and goemorphological information that has required a reevaluation of, among other things, the 'single dome' concept taught in North American universities over the past several decades.

It is the purpose of this paper to present models derived from current glacial geological information on three aspects of the Laurentide Ice Sheet: (1) its configuration and centers of outflow; (2) the latter part of its history of waxing and waning; and (3) the style of its final deterioration and retreat.

Configuration of the Laurentide Ice Sheet

The model for the configuration of the Laurentide Ice Sheet that corresponds most closely to the geological evidence presently available is that of an ice sheet comprising several centers of glacial outflow, at least during the last major glacial stage. How these independent centers interacted in space and time is not particularly clear, but evidence of their *maximum* zones of influence is unequivocal.

Abundant geomorphological evidence of ice flow directions east and west of Hudson Bay have been known for almost a century and can be mapped readily from air photographs. Drumlin and fluting orientations, esker trends, rögen and De Geer moraine orientations, and striations clearly reveal flow from two major elongate dispersal centers in Keewatin and Nouveau Québec/Labrador. Similar features are not so well developed in the high arctic or are submerged beneath marine waters, but they indicate, nevertheless, that other centers existed there, on individual arctic islands and in the Foxe and M'Clintock basins. Centers of outflow also existed on the island of Newfoundland and in Nova Scotia. Dyke *et al.* (1982) have suggested that a major late-glacial dispersal center was present over southern Hudson Bay, but in spite of their insistence that this feature was a major Wisconsin dispersal center, there is no direct evidence of its existence.

On geomorphological grounds alone, the last Laurentide Ice Sheet can be visualized as comprising at its maximum two huge ice dispersal centers, in Keewatin and Nouveau Québec/Labrador, from which ice flowed to cover most of the central part of northern North America. The ice mass formed by coalescence of glaciers flowing from these centers was surrounded by highland or marine-based ice dispersal centers on the west (Cordilleran Ice Complex), north (Arctic Islands complex and M'Clintock and Foxe Basin ice domes), and east (Newfoundland and eastern Appalachian ice flow centers). The Keewatin and Nouveau Québec/Labrador ice masses were in contact along shifting zones of confluence, depending on their relative vigor, and their peripheries were undoubtedly confluent at various times with the independent ice masses that ringed them in a semicircle from west to north to east.

Recently acquired data on dispersal patterns of distinctive rock and mineral debris (Shilts *et al.* 1979, Shilts 1980b, Hardy 1976, Hillaire-Marcel *et al.* 1980, Shilts 1982) argues strongly for this model and against a single dome model in which most of the centers mentioned above are thought to have been late-glacial features, formed as the North American ice sheet broke up late in the last glacial cycle.

The bedrock geology under the central part of the Laurentide Ice Sheet includes distinctive lithologic units that underlie large areas and produced easily traced erratics. Figure 3.1 shows the patterns, as presently known, of dispersal of late Proterozoic red sedimentary and volcanogenic rocks (Dubawnt Group of Donaldson 1965) from the Baker Lake region,[1] Proterozoic dark gray to black metagraywackes and metavolcanic rocks from the Circum-Ungava Geosyncline (Dimroth *et al.* 1970) which underlies the eastern part of Hudson Bay, and early to middle Paleozoic and Mesozoic sedimentary rocks (Sanford *et al.* 1979), which underlie most of Hudson Bay and the

---

[1] Although dispersal of Dubawnt debris is shown only eastward on Figure 3.1, the Keewatin Ice Divide lies across the Dubawnt outcrops. West and north of the divide dispersal is westward or northward, approximately as shown in the northwest corner of Figure 3.2. The latter dispersal is not shown because it is based on spot observations and has not been mapped in detail as has the pattern of eastern dispersal, which is based on quantitative analyses of many hundreds of samples (Shilts 1982).

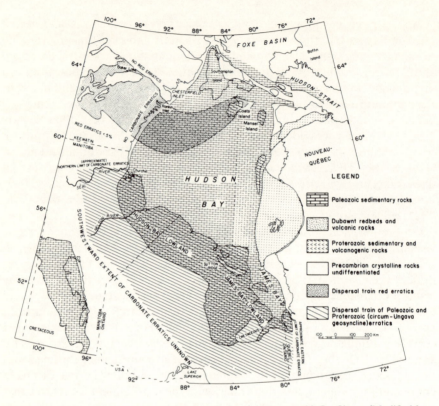

Figure 3.1 Major dispersal trains of the central part of the Laurentide Ice Sheet. (Modified from Shilts, 1980b. Copyright (c) 1980 by Macmillan Journals Ltd. Used by permission.)

lowlands west and south of the Bay. These dispersal patterns are partially the result of the integration of dispersal during each of several glacial events known to have affected the region, as deduced from stratigraphic exposures in the Hudson Bay Lowland. When possible, the patterns were drawn from data on frequencies of erratics in or derived from the youngest till. Older tills, where exposed, generally are characterized by similar pebble lithologies, mineralogy, and geochemistry in the Hudson Bay Lowlands, but recent detailed work in the Severn River drainage basin shows significant differences in frequencies of specific erratics among tills. Mineralogic studies of till and derived sediments from the bottom of Hudson Bay yield similar dispersal data (Adshead 1983, Henderson 1983a, 1983b), as do heavy mineral studies of cores collected from boreholes drilled west of Hudson Bay (Paré 1982). Detailed drift geochemical surveys (Shilts 1975, 1976, 1980a, Ridler and Shilts 1974, Klassen and Shilts 1977) west of Hudson Bay all support unidirectional (eastward to southeastward) ice flow east of the Keewatin Ice Divide. These quantitative studies are backed by hundreds of more casual observations by the author and other workers (for example, Nielsen and Dredge 1982, Hardy 1976, Dionne 1974).

The most important conclusions from the observations of glacial dispersal are as follows: (1) North of Seal River, Manitoba, no Paleozoic carbonate erratics from under Hudson Bay are found in till on land west of Hudson Bay; (2) likewise, no Paleozoic erratics are reported found in till east of Hudson Bay, except in the clayey till deposited during the late-glacial Cochrane surges; (3) no eastward displacement of erratics has been observed from Proterozoic outcrops in the eastern part of Hudson Bay; (4) both Paleozoic and Proterozoic erratics are dispersed southward and southwestward in a vast train extending from southern Hudson and James Bays as far west as western Manitoba and as far south as Lake Superior; (5) red, Proterozoic erratics form a distinctive train extending from the Baker Lake area in the District of Keewatin, at least to the mouth of Hudson Bay, a distance of over 600 kilometers.

From these dispersal data and from geomorphological evidence, the author and others (Shilts 1980b, Shilts *et al.* 1979) have attempted to reconstruct flow lines and configurations of ice dispersal centers for the central part of the last Laurentide Ice Sheet. These attempts are partially confounded by (1) the difficulty of determining when the dispersal took place with respect to the time of formation of the geomorphic features, (2) the fact that much of the area is hidden beneath Hudson Bay, (3) the demonstrable shifting of position of centers of outflow, and (4) the shifting positions of zones of confluence of ice from the various centers of outflow. These difficulties notwithstanding, preliminary maps have been drawn of possible configurations of flow at some times during the last major glaciation. These 'cartoon' reconstructions are based largely on inferences about the areas influenced by the two major dispersal centers, inferences drawn from dispersal data (Figs. 3.2 and 3.3) and from knowledge of the sequence of events that occurred as the ice sheet retreated. Figure 3.2 is thought to represent flow patterns at some time at or near the last glacial maximum. Figure 3.3 represents a possible flow scenario when the Keewatin ice center was so active that it displaced Labrador ice in much of Hudson Bay, possibly near the end of the last glaciation. It accounts for southeastward dispersal and southeastward oriented striae observed here and there throughout the Lowlands. The zone of confluence (which is *not* a zone of opposed flow as some researchers have inferred) probably shifted throughout the Hudson Bay basin from Seal River, just north of where the zone is depicted on Figure 3.2 to near the Quebec-Ontario border as depicted on Figure 3.3. The extent to which one ice mass displaced the other depended on their relative rates of nourishment and resulting vigor.

Figure 3.2 has provoked some controversy because of the apparent east-west assymetry of the flow depicted from the Nouveau Québec/Labrador dispersal center (Dyke *et al.* 1982, Denton 1980, personal communication). Dyke *et al.* (1982) have proposed that another center of glacial outflow, the Hudson Dome, existed over the southwest part of Hudson Bay, largely over an area presently covered by water. They thought that the westward dispersal of eastern Hudson Bay Proterozoic erratics and Paleozoic erratics was accomplished by reworking of already deposited debris by ice from the Hudson

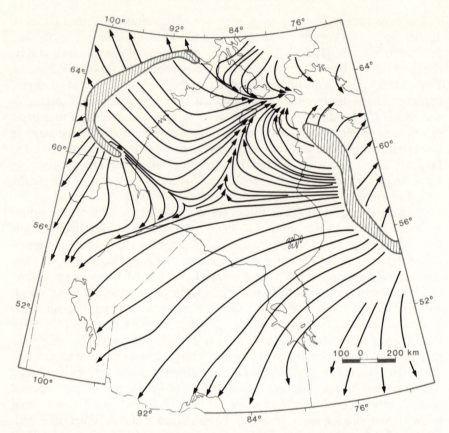

Figure 3.2 Possible ice flow configuration during maximum development of Laurentide Ice Sheet. Note zones of *confluence* of Keewatin and Labrador Ice Sheets in Hudson Bay. Zones of confluence shifted depending on vigor and location of the two major ice dispersal centers. Arrows based on dispersal and geomorphic data. (Modified from Shilts 1980b. Copyright (c) 1980 by Macmillan Journals Ltd. Used by permission).

Dome. Although their model seems to remedy the assymetry problem, there is no direct evidence for its existence, and, in fact, McDonald (1969)[1] specifically rejected the possibility of Tyrrell's (1914) Patrician ice dome, a feature Dyke *et al*. (1983) say is similar in configuration to their Hudson Dome. Isostatic recovery patterns, gravity anomalies, the configuration of the Cochrane readvance, and the configuration of end moraines in Quebec and Manitoba lend some support to the existence of a Hudson Dome, but can be explained equally well, in my opinion, by reconstructions such as that depicted by Figure 3.3.

In any case, whether the Hudson Dome existed or not, most of those formerly or presently engaged in Quaternary research in the Canadian arctic

---

[1] '... bedrock striations, striated boulder pavements, fabric studies, ice-flow features exposed at the surface, and pebble counts indicate ice flow from northeast, north, and northwest. No acceptable evidence was found to support the contentions of Tyrrell ... that "Patrician" glaciers flowed northward across the region.' (McDonald 1969, p. 91).

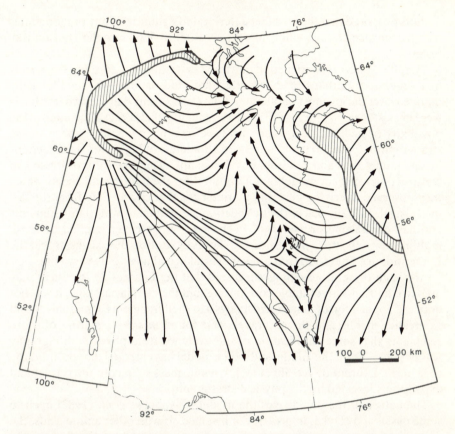

Figure 3.3 Possible ice flow configuration early or late in development of Laurentide Ice Sheet. There is dispersal and striation evidence for this configuration in Hudson Bay Lowland. Note how zones of confluence have shifted from figure 2.

and subarctic recognize the Laurentide Ice Sheet as a complex of two or more ice domes which, when confluent, formed a continuous glacier cover over the northcentral part of North America (Shilts *et al.* 1979, Andrews and Miller 1979, Hillaire-Marcel *et al.* 1980, Shilts 1980b, Prest, 1983, Tyrrell 1898, 1914). The single dome model is increasingly less attractive to Quaternary geologists as field evidence supporting the multi-dome model accumulates.

Model for glacial and non-glacial events affecting the center of the Laurentide Ice Sheet

Andrews *et al.* (1982) have recently proposed a new and controversial model for the sequence of glacial events in the Hudson Bay region. Rather than each glacial stage being represented by continuous ice cover over the Bay, with only the unstable outer margins of the Laurentide Ice Sheet fluctuating, these authors proposed that the ice sheet deteriorated significantly one or more times within each glacial stage, allowing marine water to flood into the Hudson Bay basin.

Tyrrell's (1898) reconstructions of the Keewatin and Labrador Ice Sheets first suggested, perhaps inadvertently, that Hudson Bay was inundated by marine water during most or all of the time (Wisconsin) that the two ice sheets were in existence. Skinner (1973) and McDonald (1969, 1971) also hinted at the possibility of such a model when they tried to explain physical aspects of the stratigraphy exposed in the hundreds of kilometers of high river-bank sections in the James Bay and Hudson Bay Lowlands. They were hard-pressed to construct an ice sheet model that accounted for the multiple till sheets which they observed overlying peat and marine beds that clearly date from the last interglacial. In many places the till sheets are separated by waterlaid sediments and had contrasting provenance. Each break in ice cover implied by these multiple units, located virtually on the shores of Hudson Bay, suggests that the Laurentide Ice Sheet suffered a severe shrinkage almost to its very core.

Although these observations suggested that the Laurentide Ice Sheet may not have had a massive, stable core throughout the last glacial stage, it was not until the application of amino acid analysis to shell fragments commonly found as erratics in tills and fluvial gravels and found *in situ* in marine offshore sediments that some clarification of the evidence from physical stratigraphy was possible. From the animo acid data a model has emerged that portrays the core of the Laurentide Ice Sheet as a dynamic mass subject to the retreat and expansion suggested by the physical stratigraphy.

The nature of the data leaves the model of Andrews *et al.* (1982) open to some discussion (Dyke, in press). For instance, as with other amino acids, the epimerization rate[1] for isoleucine, the amino acid on which ratios determined at the University of Colorado are based, is dependent on temperature history as well as time. Fluctuations in temperature since burial may render relative or absolute age estimates invalid. Nevertheless, we have argued that near-surface sediments in the Hudson Bay Lowlands, which now lie largely within the zone of discontinuous permafrost with a mean annual air temperature near 0°C, would not have been subjected to wide ranges in temperature during the Quaternary, regardless of whether a site was covered by ice or water or was exposed to the air. If this assumption is correct, then it should be possible to compare ratios (total aIle:Ile) of shells from various stratigraphic units in a relative sense and should even be possible to calculate *rough* absolute ages from the kinetics of amino acid racemization (Miller & Hare 1980, p. 431–3).

When the above reasoning was applied to the amino acid ratios of shells collected from various stratigraphic units throughout the Lowlands, we

[1] Epimerization is the process by which protein amino acid molecules (such as isoleucine) containing two chiral carbon atoms are converted gradually to 'mirror image', non-protein amino acids (D-alloisoleucine) after an organism dies. The ratio, aIle:Ile, of D-alloisoleucine to its protein diasteriomer, L-isoleucine, is a measure of the extent of epimerization of isoleucine and is the primary geochronological index used in the amino acid laboratory at the University of Colorado (Andrews *et al.* 1983, p. 23, Rutter *et al*, 1979).

expected to observe two groups of ratios representing interglacial deposits of the Bell Sea (Skinner 1973) and postglacial deposits of the Tyrrell Sea (Lee 1960). The ratios for the two marine deposits were established by analyzing *in situ* shells collected from marine beds known to have been deposited during these two events. We found, however, that in addition to numerous shells with ratios similar to the Bell Sea ratios, at least one and possibly two groups of shells had intermediate ratios, particularly shells collected from the younger tills in two-till sequences. Although there is considerable scatter of the data due both to natural causes and to changes in laboratory technique, the presence of shells of intermediate ratios is statistically confirmed. In addition, *in situ* variation tests (Fig.3.4) and comparison of the data with known stratigraphic relations (Andrews *et al.* 1982 p. 28, 29) leave little doubt that the intermediate groups of ratios are real.

If the intermediate amino acid ratios are reliable indicators of age, as we think, Hudson Bay must have been open at some time between deposition of last interglacial Bell Sea sediments and post-glacial Tyrrell Sea sediments. These intermediate ratios have been found not only in tills, but in low-altitude fluvial gravels and marine sediments that separate two tills that postdate interglacial (?) peats in the Fawn–Severn–Hayes River basins near the Manitoba–Ontario border in the Lowland. The physical presence of fluvial gravels separating what are interpreted to be two post-interglacial (Wisconsin) tills and the draping of the upper of these two tills over fluvial channels cut in the lower (P. Wyatt, pers. comm. 1983) alone require at least a partially ice-free Hudson Bay at some time during the Wisconsin Stage, regardless of what the amino acid data indicate. This is because during any ice-free interval in the Hudson Bay Lowlands the land would be flooded by a southward draining proglacial lake if ice blocked the connection of the Hudson Bay depression to the sea. No fluvial depositional or erosional phenomena could form and no weathering could take place at the bottom of a lake.

Having established the high probability of at least one mid-Wisconsin marine event in Hudson Bay, both on physical and amino-stratigraphic evidence, we used amino acid kinetics to estimate a *rough* age for the most prominent of the intermediate amino acid groups, the one associated with the presumed mid-Wisconsin fluvial events (Andrews *et al.* 1982 p. 26–9). As stated above, we assumed that the shells had had an integrated thermal history close to 0°C. Based on these assumptions, the intermediate group was found to have a mean age of about 76 ka, probably with a ± error of several thousand years. This is similar to the estimated age of the St. Pierre interstadial of the Great Lakes–St. Lawrence Lowlands. It suggests that the St. Pierre, and possibly other interstadials, were not just confined to the south, but represented a major retreat that opened up the very heart of the Laurentide Ice Sheet to marine inundation.

If the model described above is correct, the Laurentide Ice Sheet suffered severe shrinkage one or more times during the Wisconsin. During each shrinkage the Hudson Bay depression was at least partially inundated by marine water. The model stands in sharp contrast to the conventional view that

ADAM CREEK SECTION

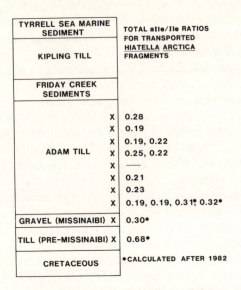

	TOTAL aIle/Ile RATIOS FOR TRANSPORTED HIATELLA ARCTICA FRAGMENTS
TYRRELL SEA MARINE SEDIMENT	
KIPLING TILL	
FRIDAY CREEK SEDIMENTS	
ADAM TILL	X 0.28
	X 0.19
	X 0.19, 0.22
	X 0.25, 0.22
	X —
	X 0.21
	X 0.23
	X 0.19, 0.19, 0.31*, 0.32*
GRAVEL (MISSINAIBI) X	0.30*
TILL (PRE-MISSINAIBI) X	0.68*
CRETACEOUS	*CALCULATED AFTER 1982

Figure 3.4 Amino acid (total aIle:Ile) ratios from *Hiatella arctica* fragments in profile samples of Adam Till at its type section, mouth of Adam Creek, Ontario (Skinner, 1973). Note minimal in-site variability, comparable ratios between till and underlying Missinaibi (interglacial?) gravel, and sharply higher ratio in pre-Missinaibi till.

the central part of the Laurentide Ice Sheet was more-or-less stable through each glacial stage.

Model for disintegration of Laurentide Ice Sheet

In the course of compiling a map of the surficial geology of southern District of Keewatin, an area of approximately 275 000 km^2 (about 90% of the combined area of New England and New York), data bearing on several possible models of deglaciation were examined. Among the most striking geomorphic features of this region are the massive esker systems which radiate like the spokes of a wheel from the region of the Keewatin Ice Divide (Lee *et al.* 1957) (Fig. 3.5). It is my opinion that a model that explains both regional esker distribution and esker and associated glaciofluvial sedimentation can provide the key to defining the model for the conditions under which much of the Laurentide Ice Sheet disintegrated.

Esker systems
In order to interpret the large-scale ice-sheet hydrology suggested by patterns of distribution of esker deposits and associated meltwater features, it is necessary to develop a model for drainage of the decaying ice sheet. The eskers

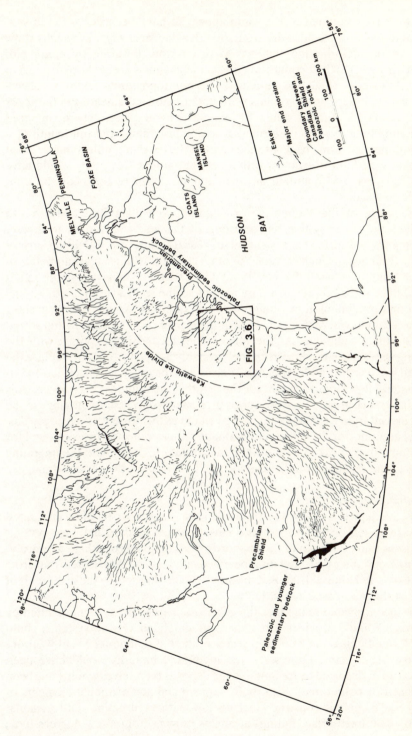

Figure 3.5. Esker systems radiating from Keewatin Ice Divide, site of last remnants of Laurentide Ice Sheet west of Hudson Bay.

that we have examined in Keewatin appear to have been deposited in sub-glacial tunnels. We deduce this from the following facts: (1) they usually show none of the deformation that would result from lowering by melting of underlying ice; (2) their connecting meltwater channels are cut into till down to bedrock; (3) they climb and descend over major topographic features, suggesting that their depositing streams were under hydrostatic head. On the other hand, it may be unreasonable to assume that an esker, several hundred kilometers long, was being formed synchronously along the full length of a basal tunnel extending back into the heart of the ice sheet; unless the ice sheet remnant was thin throughout, with a very flat gradient, it surely would have been too plastic to maintain an open tunnel anywhere but near the thin, retreating edge.

It seems to me that the most reasonable way to explain these features is to consider the esker system as consisting of two parts—a supraglacial river draining the ice surface and a short, subglacial tunnel into which the surface streams plunged through crevasses or other conduits within a few kilometers of the ice front (St. Onge 1984, Ives 1967). Once such a system was established, it would have retreated headward by melting as the ice front retreated. At any given time, sedimentation would only be taking place in the relatively short tunnel segment where the basal load of the glacier could be tapped and sorted into stratified sediment. As the tunnel sides melted from contact with the flowing water, ice lateral to the tunnel probably flowed toward it to replace the wasted ice, bringing additional basal debris into the esker system (Repo 1954). This lateral flow would account for the fact that the sediment mass in eskers is generally much greater than the debris presumed to have been contained in the ice removed from a tunnel of the same volume as the present esker ridge. This model accounts for the integrated drainage pattern reflected by eskers and for the bifurcation of eskers into lower order tributaries upstream; the migrating heads of the segments would have followed the trace of the surface drainage.

Distribution of eskers

Figure 3.6 depicts a major Keewatin esker that has the aspect of an integrated drainage system which corresponds closely to the modern drainage system. Almost nothing is known of the offshore extension of this or other eastwardly-flowing Keewatin eskers that disappear beneath Hudson Bay. Eskers that cross both Coats and Mansel Islands at the mouth of the Bay may be the distal ends of some of the land-based systems.

It is inconceivable to me that the esker drainage patterns demonstrated by Figures 3.5 and 3.6, which are classic but simplified Horton systems, could have formed unless the Keewatin sector of the Laurentide Ice Sheet was very inactive or stagnant. Otherwise, the integrated patterns would have been broken up or distorted by ice flow and the deposits themselves would have been deformed or obliterated. Also, both depositional and erosional elements of many eskers cut both constructional ice-flow features (drumlins) and erosional ice-flow features (striae, fluting) at angles ranging from a few degrees to as

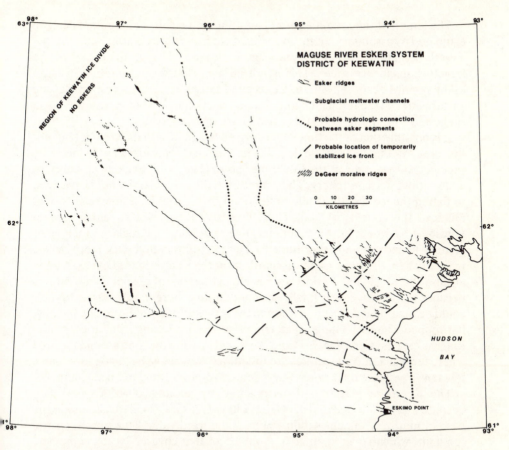

Figure 3.6 Detail of the Maguse River esker system, the location of which is shown on Figure 3.5. Note clusters of short eskers between trunk streams along belts where retreat of ice front, which was standing in >100 m of marine water of Tyrrell Sea, is inferred to have slowed temporarily.

much as 90°, a situation that is difficult to reconcile with vigorously moving ice. The esker swarm radiating from the Keewatin Ice Divide has definite down-ice terminations, which in many places are marked by major end moraine systems. End moraines are rare within the swarm. From these observations I conclude that the ice in which the eskers were deposited had, for all intents and purposes, ceased to flow. In the case of the Keewatin Ice Sheet, the dimensions of the sheet when ice flow ceased were immense, the stagnant remnant having a diameter of more than 1300 km (Fig. 3.5). By analogy, a slightly smaller remnant would have covered Labrador/Nouveau-Québec and minor satellite remnants were probably scattered throughout the arctic. In some cases, such as the Keewatin and Labrador-Quebec remnants, the remnant ice shrank back approximately to where I infer that the ice sheets originated; in others, notably on Melville Peninsula (Sim 1960), the configurations of the remnants were artifacts of the pattern and style of stagnation.

The length of time required for dissipation of the Keewatin remnant can be estimated by assuming that the invasion of Hudson Bay by marine waters of the Tyrrell Sea about 7800–7900 years ago took place through the center of the largely stagnant ice sheet. This is not an unreasonable assumption because esker systems occur seaward to Coats and Mansel Islands, suggesting that the eastern edge of the remnant approximated the initial route of marine incursion suggested by Skinner (1973), Hardy (1976), and others.

It is possible that at the time of the Tyrrell Sea incursion the volume, but not the area of the Laurentide Ice Sheet had been considerably reduced by repeated surges, such as the Cochrane surges (Hardy 1976) along its southern and southwestern periphery. The net result of the volume loss would have been to flatten the ice sheet's profile to the point that it could not flow readily over much of the area it covered. Thus, extensive ice masses on either side of Hudson Bay would have become inactive, with large supraglacial meltwater systems terminating in esker tunnels. The ice front melted back more or less regularly toward its geographical center, near the Keewatin Ice Divide. A zone around the ice divide 100 km wide is almost devoid of glaciofluvial features because the ice cover was finally so thin that no coherent channels or tunnels could be maintained. The transition to these conditions is indicated by the low, flat-topped channel fillings that comprise the upstream tributaries of the Maguse esker system (Fig. 3.7) and are found in a narrow zone around the area where few glaciofluvial features are found. Ice could have been no more than a few tens of meters thick when these deposits were formed in open channels.

The Keewatin glacier had disappeared by about 6000–6500 years BP, because marine shells dated at 6600 ± 230 yr BP (GSC-1434) are found at the base of marine sediment within 70 km of the ice divide. Thus, the Keewatin remnant, which covered an area about 1/5 of that covered by the composite Laurentide Ice Sheet at its maximum extent, persisted for a maximum of only 1400–2000 years, was largely stagnant, and had an average rate of ice-front retreat of about 325–450 m/yr.

Conclusions

This necessarily brief discussion of various models relating to the growth and development of the last, and presumably, earlier Laurentide Ice Sheet(s) demonstrates that new concepts, mostly stemming from research in remote, logistically difficult areas, must be considered in any synthesis of the geologic history of North America over the last million years. Alternative and classic ice sheet models, particularly those derived from the sparse subarctic and arctic data base that existed before about 1975, must be re-examined. Specifically, the number and areas of influence of major dispersal centers must accommodate the known geologic facts. I believe that at least two major centers of outflow existed, one on either side of Hudson Bay, throughout the existence of each Laurentide Ice Sheet and that several other major and minor dispersal centers

Figure 3.7 Low sun angle photo of flat-topped esker remnants in headwater region of Maguse River esker. At time of deposition ice was so thin that tunnels could not be formed and sediment was deposited on bedrock exposed in shallow ice channels. (Geol. Surv. Can. Photo No. 203534-y).

existed as satellites around the edges of their ice sheds. Accurate concepts of the history and configuration of these centers are prerequisite for subsequent modeling of paleoclimate, ice sheet dynamics, crustal and mantle rheology (as deduced from isostatic responses), and for various types of applied research, including mineral prospecting, oceanography of the continental shelves and epieric seas, environmental geochemistry, etc.

In addition to supporting models of the configuration of ice centers and associated flow, these data challenge the very concepts of glacial stages and the classical North American glacial stratigraphy. Study of physical stratigraphy supported by amino acid data of a small proportion of the hundreds of kilometers of stratigraphic sections exposed in the Hudson Bay Lowland has led to the conclusion that the Laurentide Ice Sheet decayed through its geographical 'heart' during what has been considered traditionally as the Wisconsin Stage. During this decay, marine waters invaded Hudson Bay, and at least some parts of its shores were ice free. The most likely age of this decay was a few thousand years on either side of 76 ka, but there is some suggestion of a similar decay somewhat later, perhaps in the 30–40 ka range (Andrews *et al.* 1982).

Lastly, a model for some aspects of the final disintegration of the Keewatin sector, and by analogy for other sectors, of the Laurentide Ice Sheet has been described, based largely on the patterns of late-glacial glaciofluvial sedimentation. Assuming that the integrated drainage patterns of the Keewatin-sector eskers are indicative of sluggishly flowing or stagnant ice, an old but well-substantiated concept, it can be argued that a very large part, 1/2 to 2/3, of the Laurentide Ice Sheet decayed by backwasting and downwasting in about 2000 years, starting 8000 years ago. Although the area covered by this ice was large, the ice must have been much thinner than it was at the glacial maximum. The eskers are a reflection of the supraglacial drainage developed on this sheet. They represent deposition of reworked basal glacial debris in tunnels extending a few kilometers back from the ice front to where the supraglacial drainage plunged to the glacier's base through crevasses or other conduits. The conduits and tunnels were extended headward as the ice front receded by melting up-stream along the courses of the supraglacial trunk and tributary streams.

Finally, it should be stressed that the models presented here are based on geological and geomorphological data and stand in contrast to some glaciological models that are currently popular. The glacial geology of the remote regions in which the bulk of the Laurentide Ice Sheet was located has only been studied in reconnaissance fashion since the advent of helicopter surveys and good air photographs in the early 1950s, with 'detailed' studies being carried out from the late 1960s to the present. Thus, our knowledge of the Canadian arctic and the research effort expended on it are probably comparable with or even less than what is known or what has been expended in the Antarctic. Nevertheless, although much critical geological information has yet to be collected, and the models suggested or accepted today are bound to change as a result, all conceptual models must accommodate the geologic facts as we presently know them. If a glaciological principle is violated by a model based on the geological facts, then perhaps the assumptions on which the principle is based should be examined, rather than the facts themselves, or the model. Two areas where this suggestion is particularly applicable are in the controversy about the apparent assymetry of the flow from the Labrador-Nouveau Québec ice dispersal center (Dyke *et al.* 1982, Denton, pers. comm. 1980) and in the suggestion by Denton and Hughes (1981) and Hughes *et al.* (1985) that much of the dispersal pattern of erratics around Hudson Bay is related to late-glacial ice streams. In the first case, the dispersal trends of east Hudson Bay (Proterozoic) erratics clearly require a real assymetry and do not lend any support to the possibility of another dispersal center over southern Hudson Bay. In the second case, the widespread late-glacial stagnation of the Keewatin sector of the ice sheet (as well as the flat topography on which it rested) argues against the presence of late glacial ice streams.

Acknowledgments

R. N. W. DiLabio and F. H. Muller made helpful comments on early versions of this manuscript. Most of their suggestions have been incorporated into the manuscript, but the author assumes full responsibility for conclusions drawn.

References

Adshead, J. D. 1983. Hudson Bay river sediments and regional glaciation: III. Implications of mineralogical studies for Wisconsinan and earlier iceflow patterns. *Can. J. Earth Sci.* **20**, 313–21.

Andrews, J. T. and G. H. Miller. 1979. Glacial erosion and ice sheet divides, northeastern Laurentide Ice Sheet, on the basis of the distribution of limestone erratics. *Geology* **7**, 592–6.

Andrews, J.T., W. W. Shilts and G. H. Miller. 1983. Multiple deglaciations of the Hudson Bay Lowlands, Canada, since deposition of the Missinaibi (Last-interglacial?) formation. *Quat. Res.* **19**, 18–37.

Denton, G. H. and T. J. Hughes (eds.). 1981. *The last great ice sheets*. New York: John Wiley.

Dimroth, E., W. R. A. Baragar, R. Bergeron and G. D. Jackson 1970. The filling of the circum-Ungava geosyncline. In *Symposium on basins and geosynclines of the Canadian shield*, A. J. Baer (ed.), 45–142. Geol. Surv. Can. Paper 70–40.

Dionne, J. C. 1974. The eastward transport of erratics in James Bay area, Québec. *Revue Géogr. Montréal* **28**, 453–7.

Donaldson, J. A. 1965. *The Dubawnt Group, Districts of Keewatin and MacKenzie*. Geol. Surv. Can., Pap. no. 64–20. Ottawa.

Dyke, A. S., L. A. Dredge and J.-S. Vincent 1982. Configuration and dynamics of the Laurentide Ice Sheet during the Late Wisconsin maximum. *Géog. phys. Quat.* **36**, 5–14.

Dyke, A. S., L. A. Dredge and J.-S. Vincent 1983. Canada's last great ice sheet. *Geos—Energy, Mines Resour. Can.* **12**, no. 4, 6–9.

Dyke, A. S. (in press). Multiple deglaciations of the Hudson Bay Lowlands, Canada, since deposition of the Missinaibi (Last-interglacial?) Formation: Discussion. *Quat. Res.*

Flint, R. F. 1943. Growth of the North American ice sheet during the Wisconsin age. *Bull. Geol Soc. Am.* **54**, 325–62.

Hardy, L. 1976. *Contribution à l'étude géomorphologique de la portion québécoise des basses terres de la baie de James*. Ph.D. thesis, McGill Univ., Montreal.

Henderson, P. J. 1983a. A study of the heavy mineral distribution in the bottom sediments of Hudson Bay. In *Current Research*, Geol Surv. Can. Pap. no. 83–1A, 347–51. Ottawa.

Henderson, P. J. 1983b. A study of the heavy mineral distribution in the bottom sediments of Hudson Bay: reply. In *Current Research*, Geol Surv. Can. Pap. no. 83–1B, 435–6. Ottawa.

Hillaire-Marcel, C., D. R. Grant and J.-S. Vincent 1980. Comment and reply on 'Keewatin ice sheet-reevaluation of the traditional concept of the Laurentide Ice Sheet' and 'Glacial erosion and ice sheet divides, northeastern Laurentide ice sheet, on the basis of the distribution of limestone erratics'. *Geology* **8**, 466–8.

Hughes, T., G. H. Denton and M.G. Grosswald 1977. Was there a late Würm Arctic ice sheet? *Nature* **266**, 596–602.

Hughes, T. J., G. H. Denton and J. L. Fastook 1985. The Antarctic Ice Sheet: an analog for northern hemisphere paleo-ice sheets? In *Models in geomorphology*, M. J. Woldenberg (ed.), 93–117. Boston: Allen & Unwin.

Ives, J. D. 1967. Glacier terminal and lateral features in northeast Baffin Island: illustrations with descriptive notes. *Geogr. Bull.* **9**, 106–14.

Klassen, R. A. and W. W. Shilts 1977. Glacial dispersal of uranium in the District of Keewatin, Canada. In *Prospecting in Areas of Glaciated Terrain*, Michael J. Jones (ed.), 80–88. London: Inst. of Mining and Metallurgy.

Lee, H. A., B. G. Craig and J. G. Fyles 1957. Keewatin ice divide. (abs.) *Bull. Geol. Soc. Am.* **68**, 1760–1.

Lee, H. A. 1960. Late glacial and post-glacial Hudson Bay sea episode. *Science* **131**, 1609–1611.

McDonald, B. C. 1969. Glacial and interglacial stratigraphy, Hudson Bay Lowland. In *Earth science symposium on Hudson Bay*, P. J. Hood (ed.), 78–9. Geol Surv. Can. Pap. no. 68–53. Ottawa.

McDonald, B. C. 1971. Late Quaternary stratigraphy and deglaciation in eastern Canada. In *The late Cenozoic glacial ages*, K. Turekian (ed.), 331–53. New Haven, Conn.: Yale Univ. Press.

Miller, G. H. and P. E. Hare 1980. Amino acid geochronology: integrity of the carbonate matrix and potential of molluscan fossils. In *Biogeochemistry of amino acids*. P. E. Hare, T. C. Hoering, and K. King, Jr. (eds). 415–43. New York: John Wiley.

Nielsen, E. and L. Dredge 1982. Quaternary stratigraphy and geomorphology of a part of the lower Nelson River. *Geol. Assoc. Can.* Winnipeg Section; Field trip no. 5, 1982.

Paré, D. 1982. *Application of heavy mineral analysis to problems of till provenance along a transect from Longlac, Ontario to Somerset Island.* M.A. thesis, Carleton Univ., Ottawa.

Prest, V. K. 1970. Quaternary geology of Canada. In *Geology and economic minerals of Canada*, R. J. W. Douglas (ed.), 676–758, Geol Surv. Can. Econ. Geol. Series, No. 1, 5th ed. Ottawa.

Prest V. K. In press. The Late Wisconsinan glacier complex. In *Quaternary stratigraphy of Canada—a Canadian contribution to IGCP Project 24*, R. J. Fulton (ed.), Geol Surv. Can. Paper 84–10, Map 1584A.

Repo, R. 1954. Om forhallandet mellan rafflor och asar (On the relationship between striae and eskers). *Geologi* **6**, no. 5, 45.

Ridler, R. H. and W. W. Shilts 1974. *Exploration for Archean polymetallic sulphide deposits in permafrost terrains: an integrated geological/geochemical technique, Kaminak Lake Area, District of Keewatin.* Geol Surv. Can. Pap. no. 73–34. Ottawa.

Rutter, N. W., R. J. Crawford and R. D. Hamilton 1979. Dating methods of Pleistocene deposits and their problems: IV. amino acid racemization dating. *Geosci. Can.* **6**, no. 3, 122–8.

St. Onge, D. A. 1974. Surficial deposits of the Redrock Lake area, District of MacKenzie. In *Current Research* Geol Surv. Can. Pap. no. 84–1A, 271–7. Ottawa.

Sanford, B. V., A. C. Grant, J. A. Wade and M. S. Barss 1979. *Geology of eastern Canada and adjacent areas.* Geol. Surv. Can. Map 1401A, 4 sheets. Ottawa.

Shilts, W. W. 1975. Principles of geochemical exploration for sulphide deposits using shallow samples of glacial drift. *Can. Min. Metall. Bull.* **68**, no. 757, 73–80.

Shilts, W. W. 1976. Mineral exploration and till. In *Glacial Till*, R. F. Legget (ed.), 205–24. Roy. Soc. Can. Spec. Publn no. 12. Ottawa.

Shilts, W. W., C. M. Cunningham and C. A. Kaszycki 1979. Keewatin ice sheet—re-evaluation of the traditional concepts of the Laurentide Ice sheet. *Geology* **7**, 537–41.

Shilts, W. W. 1980a. Geochemical profile of till from Longlac, Ontario to Somerset Island. *Can. Min. Metall. Bull.* **73**, no. 822, 85–94.

Shilts, W. W. 1980b. Flow patterns in the central North American ice sheet. *Nature* **286**, 213–18.

Shilts, W. W. 1982. Quaternary evolution of the Hudson/James Bay region. *Le Naturaliste Canadien* **109**, 309–32.

Sim, V. W. 1960. A preliminary account of late 'Wisconsin' Glaciation in Melville Peninsula, Northwest Territories. *Can. Geogr.* **17**, 21–4.

Skinner, R. G. 1973. *Quaternary stratigraphy of the Moose River basin, Ontario.* Bull. Geol Surv. Can. no. 225.

Tyrrell, J. B. 1898. The glaciation of north central Canada. *J. Geol.* **6**, 147–60.

Tyrrell, J. B. 1914. The Patrician glacier south of Hudson Bay. In *Congrés Géol. Internat., Compte-Rendus de la XIIe Session*, Canada. 523–37.

4

Patterns of glacial erosion and deposition around Cumberland Sound, Frobisher Bay and Hudson Strait, and the location of ice streams in the Eastern Canadian Arctic

J. T. Andrews, J. A. Stravers, and G. H. Miller

Introduction

In recent years there has been active debate on a number of problems associated with the glacial geology of the Canadian Arctic. These debates have included the extent and timing of glaciation, the nature and amount of glacial erosion, and oceanographic and atmospheric conditions during periods of glaciation and deglaciation. Central to many of these discussions is the origin of the large sounds and straits that fringe the eastern Canadian Arctic and connect the land glacial record with the deep-sea marine evidence from the North Atlantic. It is the purpose of this paper to examine the bedrock and glacial geology of Cumberland Sound, Frobisher Bay, and Hudson Strait (Fig. 4.1 and

Table 4.1 Areas within the different drainage basins (see Figure 4.1).

Location	Area(km^2)—maximum model	Area(km^2)—minimum model
Cumberland Sound	177 000	39 400
Frobisher Bay	130 000	43 000
Hudson Strait/Bay	1 159 000	1 453 750
Drainage Hudson Bay/ Foxe Basin (dashed line, Fig. 4.1)	3 100 000	

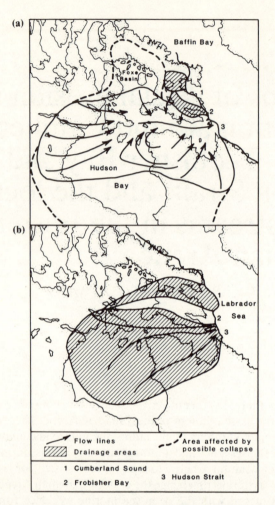

Figure 4.1 (a) Ice flow and drainage basins for Hudson Strait, Frobisher Bay, and Cumberland Sound with a 'minimum' ice sheet model. (b) Ice flow and drainage basins for a 'maximum' ice sheet reconstruction (after Denton & Hughes 1981). Table 4.1 lists the drainage basin size. The broken line represents the area that might be affected by a major collapse of the ice sheet and drainage through Hudson Strait.

Table 4.1) in order to ascertain if there are any common threads in the style of glaciation or the tectonic setting of these large bathymetric features.

Hudson Strait must concern anyone interested in the dynamics of the Laurentide Ice Sheet (e.g. Falconer *et al.* 1965, Blake 1966, Falconer 1969, Sugden 1977, Denton & Hughes 1981); in addition the glacial history of Hudson Strait/Hudson Bay is central in attempts to link the terrestrial glacial record with specific events in the North Atlantic (Ruddiman & McIntyre 1981, Fillon & Duplessy 1980). We suggest that the presence of Hudson Strait, which drained a large proportion of the interior of the Laurentide Ice Sheet (Fig. 4.1

and Table 4.1), was a primary reason why Denton and Hughes (1981) invoked an interrelated and dynamic glaciological system in the region of the Labrador Sea and Baffin Bay which consisted of ice shelves/ice streams/ and ice domes. In their model the shelves are emplaced largely to support the ice dome over Hudson Bay, which would collapse by rapid ice discharge through Hudson Strait except for the buttressing effect of the ice shelves (Weertman 1974, Thomas & Bentley 1978, Denton & Hughes 1981, Andrews 1982). Ruddiman and McIntyre (1981) have suggested that 50% by volume of the Laurentide Ice Sheet was discharged through major ice streams into the North Atlantic between 16 000 and 13 000 BP. *If* the Laurentide Ice Sheet had a late Wisconsin central dome over Hudson Bay, and *if* collapse occurred, then the area potentially affected by rapid discharge through Hudson Strait is enormous (Table 4.1). The notion of ice sheet collapse and the migration of ice divides is not new (Flint 1943, Falconer *et al.* 1965, Prest 1970, Andrews & Peltier 1976). However, we ask: What is the nature of the glacial evidence from within and around Hudson Strait? Given the central importance of Hudson Strait in any theoretical reconstruction of the ice sheet, there have been little field data available for verification or falsification of the various models, i.e. the single domed or multidomed model (Denton & Hughes 1981, Shilts 1980, Dyke, Dredge, & Vincent 1982). Even now the available data (on which we will report later in this paper) is sparse but these data are sufficient to pose serious questions about our conventional view of the glaciological situation in Hudson Strait.

Objectives of the paper
We have two main objectives in this paper: (1) to examine the evidence around and within some of the major sounds and troughs (e.g. Cumberland Sound, Frobisher Bay, and Hudson Strait) for their origin(s); and (2) to analyze evidence from Landsat imagery, erratics, and striations for the location of discrete ice streams in this section of the Laurentide Ice Sheet. Because of increasing offshore seismic exploration and drilling the bedrock geology of the area is rapidly becoming better known (Fig. 4.2). It is our contention that a major reason for many of the large bathymetric troughs must be sought in the tectonic style that affected the area from the Cretaceous to the present; thus next we briefly review this evidence.

Cretaceous to present geological history and the role of tectonics?

Various tectonic settings for the opening of Baffin Bay (Fig. 4.2) and adjacent embayments have been postulated. Kerr (1967) suggested that Baffin Bay is a submerged continental rift valley, however, LePichon *et al.* (1971) argued that sea floor spreading along a northern extension of the Labrador Sea mid ocean ridge is responsible for the opening of the bay. Seismic studies (Barrett *et al.* 1971, Keen & Barrett 1972, Keen *et al.* 1974) have demonstrated that oceanic

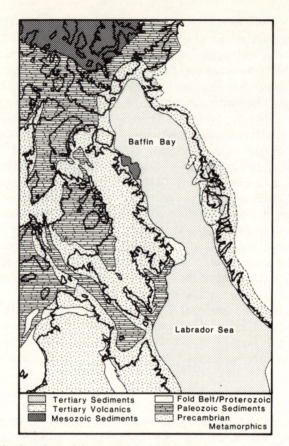

Figure 4.2   Bedrock geology of the eastern Canadian Arctic.

crust underlies thick sedimentary strata on the basin floor (200 m depth) while the shallower Davis Strait 'sill' is underlain by Cretaceous continental crust (Grant 1975).

The original opening of the Labrador Sea is believed to have occurred in the late Cretaceous (Srivastava 1978). Sea floor spreading extending northward into Baffin Bay took place between the early Paleocene and early Oligocene (Keen *et al.* 1972).

Some geologic features related to the opening of Baffin Bay as well as ancient Precambrian structural features are important to the physiography and geology of the surrounding embayments. Major faulting and graben formation along the northeastern coast of Baffin Island occurred prior to, or during the opening of, Baffin Bay (Srivastava 1978). Much of the continental shelf is underlain by late Cretaceous(?)–early Tertiary sediments that are locally up to 6 km thick (Jackson *et al.* 1977) and fill the down-faulted structures and grabens.

Farther north, Lancaster Sound (Fig. 4.2) is bounded by late Cretaceous-early Tertiary fault systems (Kerr 1980). The faults follow structural trends in the crystalline basement. A thick sequence of Proterozoic rocks underlie (at >6 km depth) the floor of the sound. These rocks are overlain by Paleozoic carbonate rocks and a thick wedge of Cretaceous/Tertiary sediments.

The data suggest that the tectonic features related to the opening of Baffin Bay closely follow ancient Precambrian structures. Unfortunately, no data are available for the Precambrian basement rocks of Cumberland Sound, Frobisher Bay, or Hudson Strait. However, it is possible that they may share some of the structural features common to Lancaster Sound. The existence of Paleozoic rocks underlying the major bays of southeastern Baffin Island, and the adjacent continental shelf, suggests that their original extent was much greater than at present. Significant thicknesses of these rocks may have been present over Baffin Island as suggested by scattered outcrops along the inland extension of the Frobisher Bay half-graben and in Foxe Basin.

Along the northern coast of Cumberland Peninsula scattered outcrops of Paleocene basalts overlie Precambrian crystalline rocks (Clarke & Upton 1971). If Paleozoic rocks had been present they were eroded from the land by this time (MacLean *et al.* 1977). Thus it is possible that the Paleozoic limestones were preserved in the structural depressions of the Precambrian basement, which may align roughly with Frobisher Bay, Cumberland Sound, and Hudson Strait. Cretaceous/Tertiary graben formation of the sounds then followed these ancient structural lineaments. The present day physiography owes much to tectonic features.

Glacial geology of Cumberland Sound, Frobisher Bay, and Hudson Strait

Our analysis of the role and importance of glacial erosion on the size and form of these troughs is based on field mapping, bathymetry, seismic stratigraphy, and bedrock geology. An overall framework is provided by the mapping of the glacial landforms from Landsat 1:1000 000 imagery (Fig. 4.3) and from consideration of lake density based on topographic maps of the same scale. This work represents an amplification of the original glacial mapping carried out by Sugden (1977, 1978, Andrews, Clark, & Stravers, in press).

Cumberland Sound
In Figure 4.4 the heavy dotted symbol represents areas that have been heavily ice scoured. Such areas are largely devoid of till, bedrock crops out at the surface, and the landscape is dominated by myriads of small lakes occurring in bedrock depressions excavated along lines of structural weakness that have been exploited by glacial erosion (Fig. 4.4). The bedrock thus shows evidence of extensive glacial scouring; from a basal thermal viewpoint this landscape unit (*D* on Fig. 4.3) indicates that the ice sheet was either warm based (i.e. at the pressure melting point) or more probably warm-freezing (Sugden 1978).

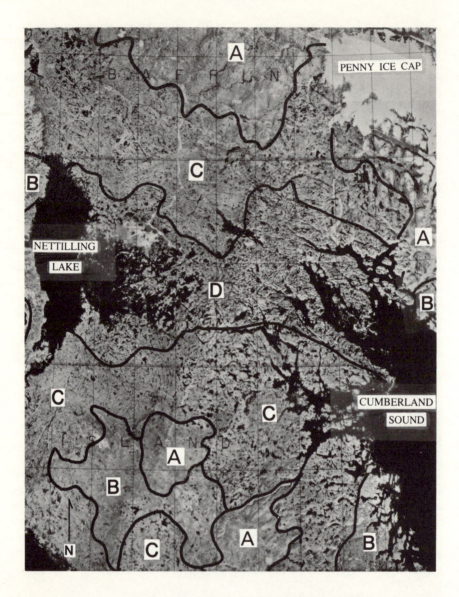

Figure 4.3 Landsat image of the head of Cumberland Sound showing the boundaries between different intensities of glacial scouring as reflected by different areas of lake density. A is 0–5% lake area, B is 5–10%, C is 15–25% and D is greater than 30%. Compare to Figure 4.4.

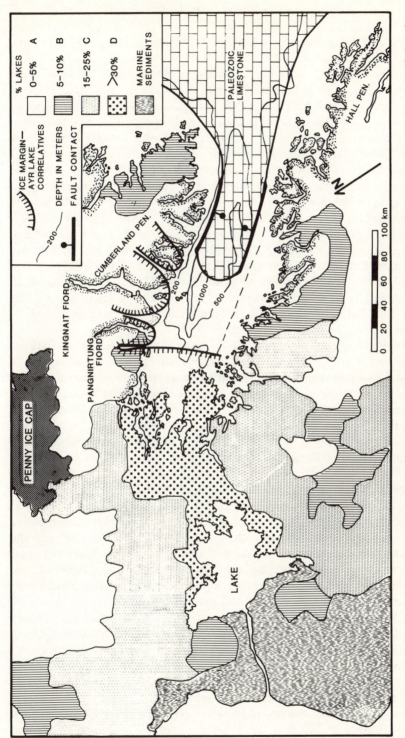

Figure 4.4 Intensity of glacial scouring around Cumberland Sound as shown by lake density and in addition showing the bathymetry and extent of Paleozoic limestone in the floor of the trough. Margin of the glacial ice ca. 100 000 BP during the Ayr Lake stade is sketched. (Note position of North; compare to Fig. 4.3).

Statistical analysis of about 550 grid points (Andrews *et al.* in press) shows that the intensity of glacial erosion (as expressed by lake density) is a function of elevation and distance from the east coast of Baffin Island. Both of these measured variables are indirect measures of ice thickness. The contact between high/moderate lake density lies rather uniformly between 350–410 m a.s.l.. This again suggests that these landscapes reflect long-term (?) stable conditions at the ice/bedrock interface (cf. Boulton 1982) with lower elevations, on average, showing more intensive evidence of glacial scouring.

Figures 4.3 and 4.4 represent a Landsat image of the head of Cumberland Sound and a map of the glacial geology and bathymetry. Cumberland Sound is 250 km long by 90 km wide. The sound is about 200 m deep from the head to near Pangnirtung Fiord; from here depth increases rapidly seaward to 1100 m off Kingnait Fiord (Fig. 4.4). Depths then slowly rise eastward toward the shelf and off Cumberland Sound the sill between the inner basin and the offshore deep ocean lies between 400 and 200 m. Seismic mapping indicates that stratified rocks, identified as Paleozoic limestones (Grant 1975, Sanford *et al.* 1979), extend to the inner basin nearly to the deepest portion (Fig. 4.4). The contact of the Paleozoic and flanking Precambrian granites and granite gneisses is shown as being faulted.

The thickness of the Paleozoic rocks and the nature of their contact with the Precambrian is not yet known. Two possible alternatives are illustrated in Figure 4.5 and it is clearly important at some future date to ascertain which situation is the correct one. In Figure 4.5A the Cumberland Sound trough is envisaged as being fault bounded near the inner part of the shelf. MacLean and Falconer (1979) have noted evidence for a major coastal fault farther north along the eastern coast of Baffin Island and Figure 4.5A suggests that the shelf off Cumberland Sound may be a structural high through either faulting or a simple fold. Alternatively, Figure 4.5B suggests a considerable thickness of Paleozoic sediment in an ancestral sound with glacial erosion (?) being one possible candidate for the removal of several 100 m of rock.

Is Cumberland Sound glacially eroded? Figure 4.4 shows that intensive glacial scour has taken place along the low saddle between Nettling Lake and the head of the sound (Fig. 4.3). This zone of intense scour is flanked by areas of moderate to low lake density and the uplands of Cumberland Peninsula and Hall peninsula are only lightly affected by glacial scour. The mapping in Figure 4.3 is amply supported by 1: 50 000 air photo mapping (Dyke, Andrews, & Miller 1982). Figure 4.3 thus suggests that an ice stream may have moved toward the head of Cumberland Sound. Because of the flanking uplands, Cumberland Sound would be the locus of converging ice flow trajectories and hence rapid ice velocities (see Andrews 1980). However, the zone of heavy aerial scour stops abruptly at the head of the sound and eastward light scouring occurs part way along both sides of the sound.

Figure 4.4 shows the margins of the ice sheet and local glaciers during the Ayr Lake stadial (Miller *et al.* 1977, Dyke, Andrews, & Miller 1982). This margin probably corresponds to an early Foxe (Wisconsin) advance broadly

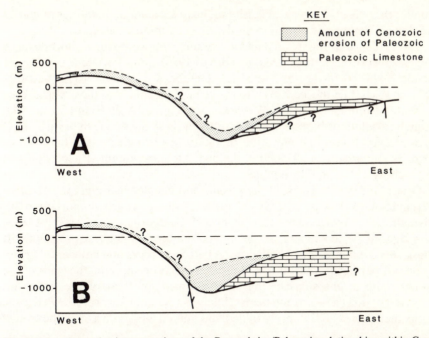

Figure 4.5 Alternative interpretations of the Precambrian/Paleozoic relationships within Cumberland Sound and the importance for models of glacial erosion. The sketches are drawn along the Cumberland Sound trough (see Fig. 4.4).

occurring sometime within marine isotope stage 5. During the late Foxe Glaciation the Laurentide Ice Sheet and the local glaciers were well behind the margins shown on Figure 4.4 (Birkeland 1978, Locke 1979, Bockheim 1980, Dyke 1979, Dyke, Andrews, & Miller 1982). Thus field mapping does not permit the extension of glacial ice farther than that shown on Figure 4.4 for about the last 100 ka. Outside the limits of the Duval moraines (Dyke, Andrews, & Miller 1982) there is some evidence for glaciation, but along the northern coast of the sound it consists of local cirque and valley glacial landforms and deposits with no evidence that these features have been overrun by Laurentide ice.

In contradistinction to Frobisher Bay and Hudson Strait (see below) the outer coast and islands of Cumberland Sound *do not* have a cover of limestone-rich till nor do they bear limestone erratics (Andrews & Miller 1979). Thus if outer and middle Cumberland Sound were glaciated, and if substantial glacial erosion occurred, then this occurred prior to the last glaciation (Fig. 4.4), and indeed may have occurred prior to the development of the present alpine landforms of outer Cumberland Peninsula (Sugden 1978, Dyke, Andrews, & Miller 1982). Estimates of rates of local cirque erosion (Andrews 1972, Anderson 1978) suggest that the local alpine glacial landforms originated ±3 million years ago! Thus given the available evidence it is not clear when and how glacial erosion of Cumberland Sound occurred. However, if Figure 4.5A applies there is no longer a necessity to move substantial quantities of glacier ice down Cumberland Sound to erode the deep middle basin.

The evidence in Figures 4.3 and 4.4 requires an ice divide a short distance (50 km?) west of Nettling Lake (Andrews 1980) (Fig. 4.1A & B). If the ice divide is shifted farther westward (cf. Denton & Hughes 1981) it becomes difficult to confine the ice stream to the head of the Sound as the drainage area increases from 39.4×10^3 to $177 \times 10^3 \, km^2$ (Fig. 4.1A & B and Table 4.1). Rough calculations based on a net accumulation at the divide of 0.8 m/yr H_2O and an equilibrium line elevation of 1000 m a.s.l. indicate that the average velocity increased from 0 km/yr at the divide to a maximum of 1 to 5 km/yr in the vicinity of the equilibrium line.

Grant (1975) has discussed the seismic and magnetic stratigraphy of Cumberland Sound. Two profiles (A–A' and B–B' of Fig. 4.6) are important to this discussion. A–A' is a track taken near outer Cumberland Sound (Grant 1975) at a right angle to the axis of the sound (see Fig. 4.7 for track location). It illustrates the fault contact between the Paleozoic rocks and adjacent Precambrian. Of particular interest is that these contacts occur along the walls of the trough. The profile suggests that excavation of the floor of the sound has occurred, with glacial erosion being a likely candidate. However, the Paleozoic rocks themselves are intensely folded and faulted along the axis of the trough

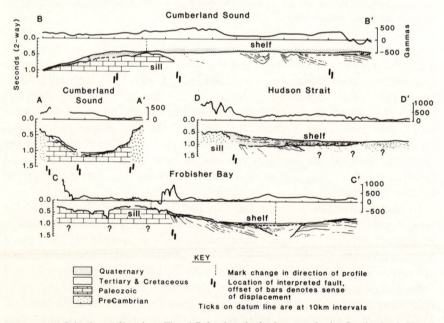

Figure 4.6 Seismic profiles (see Fig. 4.7 for location) along tracks in Cumberland Sound, Frobisher Bay, and Hudson Strait (from Grant 1975. Used by permission of the Canadian Society of Petroleum Geologists). The interpretation of the bedrock is based on the magnetic records and interpretation of the seismic records.

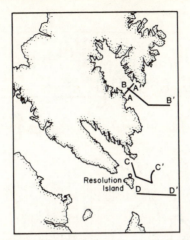

Figure 4.7 Location of the tracks illustrated as Figure 4.6. (From Grant 1975. Used by permission of the Canadian Society of Petroleum Geologists).

and thus the total relief of the trough may be due strictly to graben style tectonics.

Profile B–B′ (Fig. 4.6) is a track along the axis of the sound (Grant 1975), across the sill, and onto the shelf. The sill is underlain by deformed Paleozoic limestones with a prominent uplifted fault block located along the shallowest part of the sill and inner shelf. Glacial excavation of Cumberland Sound cannot be discounted; however, it seems rather fortuitous that the sill is 'bracketed' by bedrock structural features. Seaward from the sill, folded Tertiary and Cretaceous 'coastal plain' sediments (Grant 1975) are either truncated against the sill or exhibit an onlapping relationship. The profiles clearly show that the physiography of Cumberland Sound owes much to Tertiary or older tectonism. A strictly glacial origin for the bathymetric features is highly unlikely.

Frobisher Bay

Frobisher Bay is a half-graben with a projected fault running close to and parallel with the south coast. Paleozoic limestones again crop out along the floor of the bay (Fig. 4.8) and extend seaward where they are in fault contact with Tertiary rocks at the outer sill edge. Farther west a series of fault blocks have preserved small outliers of Paleozoic limestone all the way from the head of Frobisher Bay westward to the edge of the main contact at Amadjuak Lake (Fig. 4.8) where the carbonate rocks crop out under present Foxe Basin and merge toward the Paleozoic terrain within and beneath Hudson Bay (Sanford *et al.* 1979, Grant & Manchester 1970) (Fig. 4.2).

Frobisher Bay is flanked to the north by the uplands of Hall Peninsula, whereas to the south the land rises abruptly to the Everett Mountains and the Meta Incognita Peninsula. Glacial geological mapping of the area has

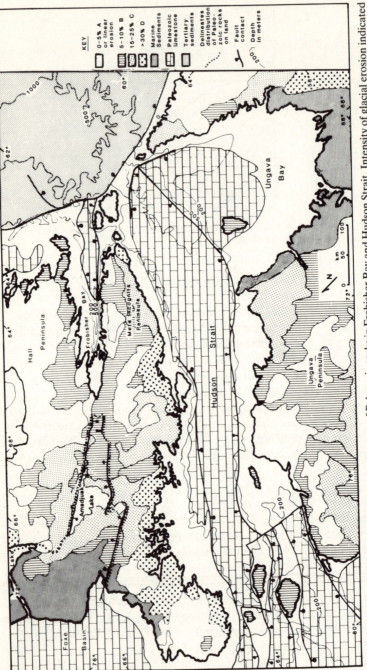

KEY

0-5% A or linear erosion

5-10% B

15-25% C

>30% D

Marine Sediments

Paleozoic limestone

Tertiary sediments

.... Delineates distribution of Paleozoic rocks on land

Fault contact

⌒200⌒ Depth in meters

Figure 4.8 Glacial erosion, bathymetry, and extent of Paleozoic rocks in Frobisher Bay and Hudson Strait. Intensity of glacial erosion indicated by lake density (%).

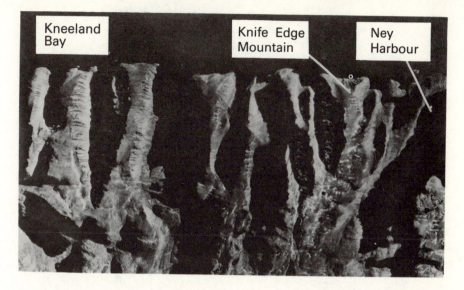

Figure 4.9 Air photomosaic of a section of the faulted escarpment of the northern face of Meta Incognita Peninsula, Frobisher Bay.

employed Landsat imagery and black and white air photos (Muller 1980, Colvill 1982, Lind 1983) (e.g. Fig. 4.9).

The extent and degree of aerial scour along the coast of Frobisher Bay is radically different from that observed in Cumberland Sound (Figs. 4.3 & 4.8). Moderate scouring (15–25% lake cover) is found along much of the north coast but rapidly drops to between 0–5% on the Hall Peninsula. Along the south coast the steep, faulted coast rises abruptly to about 600 m and the scarp face is eroded by a number of emerged and submerged cirque basins (Fig. 4.9). Behind this scarp, in the vicinity of the present Terra Nivea and Grinnell Ice Caps, there is little evidence for glacial scour. However, the Everett Mountain plateau shows moderate to high scour (Muller 1980, Colvill 1982) (Fig. 4.9). The deep trough of Frobisher Bay (Fig. 4.9) hugs the south coast and reaches depths of between 450 and 600 m. The bay shallows rapidly in the inner third and within the line of islands depths of less than 200 m prevail.

Research on outer Frobisher Bay indicates that the tills are heavily charged with carbonate (Miller 1980) (Miller 1982, Fig. 4.10). In contrast the inner third of the bay limestone erratics are present but rare (Andrews & Miller 1979, Colvill 1982, Lind 1983) and there is little carbonate within the till matrix. Glacial marine sediments from 500 m ± depth in the outer third of the bay are between 20–50% carbonate by weight.

Extensive radiocarbon dating of shell bearing marine, glacial marine, and till sediments indicate that grounded ice extended toward the outer part of

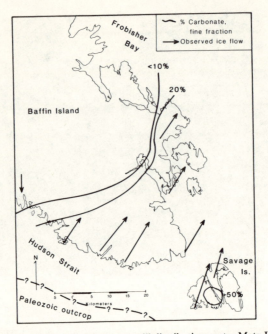

Figure 4.10   Ice flow indicators and carbonate till distribution, outer Meta Incognita Peninsula.

Frobisher Bay between 10 and 11 ka BP (Miller 1980). This suggestion is also supported by faunal analysis of marine cores within the bay (Osterman 1982, Osterman & Andrews in press). Thus the carbonate-bearing tills, the extensive carbonate content of the marine sediments (Osterman 1982, Osterman & Andrews in press), and the extent of late Foxe glaciation (Miller 1980, Osterman 1982) indicate that glacial erosion of the Paleozoic outcrop within the Frobisher Bay trough has occurred in recent time. However, there is considerable evidence (Osterman 1982) that grounded glacier ice did not extend across the deep trough of Frobisher Bay and Osterman (1982) has proposed that a floating ice shelf extended over a considerable part of the bay.

High resolution 'Huntec' acoustic and seismic profiles (MacLean & Falconer 1979) and seismic and magnetic survey data (Grant 1975) have been collected for central and outer Frobisher Bay. They show that a situation analogous to that in Cumberland Sound occurs here (Fig. 4.6), although on a smaller scale. Late Foxe stratified sediments, and possibly till, overlie the Precambrian basement within the deep trough of Frobisher Bay. The surface of the basement shows moderately smooth relief (approximately 100 m) suggesting possible glacial sculpting. However, since these are two-dimensional profiles, it is difficult to extrapolate to the third dimension. Locally, the basement is so sharply incised that ancient fluvial incision or faulting seem more likely than glacial erosion.

The Frobisher Bay sill is roughly marked by a line between the outermost islands on either side of the mouth of the bay. Profile C–C' (Fig. 4.6) includes part of the sill (see Fig. 4.7 for location of track). It consists of crystalline basement rocks and overlying Paleozoic strata. Eastward toward the shelf, this ridge is in fault contact with an onlapping sequence of Cretaceous–Tertiary sediments (similar to the relationship in outer Cumberland Sound). A key feature is that the onlapping sediments appear to be truncated along the fault, with no Tertiary sediments on the uplifted block. Furthermore, Paleozoic strata are preserved along this structural high. These data suggest that the Tertiary sediments have been eroded from the crest of the sill but that erosion has not been so intense as to strip away the Paleozoic rocks. Alternatively, the absence of the Tertiary sediments on the ridge could be due strictly to structural relationships and the pattern of Tertiary sedimentation. Nonetheless, the question still remains: Is the relief on the surface of the Paleozoic rocks from the sill toward the inner part of the bay due totally to glacial excavation?

Toward the outer part of the deep trough, flat lying Paleozic limestones crop out continuously. Bedding planes form the surface of the outcrop and there is no evidence for glacial gouging of this relatively soft rock. Further out toward the sill these strata are cut by a series of normal faults that parallel the Frobisher Bay escarpment (Fig. 4.5). These faults appear to offset Quaternary sediments. The topography is controlled exclusively by the surface (parallel seismic reflectors = bedding planes) and escarpments of the fault blocks. The crests of the fault blocks are very sharp and create the impression that no glacial scouring has occurred. The data suggest that Osterman's (1982) interpretation of a floating ice shelf over the deep trough is valid, and indeed may have been the style of glaciation throughout much of the Quaternary. If this is the case the limestone erratics from the tills of outer Frobisher Bay were derived from the shallower parts of the floor of the bay, or possibly were carried in by Hudson Strait ice as it was deflected northward across the southern tip of Meta Incognita Peninsula.

Hudson Strait

The bedrock geology and bathymetry of Hudson Strait is complex. A narrow sill 200 m deep extends from the Resolution Islands (Fig. 4.7) southward to Labrador. Based on his analysis of seismic records, Fillon (1980) suggested that the inner sill is primarily composed of Quaternary sediments. On the other hand, Grant's (1975) seismic and magnetic survey of the outer sill and adjacent continental shelf show that this sill is clearly underlain by an uplifted block of Precambrian rock (profile D–D', Fig. 4.6). Westward from the sill most of the floor of Hudson Strait is 300–450 m deep with some small basins in eastern Ungava Bay and near the islands at the head of the strait (Fig. 4.8).

Despite the overriding importance of Hudson Strait to the glacial history of the Laurentide Ice Sheet there are hundreds of kilometers of coastline in Ungava, Labrador, Baffin Island, and the outer islands that are totally unvisited! However, recent work by Miller, Stravers, Osterman, and Clark (University of Colorado, unpublished), Lauriol and Gray (1982), and Gray and Lauriol (1982), has started to shed some light on the history of the region.

In addition, Fillon (1978), Fillon and Harmes (1982), have contributed data on the problem of the dynamics of the ice margin within outer Hudson Strait and adjacent shelves.

Much of the floor of Hudson Strait and Ungava Bay is floored with sedimentary rocks of Silurian to Ordovician age (Fig. 4.8) (Sanford *et al*. 1979, Grant & Manchester 1970). The deeper portions of Hudson Strait are floored by sedimentary rocks with the exception of the Ungava Trough which is mapped as cutting into the Precambrian shield terrain.

Landsat mapping (Sugden 1977, 1978) (Fig. 4.8) indicates that much of the northern coast of Hudson Strait is moderately to heavily ice-scoured. Heavy ice scouring is also evident on the outlying Resolution Islands and along parts of the Ungava Peninsula. However, the uplands of that peninsula are only lightly affected by scouring and we suggest that the ice there was thin and/or cold based. The islands at the head of the strait are moderately scoured. Gray and Lauriol (1982, p. 117) identify evidence for a radial outflow from the Ungava Peninsula for '. . . several millenia prior to 8,000 yrs. B.P.' Taylor (1974, p. 2) also noted in reference to Ungava Peninsula: 'Glacial ice chiefly flowed outward to the sea from the center of the map-area, but local evidence of what is probably an earlier southern flow in the central part of the map-area is present.'

At first glance the pattern of glacial scour (Fig. 4.8) would appear to suggest that a rapidly moving, warm-based ice-stream existed in Hudson Strait and extended its influence inland along the coast of southern Baffin Island and seaward to the outer islands. However, the problem with the simple acceptance of this notion is that mapping of local patterns of striations and erratic trains indicates the following:

(1) In the vicinity of Lake Harbour, southern Baffin Island, all striations and large rock bedforms indicate movement north to south with no evidence for cross-cutting striations. In fact most of the northern shore of Hudson Strait is characterized by southward ice flow directional indicators (Blake 1966, Prest *et al*. 1968). Limestone drift (Clark unpublished) occurs in a thin belt along the Hudson Strait coast near Lake Harbour.

(2) In the vicinity of the outer SE tip of Baffin Island and eastward to the Savage Islands (Fig. 4.8) the striation and carbonate erratic evidence shows a movement *toward* N 20° E (Miller & Hearty, in preparation, Osterman *et al*. in press) that extended to elevations of at least 400 m in the interior of Meta Incognita Peninsula (Fig. 4.10).

(3) To the west of the area noted in (2) striations and the lack of carbonate erratics suggest that ice moved from the north *toward* the coast of Hudson Strait. Again, in neither area (2) nor (3) was there evidence of crossing striations that could have been used to support a model of a major ice flow WNW–ESE along Hudson Strait toward the North Atlantic.

Evidence from ^{14}C dated shells (Miller unpublished, Osterman in press,

Andrews & Short in press) suggest that glacial flow toward N 20° E occurred during the late Foxe Glaciation and may cover the interval 18–9 ka BP.

Two explanations could be advanced as to why the ice flow was toward the north-northeast on southeast Baffin Island during the late Foxe. The first scenario (Fig. 4.11A) considers that the reason is related to the dominance of flow from the Labrador–Ungava ice divide. Thus ice moves northward down Ungava Bay and is confined to this route by the N–S bathymetric ridge south of the Resolution Islands, and possibly a fringing ice shelf that existed to the east. Crucial to this or another theory would be the record of ice movement on the Button Islands off the northern tip of Labrador. The second suggestion (Fig. 4.11B) accepts the argument of Fillon and Harmes (1982) that a grounded ice shelf existed on the Labrador banks as recently as 10 ka BP. Such a feature may have extended further north along the Baffin Island shelf where indeed diamictons can be recognized in the seismic record (MacLean in press). The flow patterns of Figure 4.11B could fit an ice stream/ice shelf flow-line model (Hughes 1975). To decide between Figures 4.11A or 4.11B it will be critical to determine if the source of the ice impinging on outer northern Hudson Strait was from the southwest or north-northwest. Both Figures 4.11A and 4.11B may have validity. Further, we suggest that the *relative* depths of Ungava Bay and inner Hudson Strait, with a fault contact between, argue strongly for the major control on water depth to be associated with tectonic style and *not* intensity of glacial erosion.

There is, however, evidence for glacial excavation in the area adjacent to either side of the Hudson Strait sill. It is delineated roughly by the 500 m isobath (Fig. 4.8). No bedrock information is available for the area just inside the sill. However, a seismic track from the sill seaward across the continental shelf (profile D–D' Figs. 4.6 & 4.7) was taken by Grant (1975). Of particular interest is the relationship between the Tertiary sediments and Precambrian crystalline rocks. Three interpretations are possible: (1) the truncation of the subparallel reflectors (bedding planes?) on the surface of the Tertiary sediments and the topographic relief (ca. 200 m) on the pre-Quaternary acoustic strata may be due to glacial excavation; (2) the relief on the surface of the pre-Quaternary rocks could be due to displacement across the inferred fault contact; (3) Pre-Pleistocene fluvial erosion (Bornhold *et al.* 1976, Fortier & Morley 1956, Pelletier 1969) may be responsible for the apparent removal of some 200 m of Tertiary/Cretaceous sediments.

The observed ice flow in outer Hudson Strait (Fig. 4.10) may present a major problem to proponents of the downdraw theory of the Laurentide Ice Sheet. The field evidence, in support of a long-continued flow across southeast Baffin Island from the south which is not disrupted until ca. 9 ka BP (Miller unpublished), is supported by Quinlan's (1981 in press) reconstruction of the thickness and extent of the Laurentide Ice Sheet in this region based on an inverse theory of glacial isostatic response (Peltier 1976, Quinlan 1981). The reconstructions show an essentially steady-state ice sheet extended over the region between 18–8 ka BP, which then rapidly retreats.

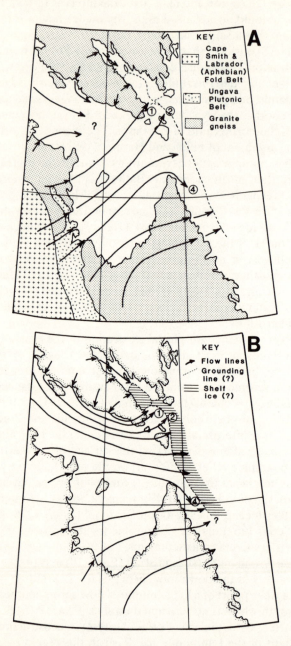

Figure 4.11 Alternative reconstructions of the flow regime within Hudson Strait/southern Baffin Island. Either would account for the N/NE trending striations on the extreme eastern end of Meta Incognita Peninsula, but only Figure (A) accounts for the lithologies from the Ungava Peninsula being present on Southern Baffin Island.

Summary
Our analysis of the glacial geological evidence from Cumberland Sound (Fig. 4.3), Frobisher Bay (Fig. 4.8), and Hudson Strait (Fig. 4.8) indicates that each of these large geographic features has experienced a different history of Cenozoic glaciation. They are all obviously tectonic features of considerable magnitude that probably owe much to early Cenozoic/late Mesozoic sea-floor spreading and isostatic uplift and rifting. The bedrock geology of these troughs does not *demand* significant glacial erosion. The fact that glacial erosion of the Paleozoic surface occurred in Frobisher Bay and Hudson Strait is demonstrated by the high detrital carbonate content of both tills and glacial marine sediments; however, we cannot demonstrate that the same process affected Cumberland Sound, and if it did, those events occurred prior to ca. 100 ka BP (Fig. 4.7, cf. Dyke, Andrews, & Miller 1982).

Denton and Hughes (1981) place great weight on the bathymetry of these troughs and argue that sills and the deep inner basins are obvious products of glacial erosion. In making such statements these authors are following most other workers, however, we remind our readers that until the last decade there have been few definitive studies on the bedrock/tectonics/bathymetry associations of troughs and fjords. We contend that virtually all arguments on the glacial origin of such features lack facts, and lack any rigorous analysis. Our analysis of three major troughs indicates that although they exhibit substantial differences in glacial style, they have as a common denominator the importance of probable Cenozoic faulting and rifting. Kerr's (1980) analysis suggests that these conclusions can be extended northward to many of the large channels in the Canadian Arctic Archipelago. Dyke (1982) and Dyke, Dredge, & Vincent (1982) have documented, as we have, that in several instances the flow of the Laurentide Ice Sheet was *across* major troughs.

Ice streams in the northeastern Laurentide Ice Sheet

For our purpose we define an ice stream, or rather we reconstruct ice streams, on the basis of observable glacial geological evidence. Thus we do not consider *a priori* that ice streams can be located along structural/bathymetric lows. We do, however, feel that areas of warm-based freezing ice and areas of probable rapid flow can be identified on the basis of: (1) areas of intense to moderate areal scour; (2) areas of distinct till composition derived from known sources, specifically the location of carbonate-rich tills; and (3) areas of extensive striations and bedrock molding. Dyke, Dredge, & Vincent (1982) used some of these same lines of evidence to present a reconstruction of the Laurentide Ice Sheet. This reconstruction has many similarities to the 'minimum' ice extent model of Denton and Hughes (1981).

Our own model (Fig. 4.12) differs from that of Dyke, Dredge, & Vincent (1982) mainly in locating the ice divide closer to the coast of western Baffin Island. The reason for this is the absence of carbonate drift at the head of

Cumberland Sound. If ice had flowed from Foxe Basin we would expect a swath of limestone-rich till to extend from Foxe Basin eastward (see Figs. 4.1 & 4.2 and Table 4.1). The pattern of areal scour and the presence of extensive limestone erratics in Home Bay (Andrews *et al.* 1970) suggests ice stream *F* (Fig. 4.12). In a similar fashion, Blackadar's observations (1967) of striations and up to 100 m of carbonate-rich till south of Mingo Lake, implies the presence of ice stream C heading toward Frobisher Bay. An ice dome or divide over Amadjuak Lake appears to be called for (Blake 1966, Colvill 1982, Dyke, Dredge, & Vincent 1982). However, the major feature of Figure 4.12 and most other reconstructions is the situation in Hudson Strait. We show two alternatives (Fig. 4.11). In the first an ice stream moves out of Ungava Bay northward across the southern tip of Baffin Island; in the second an ice stream flowing west to east in Hudson Strait is deflected northward, across Baffin Island, as the ice stream reaches the ice shelf (e.g. Hughes 1975, Fig. 16, Denton & Hughes 1981). In the latter case flow lines in the southern Hudson ice stream strait would be deflected southward across northern Labrador. Loken (1964) has in fact recorded northwest to southeast striations on an island off northernmost Labrador.

The difference in these two scenarios (Fig. 4.11) is probably *the* most significant problem confronting glacial geologists in the eastern Canadian Arctic, and indeed, is probably *the* most critical problem in current efforts to model the late glacial history of the Laurentide Ice Sheet. In this context the till

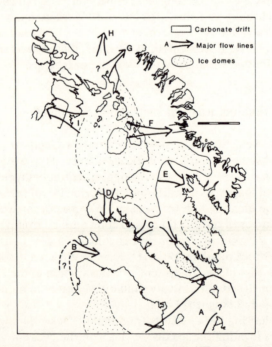

Figure 4.12 Suggested location of major ice streams in the NE sector, Laurentide Ice Sheet.

mineralogy of the noncarbonate fraction of sediments on southeast Baffin Island is critical. Volcanic and metamorphic rocks of the Circum-Ungava foldbelts would be picked up by ice along either flowline 1 or 2 (Fig. 4.11B). In fact, Bell (1898) noted the presence of dark volcanic erratics on the southern tip of Nottingham Island, and Shilts (1980, 1982) has also used them as a diagnostic erratic within Hudson Bay. In our reconstruction (Fig. 4.11A) flowlines 1 and 2 would largely consist of local Precambrian rocks and carbonates from the floor of Hudson Strait. However, they would both probably contain a high proportion of dark volcanic and/or metamorphic clasts from the Circum Labrador–Ungavafold belt. In contrast, according to model Figure 4.11B, these clasts would be unlikely to cross Hudson Strait and end up in tills on southeast Baffin Island.

Analysis of thin sections of erratic pebbles from southern Baffin Island suggest that three distinctive lithologies found in tills from the tip of the Meta Incognita Peninsula are very likely derived from *south* of Hudson Strait (i.e. Fig. 4.11A). Pyroxene-bearing gneiss (granulite facies) from northern Ungava Peninsula, dark banded slates, and metabasalts presumably derived from the Cape Smith foldbelt or the Labrador Trough (Hynes & Francis in prep., Westra 1978, G. Albino & S. A. Morse pers. comm.) are all present.

A major problem in ice sheet reconstructions (cf. Andrews 1982) is that each reconstruction might merely represent a 'snapshot' in time and that as the ice sheets wax and wane several configurations occur. Indeed this situation must be expected and tested using several lines of evidence. A major factor in the interpretation of the evidence from outer Hudson Strait is the length of the flowlines from different proposed ice divides (Fig. 4.1). In the multi-dome model the distance to the mouth of the Hudson Strait from the different divides is as follows:

central Labrador, 1000 km; central Ungava, 600 km; and Keewatin, 1600 km.

It follows from these measurements, especially in view of the probable higher activity on climatic grounds of the Labrador/Ungava center(s) (Williams 1979), that outer Hudson Strait would be initially influenced by ice flowing off Labrador–Ungava, into Ungava Bay and then across to southern Baffin Island (Figs. 4.11 & 4.12). The flow line from Keewatin is 600–1000 km *longer* than the more direct route from the high plateaux of Labrador–Ungava. However, as Figure 4.11A shows we are unsure of the interrelationships between Keewatin and Labrador ice in this model, whereas the situation is simpler in Figure 4.11B. The question that can be posed is: Are the two models mutually exclusive, or could the glaciological conditions of Figure 4.11A be a precursor of Figure 4.11B? Tests for this question must be sought in the offshore mineralogy of marine sediments.

Acknowledgments

Our research on the history and glacial geology of the Laurentide Ice Sheet has been supported by the National Science Foundation, specifically under grants to Andrews and Miller (EAR–79–26061) and (EAR–81–21296) and to Andrews and Osterman (DPP–81–116048). Dr. Parker Calkin read a version of this paper and made valuable suggestions for improving the text and figures for which we thank him!

References

Anderson, L. W. 1978. Cirque glacier erosion rates and characteristics of neoglacial tills, Pangnirtung Fiord area, Baffin Island, N.W.T., Canada. *Arctic & Alpine Res.* **10**, 749–60.

Andrews, J. T. 1972. Glacial power, mass balances, velocities, and erosion potential. *Zeitschrift Geomorph.* Supp. Bd. **13**, 1–17.

Andrews, J. T. 1980. Progress in relative sea level and ice sheet reconstructions, Baffin Island, N.W.T., for the last 125 000 years. In *Earth Rheology, Isostasy and Eustasy*, N-A. Mörner (ed.) 175–200. New York: Wiley.

Andrews, J. T. 1982. On the reconstruction of Pleistocene ice sheets: a review. *Quat. Sci. Rev.* **1**, 1–30.

Andrews, J. T., J. T. Buckley and J. H. England 1970. Late-glacial chronology and glacio-isostatic recovery, Home Bay, east Baffin Island. *Geol. Soc. Am. Bull.* **81**, 1123–48.

Andrews, J. T., P. U. Clark, and J. A. Stravers, in press. The pattern of glacial erosion across the eastern Canadian Arctic. In *Quaternary studies on Baffin Island, Baffin Bay and West Greenland*, J. T. Andrews and M. L. Andrews, (eds). Oxford: Pergamon.

Andrews, J. T. and G. Falconer 1969. Late glacial and postglacial history and emergence of the Ottawa Islands, Hudson Bay, N.W.T., evidence on the deglaciation of Hudson Bay. *Can. J. Earth Sci.* **6**, 1263–76.

Andrews, J. T. and G. H. Miller 1979. Glacial erosion and ice sheet divides, northeastern Laurentide ice sheet, on the basis of the distribution of limestone erratics. *Geology* **7**, 592–6.

Andrews, J. T. and W. R. Peltier 1976. Collapse of the Hudson Bay ice sheet. *Geology* **4**, 73–5.

Andrews, J. T. and S. K. Short, in press. *Radiocarbon Date List V from Baffin Island; Radiocarbon Date List II from Labrador*. Institute of Arctic and Alpine Research, Occasional Paper, University of Colorado, Boulder.

Barrett, P. J., C. E. Keen, K. S. Manchester, and D. I. Ross 1971. Baffin Bay – an ocean. *Nature* **229**, 551–2.

Bell, R. 1898. Report of an exploration on the north side of Hudson Strait, Canada. *Geol. Surv. Canada Ann. Rep.* XI, Part M.

Birkeland, P. W. 1978. Soil development as an indication of relative age of Quaternary deposits, Baffin Island, N.W.T., Canada. *Arctic & Alpine Res.* **10**, 733–47.

Blackadar, R G. 1967. Geological reconnaissance, southern Baffin Island, District of Franklin. *Geol. Surv. Canada Paper* 66–47.

Blake, W. Jr. 1966. End moraines and deglacial chronology in northern Canada, with special reference to southern Baffin Island. *Geol. Surv. Canada Paper* 66–28.

Bockheim, J. G. 1980. Properties and relative age of soils of southwestern Cumberland Peninsula, Baffin Island, N.W.T., Canada. *Arctic & Alpine Res.* **11**, 289–306.

Bornhold, B. D., N. W. Finlayson, and D. Monahan 1976. Submerged drainage pattern in Barrow Strait, Canadian Arctic. *Can. J. of Earth Sci.* **3**, 305–11.

Boulton, G. S. 1982. Sedimentary and geomorphological indicators of ice sheet dynamics. McMaster Univ. Ontario, Canada: *Intern. Assn. of Sedimentologists, 11th Intern. Cong*, Abstracts *1*, 77.

Clarke, D. B. and G. J. Upton 1971. Tertiary basalts of Baffin Island, field relations and tectonic setting: *Can. J. Earth Sci.* **8**, 248–58.

Colvill, A. 1982. *Glacial landforms at the head of Frobisher Bay, Baffin Island, Canada.* M.A. thesis, University of Colorado, Boulder.

Denton, G. H. and T. J. Hughes (eds) 1981. *The last great ice sheets.* New York: Wiley.

Dyke, A. S. 1979. Glacial and sea level history of the southwestern Cumberland Peninsula, Baffin Island, N.W.T., Canada. *Arctic & Alpine Res.* **11**, 179–202.

Dyke, A. S., J. T. Andrews, and G. H. Miller 1982. Quaternary geology of Cumberland Peninsula, Baffin Island, District of Franklin. *Geol. Surv. Canada Mem.* 403.

Dyke, A. S., L. A. Dredge, and J-S Vincent 1982. Configuration and dynamics of the Laurentide ice sheet during the late Wisconsin maximum. *Geog. Phys. Quat.* **36**, 5–14.

Falconer, G., J. D. Ives, O. H. Loken and J. T. Andrews 1965. Major end moraines in eastern and central Arctic Canada. *Geogr. Bull.* **7**, 137–53.

Fillon, R. H. 1978. Glacier termini in eastern Hudson Strait. *Geol. Soc. Am. Abstracts 10*, 401.

Fillon, R. H. and J-C. Duplessy 1980. Labrador Sea bio-, tephro-, oxygen isotope stratigraphy and late Quaternary paleoceanographic trends. *Can. J. Earth Sci.* **17**, 831–54.

Fillon, R. H. and R. A. Harmes 1982. Northern Labrador shelf glacial chronology and depositional environments. *Can. J. Earth Sci.* **19**, 162–92.

Flint, R. F. 1943. Growth of the North American ice sheet during the Wisconsin Age. *Geol. Soc. Am. Bull.* **54**, 325–62.

Fortier, Y. O. and L. W. Morley 1956. Geological unity of the arctic islands, *Trans. Roy. Soc. Canada* **50** (3), 3–12.

Grant, A. C. 1975. Geophysical results from the continental margin off southern Baffin Island; In *Canada's Continental Margins and Offshore Petroleum Exploration*, C. J. Yorath, E. R. Parker, and D. J. Glass, (eds), 411–31. Canadian Society of Petroleum Geologists, Memoir 4.

Grant, A. C. and K. S. Manchester 1970. Geophysical investigations in the Ungava Bay-Hudson Strait region of northern Canada. *Can. J. Earth Sci.* **7**, 1062–76.

Gray, J. and B. Lauriol 1982. Evolution of the late Wisconsin ice sheet in Ungava: I. The morphological evidence. *Abstract Volume* **1**, 117. Moscow: IXth INQUA Congress.

Hardy, L. 1976. *Contribution a l'étude geomorphologique de la portion quebecoise des basses terres de la baie de James.* Ph.D. thesis, McGill University, Montreal.

Hughes, T. J. 1975. *Ice stability coordinated Antarctic Program. West Antarctic ice streams.* ISCAP Bulletin No. 4, University of Maine, Orono, Institute for Quaternary Studies.

Hynes, A., and D. M. Francis, In preparation. *A transect of the Early Proterozoic Cape Smith Foldbelt, New Quebec.*

Jackson, H. R., Keen, C. E., and D. L. Barrett 1977. Geophysical studies on the eastern continental margin of Baffin Bay and in Lancaster Sound. *Can. J. Earth Sci.* **14**, 1991–2001.

Keen, C. E., and D. L. Barrett 1972. Seismic refraction studies in Baffin Bay: an example of a developing ocean basin. *Roy. Astron. Soc. Geophys. J.* **30**, 253–71.

Keen, C. E., D. L. Barrett, K. S. Manchester and D. I. Ross 1972. Geophysical studies in Baffin Bay and some tectonic implications. *Can. J. Earth Sci.* **9**, 239–56.

Keen, C. E., M. J. Keen, D. I. Ross, and M. Lack 1974. Baffin Bay: small ocean basin formed by sea floor spreading. *Am. Assoc. Petrol. Geol. Bull.* **58**, 1089–1108.

Kerr, J. Wm. 1967. Nares submarine rift valley and the relative rotation of North Greenland. *Bull. Can. Petrol. Geol.* **15**, 483–520.

Kerr, J. Wm. 1980. Structural framework of Lancaster Aulacogen, Arctic Canada. *Geol. Surv. Canada Bull.* 319.

Lauriol, B. and J. Gray. 1982. Evolution of the late Wisconsin ice sheet in Ungava Quebec: II. The evidence provided by marine and lacustrine isobases. *Abstract Volume* **1**, 188. Moscow: IXth INQUA Congress.

Le Pichon, X., R. E. Houtz, C. L. Drake, and J. E. Nafe 1971. Crustal structure of the mid-ocean ridge. I. Seismic refraction measurements: *J. Geophys. Res.* **70**, 319–39.

Lind, E. K. 1983. *Sedimentology and paleoecology of the Cape Rammesberg area, Baffin Island, Canada*. M.Sc. thesis, University of Colorado, Boulder.

Locke, W. W. III 1979. Etching of hornblende grains in Arctic soils: an indicator of relative age and paleoclimate. *Quat. Res.* **11**, 197–212.

Loken, O. H. 1964. *A study of the late and postglacial changes of sea level in northernmost Labrador*. Unpublished report to the Arctic Institute of North America (mimeo).

MacLean, B. In press. Geology of the Baffin Island shelf. In *Quaternary Studies on Baffin Island, Baffin Bay, and West Greenland*, J. T. Andrews (ed.) London: George Allen & Unwin.

MacLean, B. and R. K. H. Falconer. 1979. Geological/geophysical studies in Baffin Bay and Scott Inlet-Buchan Gulf and Cape Dyer-Cumberland Sound areas of the Baffin Island shelf: In *Geol. Surv. Canada, Curr. Res. Pt B*, Paper 79–1B, 231–44.

MacLean, B., L. F. Jansa, R. K. H. Falconer, and S. P. Srivastava 1977. Ordovician strata on the southeastern Baffin Island shelf revealed by shallow drilling: *Can. J. Earth Sci.* **14**, 1925–39.

Miller, G. H. 1980. Late Foxe glaciation of southern Baffin Island, N.W.T., Canada. *Geol. Soc. Am. Bull.*, Part I, 91, 399–405.

Miller, G. H. 1982. Dynamics of the Laurentide ice sheet based on field evidence from northeastern Canada. *Abstract Volume* **1**, 222. Moscow: IXth INQUA Congress.

Miller, G. H., J. T. Andrews and S. K. Short 1977. The last interglacial/glacial cycle, Clyde Foreland, Baffin Island, N.W.T.: stratigraphy, biostratigraphy, and chronology. *Can. J. Earth Sci.* **14**, 2824–57.

Muller, D. S. 1980. *Glacial geology and Quaternary history of southeast Meta Incognita Peninsula, Baffin Island, Canada*. M.Sc. thesis, University of Colorado, Boulder.

Osterman, L. E. 1982. *Late Quaternary history of southern Baffin Island, Canada: A study of foraminifera and sediments from Frobisher Bay*. Ph.D. dissertation, University of Colorado, Boulder.

Osterman, L. E. and J. T. Andrews 1983. Changes in glacial-marine sedimentation in core HU77–159, Frobisher Bay, Baffin Island, N.W.T.: a record of proximal, distal and ice-rafting glacial-marine environments. In *Glacial-marine sedimentation*, B. E. Molina (ed.), 451–94. New York: Plenum Press.

Osterman, L. E., G. H. Miller, and J. A. Stravers. In press. Middle and Late Foxe glacial events in southern Baffin Island. In *Quaternary Studies in Baffin Island, Baffin Bay and West Greenland*, J. T. Andrews (ed.) London: George Allen & Unwin.

Pelletier, B. R. 1969. Submarine physiography, bottom sediments, and models of sediment transport in Hudson Bay. *Geol. Surv. Canada*, Paper 68–53, 100–36.

Peltier, W. R. 1976. Glacial isostatic adjustment-II: The inverse problem. *Geophys. J. Roy. Astron. Soc.* **46**, 699–706.

Prest, V. K. 1979. Quaternary geology in Canada. In *Geology and Economic Minerals in Canada*, 5th Edition, R. J. Douglas (ed.) 676–764. Ottawa: Department of Energy, Mines, and Resources.

Prest, V. K., D. R. Grant and V. N. Rampton 1968. Glacial Map of Canada. *Geol. Surv. Canada, Map* 1253A.

Quinlan, G. M. 1981. *Numerical models of postglacial relative sea level change in Atlantic Canada and the eastern Canadian Arctic*. Ph.D. dissertation, Dalhousie University, Halifax, Canada.

Ruddiman, W. F. and A. McIntyre 1981. The mode and mechanism of the last deglaciation: oceanic evidence. *Quat. Res.* **16**, 125–34.

Sanford, B. V., A. C. Grant, J. A. Wade and M. S. Barss 1979. Geology of Eastern Canada and adjacent areas. *Geol. Surv. Canada, Map* 1401A (4 sheets).

Shilts, W. W. 1980. Flow patterns in the central North American ice sheet. *Nature*, **286**, 213–18.

Shilts, W. W. 1982. Quaternary evolution of the Hudson/James Bay region. *Naturaliste Can.* **109**, 309–32.

Srivastava, S. P. 1978. Evolution of the Labrador Sea and its bearing on the early evolution of the North Atlantic. *Roy. Astron. Soc. Geophys. J.* **52**, 313–57.

Sugden, D. E. 1977. Reconstruction of the morphology, dynamics and thermal characteristics of the Laurentide ice sheet. *Arctic & Alpine Res.* **9**, 21–47.

Sugden, D. E. 1978. Glacial erosion by the Laurentide ice sheet. *J. Glaciol.* **20**, 367–92.

Taylor, F. C. 1974. Reconnaissance geology of a part of the Precambrian shield, northern Quebec and Northwest Territories. *Geol. Surv. Canada Paper* 74–21.

Thomas, R. H. and C. R. Bentley 1978. A model for Holocene retreat of the west Antarctic ice sheet. *Quat. Res.* **10**, 150–70.

Tyrrell, J. B. 1898. The glaciation of north-central Canada. *J. Geol.* **6**, 147–60.

Weertman, J. 1974. Stability of the junction of an ice sheet and an ice shelf. *J. Glaciol.* **13**, 3–11.

Westra, L. 1978. Metamorphism in the Cape Smith-Wakehano Bay area north of 61N, New Quebec. *Geol. Surv. Canada Paper* 78–10, 237–44.

Williams, L. D. 1979. An energy-balance model of potential glacierization of northern Canada. *Arctic & Alpine Res.* **11**, 443–56.

5

Forward and inverse models in sea-level studies

James A. Clark

Introduction

Models are simplified representations of the real world and provide insight into processes or improved understanding of complex interactions. In geomorphology models allow for the acceleration of time, reduction of sizes, or control of boundary conditions, depending upon the type of model. Numerical modeling, in particular, allows for the control of boundary conditions and time. It was therefore a likely choice in the study of sea-level changes on a viscoelastic Earth caused by the retreat of the ice sheets of the last ice age. The overall modeling approach was divided into two distinct phases. The first phase is called the forward calculation and it includes the steps normally considered in generating a numerical model—those of condensing the essential processes into the form of equations or algorithms to be solved on a computer and of representing continuous variables with discrete piecewise continuous functions. Previous work was predominantly concerned with this problem (e.g. Walcott 1972a, Chappell 1974, Cathles 1975, Peltier & Andrews 1976) and the method used here follows that of Farrell and Clark (1976). During this phase the glacial histories and Earth rheology were assumed to be known and the numerical model predicted sea-level curves at any location on the Earth. These predictions could then be compared to observations of sea-level change so that the numerical model could be evaluated. Foremost among the goals of the forward calculation was the reconciliation of apparently conflicting sea-level data. Regions in the southern hemisphere typically reported high mid-Holocene sea levels whereas many northern hemisphere regions reported continual transgression during the same time period (Jelgersma 1966).

The second phase is called the 'inverse calculation' and in this phase the assumptions of the forward calculation are critically and quantitatively examined in light of the ability of the predictions to fit the data. The chief goal of this phase is to use the sea-level data to reconstruct the deglacial history of the North American ice sheet of the last ice age. Once an ice sheet is found that causes the predicted sea levels to fit the observed sea-level changes, the reconstruction must be evaluated in the light of uniqueness criteria. The critical question with respect to uniqueness is: 'are there other ice-sheet reconstructions that fit the sea-level data at least as well?' This last step gives necessary information about the true state of our knowledge concerning the past ice sheet.

The main goal of this paper is to give an example of forward and inverse calculations, especially demonstrating the usefulness of inverse calculations to any modeling study.

Forward calculation

Assume the Earth to be modeled adequately by a linear spherically symmetrical viscoelastic (Maxwell) material with known variation of viscosity and elastic parameters as a function of depth. This assumption is supported by two facts: elastic waves generated by earthquakes propagate through the Earth; and slow viscous deformation of the Earth is a well-known cause of postglacial uplift (e.g. Walcott 1972). Furthermore, assume that the thickness and extent of ice sheets of the most recent ice age are known with respect to time. With these assumptions the change in sea level, anywhere on the Earth's surface, can be calculated. Actually neither the viscoelastic structure nor the ice-sheet histories are known exactly but for the purposes of the forward calculation this is not a hindrance. Errors in these assumptions are considered in subsequent sections of this paper.

Because of the assumed linearity of the Earth rheology, the crux of the problem of determining sea-level response caused by ice-sheet retreat is in finding the induced sea-level as a function of time and location caused by a 1 kg point load placed on the Earth's surface. This is because any arbitrary ice-sheet loading history can be represented by a collection of point loads and the linearity of the problem allows the solution to be the superposition of the sea-level response for each point load. It is easy to show (Farrell & Clark 1976) that, to first order, the sea-level change, s, is ϕ/g where ϕ is the gravitational potential perturbation generated by the sum of three effects:

(1) the redistribution of surface masses;
(2) the movement of the Earth's surface through the ambient gravitational potential field with gradient g, the gravitational acceleration at the Earth's surface; and,
(3) the redistribution of mass within the Earth caused by deformation of the Earth from the surface load.

The calculation of the potential response due to the point load is lengthy but numerical solutions exist (Peltier 1974, 1980) and results of these calculations have been used here.

The exact sea-level solution at position $\mathbf{r}$ and time t for any ice-sheet history is (Farrell & Clark 1976)

$$s(\mathbf{r},t) = \frac{1}{g} \int_{-\infty}^{t} [\rho_w \iint_{\text{oceans}} G(\mathbf{r}-\mathbf{r}',t-\tau)s(\mathbf{r}',\tau)d\mathbf{r}'$$

$$+\rho_I \iint_{\substack{\text{ice} \\ \text{sheet}}} G(\mathbf{r}-\mathbf{r}',t-\tau)\gamma(\mathbf{r}',\tau)d\mathbf{r}']d\tau$$

$$-k(t) \tag{5.1}$$

where: $G(\mathbf{r}-\mathbf{r}',t-\tau)$ is the potential response at position $\mathbf{r}$ and time t due to a 1 kg point load placed on the Earth's surface at position $\mathbf{r}'$ and time τ. G is called a Green's function. When $t = \tau$ the Green function is for the immediate elastic response of the Earth. When $t>\tau$, G is the viscous response of the Earth caused by the load at time $t-\tau$ in the past. γ is ice-sheet thickness which is a function of position, $\mathbf{r}$, and time, τ. ρ_w is the density of water. ρ_I is the density of ice. k is a correction term that insures the conservation of mass. Included in k is the eustatic sea-level rise.

To evaluate Equation 5.1 numerically, two simplifying assumptions are necessary. The first is that both the ice-sheet load and the sea-level load can be represented by many small discs with constant load thickness over each disc. The second assumption is that ice-sheet changes occur only at discrete times and for the model given here these times are at 1000 year intervals.

Making these assumptions, Equation 5.1 can be rewritten in a discrete matrix form for the sea-level rise occurring $L\times1000$ years after the initiation of ice-sheet retreat:

$$\mathbf{s}_L = \mathbf{A}^e\mathbf{s}_L + \mathbf{B}^e\gamma_L + \mathbf{k}_L + \sum_{l=1}^{L-1} \left[\mathbf{A}_l^v\Delta\mathbf{s}_l + \mathbf{B}_l^v\Delta\gamma_l \right] \tag{5.2}$$

Here the integrations of Equation 5.1 are represented by the product of a response matrix (upper case letters) and a load vector (lower case letters) and the elastic part of the viscoelastic Green's function (superscript e) is separated from the time-dependent viscous portion (superscript v). The first term on the right is the perturbation in sea level due to the immediate elastic response of the Earth to ocean loading. The second term is the immediate elastic sea-level response caused by ice-sheet unloading and the third term is a vector with constant terms that assures conservation of mass. The final term accounts for the slow viscous response of the Earth from *changes* in ocean and ice loads ($\Delta\mathbf{s}_l$ and $\Delta\gamma_l$, respectively) that occurred previously. The summation in this term is over all time increments from the initiation of ice retreat to the time increment just prior to that at which the sea-level result is desired. The vector $\mathbf{k}_L$ insures conservation of mass. All of its components, k_L are equal.

Mass is conserved if the total amount of mass added to the oceans as meltwater equals the amount of ice mass melted. Therefore

$$\rho_w \mathbf{d^t s}_L + \rho_w \alpha k_L = \rho_I \mathbf{e^t} \gamma_L$$

where superscript t represents transposition, k_L accounts for any discrepancy in mass balance, $\mathbf{d}$ is a vector of ocean grid areas, $\mathbf{e}$ is a vector of ice grid areas, and α is the area of the ocean.
Solving for k_L:

$$k_L = \frac{\rho_I}{\rho_w \alpha} \mathbf{e^t} \gamma_L - \frac{1}{\alpha} \mathbf{d^t s}_L$$

with the first term on the right representing the eustatic sea-level rise and the second term, which is usually small, representing a correction for deformation of the ocean floor. Define the vector

$$\mathbf{k}_L = \left(\frac{\rho_I}{\rho_w \alpha} \right) \mathbf{E^t} \gamma_L - \frac{1}{\alpha} \mathbf{D^t s}_L$$

where every row of matrix $\mathbf{E}$ is the vector $\mathbf{e}$ and every row of $\mathbf{D}$ is $\mathbf{d}$. With this definition every element of $\mathbf{k}_L$ equals k_L.
Substitution of $\mathbf{k}_L$ into Equation 5.2 yields

$$\mathbf{s}_L = \mathbf{A}^e \mathbf{s}_L + \mathbf{B}^e \gamma_L + \sum_{l=1}^{L-1} \left[\mathbf{A}_l^v \Delta \mathbf{s}_l + \mathbf{B}_l^v \Delta \gamma_l \right]$$

$$- \frac{1}{\alpha} \mathbf{D^t s}_L + \frac{\rho_I}{\alpha \rho_w} \mathbf{E^t} \gamma_L \tag{5.3}$$

Combining all terms of $\mathbf{s}_L$ gives:

$$[\mathbf{I} - \mathbf{A}^e + \frac{1}{\alpha} \mathbf{D^t}]\mathbf{s}_L = \mathbf{B}^e \gamma_L + \sum_{l=1}^{L-1} \left[\mathbf{A}_l^v \Delta \mathbf{s}_l + \mathbf{B}_l^v \Delta \gamma_l \right]$$

$$+ \frac{\rho_I}{\alpha \rho_w} \mathbf{E^t} \gamma_L \tag{5.4}$$

This equation is of the form $\mathbf{Q s}_L = \mathbf{w}$ with the vector of sea level rise $\mathbf{s}_L$ the only unknown (assuming the ice-sheet history is known) and so in theory this system of linear equations can be solved for $\mathbf{s}_L$. In practice, for the realistic ocean configuration used here, $\mathbf{Q}$ has a rank of 452. Because standard numerical methods for solving linear equations are unstable for such a large matrix, an iterative method, applied to Equation 5.3 was used. This method gave a very good approximation to s_L.

To advance another time step and determine s_{L+1}, solve Equation 5.3 again after finding $\Delta s_L = s_{L-1}$ and advancing the summation limit to L. In this way the evolution of sea level can be calculated if the ice-sheet history is known.

Results of this forward calculation for melting of the Laurentide and Fennoscandian ice sheets have been reported elsewhere (Clark *et al.* 1978, Clark 1980, Peltier *et al.* 1978) and the Antarctic ice sheet (Clark & Lingle 1977, 1979, Lingle & Clark 1980). Typical results are illustrated in Figure 5.1 for the case where the eustatic sea-level rise ceases 5000 years ago. It is clear that the sea-level rise in response to ice-sheet melting is not uniform every-where. In fact, no two sea-level curves for different regions, predicted by this viscoelastic Earth model, are identical. The success of this calculation has been in reconciling the 'high Holocene' sea-level data observed predominantly in the southern hemisphere (Zone V and VI) with the steady Holocene transgression

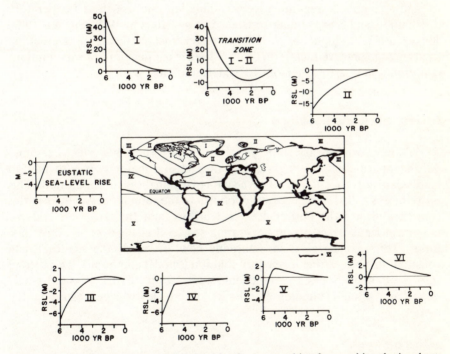

Figure 5.1 Distribution of six predicted sea-level zones resulting from melting the ice sheets 18 000–5000 years ago. Within each zone the form of the sea-level response is similar. Typical curves indicating the predicted change in sea level relative to the present geoid for each zone are included (RSL is the relative sea level). The curves show the wide variety of sea-level expressions possible despite the assumption of no change in ocean volume during the past 5000 years (no change in eustatic sea level). (From Clark & Lingle 1979. Used by permission of Academic Press).

observed in much of the northern hemisphere (Zone II). Thus, one part of the modeling procedure attained its goal—demonstration of a mechanism that can reconcile conflicting data.

Inverse calculation

Although the forms of the predicted sea-level curves were similar to those of the observed curves, the magnitude of predicted sea-level change for regions close to the glaciated regions was too large by up to a factor of three. This discrepancy in magnitude, but not form, of the curves suggested that the ice-sheet history was in error and in need of revision. Because ice-sheet thickness (especially the Laurentide ice-sheet thickness) is unknown, the sensitivity of the sea-level predictions to changes in ice-sheet thickness suggests a means for improving our knowledge about the retreat of the ice sheets.

Previous attempts to reconstruct ice-sheet thickness followed a completely independent approach from that used here. Those studies used the flow law of ice and assumptions regarding ice accumulation rate, ice-sheet temperature distribution and basal sliding resistance (e.g. Paterson 1972, Sugden 1977, Jenssen 1977, & Hughes *et al.* 1981) to reconstruct ice sheets. In the method used here the observed sea-level data are used in conjunction with assumptions concerning the Earth's rheology to achieve the same goal.

Inverse calculation: method

There are two necessary steps in solving for ice-sheet histories (i.e. time-dependent ice-sheet thickness changes). The first major step is to find an ice-sheet history that fits the observed data to within the errors of the data. This procedure is called model construction where the term *model* as used in this context indicates the desired ice-sheet history, not the computer code or numerical steps necessary in generating sea-level changes or ice-sheet histories. The second step is to determine how effectively the sea-level data constrain this best-fitting ice-history solution ('model appraisal'). This second step is often overlooked in modeling studies even though it is extremely important in understanding the limits of our knowledge regarding the model.

Model construction

There are several possible methods that could result in an ice-sheet history that fits the sea-level data. One might use a trial-and-error method where the ice-sheet history is subjectively varied until the forward calculation predicts sea-level changes sufficiently close to the observed data; or, glacial histories might be selected at random until a suitable one is accidentally found. The former method is more efficient but biased while the latter method is extremely inefficient. The approach adopted here is to use a least-squares method, where the ice-sheet history is, in addition, constrained to lie within physically plausible limits. This method is therefore highly efficient and the resulting ice-sheet history is known to be the one that best fits the data while simultaneously satisfying the constraints.

The inverse calculation can be greatly simplified because, for regions close to ice sheets the sea-level response is dominated by the effects of ice-sheet loading and the eustatic sea-level rise. Ignoring the small second order sea-level loading effects for these regions, Equation 5.3 becomes

$$s_L = \sum_{l=1}^{L} (\mathbf{B}^e + \mathbf{B}_l^v - \frac{\rho_I}{\alpha\rho_w}\mathbf{E}^t)\Delta\gamma_l \tag{5.5}$$

where it is understood that $\mathbf{B}_l^v$ equals the zero matrix if $l = L$. Equation 5.5 shows that the sea-level rise is linearly related to the incremental ice-thickness changes, $\Delta\gamma$, occurring at 1000 year intervals. Because the observed sea-level change is always measured with respect to *present* sea level, Equation 5.5 must be modified slightly so that $s_q^{o\prime}$ is found where $s^{o\prime}$ is the predicted sea-level change measured from present sea level and q is the number of 1000 year time increments before the present. The prediction $s_q^{o\prime}$ can then be compared directly to observations. Making these algebraic changes (see Clark 1980) and dropping the subscript q, which is implied, Equation 5.5 takes the form

$$s^{o\prime} = \mathbf{H}'\mathbf{m} \tag{5.6}$$

where each element of the vector $\mathbf{m}$ is the incremental change in ice thickness, $\Delta\gamma$, at a given location and time. $\mathbf{H}'$ is a matrix derived from the physical principles of viscoelasticity that relates the time-dependent ice-sheet thickness changes to the observed sea-level changes. In deriving Equation 5.6 the assumption that the ocean-loading effect is negligible compared to ice-sheet loading for regions close to the ice sheet is not mandatory (Clark 1980). An equation of a form similar to Equation 5.6 results even if ocean-loading effects are included. The resulting numerical calculation, however, is very lengthy and so this added small refinement is not included here.

Equation 5.6 is in exactly the same form as that used in a multiple-regression model. In multiple regression, however, the independent variables comprise the matrix $\mathbf{H}'$ and the unknown slopes and constant of the hyperplane that best fits the dependent variables in the least squares sense is the $\mathbf{m}$ vector. For the ice-sheet model discussed here the matrix $\mathbf{H}'$ is determined by the physics of the problem, not a measured variable, and $\mathbf{m}$ is the vector of the desired ice-thickness change. Otherwise the method of least-squares solution for $\mathbf{m}$ is identical for both.

If $\mathbf{d}'$ is the vector of observed sea-level changes our goal is to find a model vector, $\hat{\mathbf{m}}$ such that the sum of squared deviations, ϵ^2, between $\mathbf{d}'$ and $s^{o\prime}$, the predicted sea-level change, is minimized (i.e. a least-squares constraint). The hat is included over **m** because, in general, the solution will not exactly fit the data and so **m̂** is only a best estimate of **m**. Because the sea-level data have errors that may vary in magnitude, the data should be transformed so that the standard deviation of every transformed datum is unity. This procedure insures that the ice-sheet model will depend more upon accurate data than inaccurate data. To standardize the data, construct a diagonal matrix, **W**, where the elements of the diagonal are the respective standard deviations of the data (e.g. Clark 1977). In the transformed system

$$\mathbf{s}° \equiv \mathbf{W}^{-1}\mathbf{s}°'$$

$$\mathbf{H} \equiv \mathbf{W}^{-1}\mathbf{H}'$$

$$\mathbf{d} \equiv \mathbf{w}^{-1}\mathbf{d}'$$

so that Equation 5.6 becomes

$$\mathbf{s}° = \mathbf{Hm}$$

In matrix notation the least-squares approach is to minimize

$$\epsilon^2 = (\mathbf{s}°-\mathbf{d})^2 = (\mathbf{Hm}-\mathbf{d})^2.$$

The solution for minimum ϵ^2 is

$$\mathbf{\hat{m}} = (\mathbf{H}^t\mathbf{H})^{-1}\mathbf{H}^t\mathbf{d} \qquad (5.7)$$

which is the least-squares solution for a multiple-regression model. If there were no errors in the data this would be the appropriate equation to solve. However, data errors do exist and these errors will adversely affect the stability of the ice-sheet model causing unrealistically large fluctuations in ice-sheet thickness. The strict least-squares criteria can, and should, be relaxed. There are a number of ways to do this. The one employed here is to include additional constraints dictated by physically plausible limits for the ice-thickness solution. The first method is to minimize ϵ^2 under the constraint that **m^{t}m** equals some prescribed scalar. Reworded this means that an ice-thickness solution is determined which results in a good fit to the data (but not as good as the exact least-squares solution) and where the average incremental changes in ice thickness for all locations and times equals some prescribed value. In practice the ice-sheet model with the best fit to the data also has unrealistically large fluctuations in ice-sheet thickness. To minimize these fluctuations solve a slightly augmented form of Equation 5.7.

$$\mathbf{\hat{m}} = (\mathbf{H}^t\mathbf{H}+\beta\mathbf{I})^{-1}\mathbf{H}^t\mathbf{d}$$

where β is an arbitrary scalar and **I** is the identity matrix. This equation imposes the **m^{t}m** constraint upon the solution (Clark 1977) where β is a 'dial' which can be changed at will. As β increases the average incremental ice-thickness change ($\mathbf{m^t m}$) decreases but the fit to the data becomes worse. There is therefore a trade-off between stability of the solution and fit to the data.

These constraints are 'soft' in that ice-thickness changes at individual locations may exceed the average thickness change controlled by the 'β dial.' In some instances it is desirable to impose 'hard' constraints upon the solution so that a given class of solutions is never permissible. An example of such a constraint was used in this study because it is physically impossible for total ice-sheet thickness ever to become negative. This constraint is, of course, physically justified and it greatly limits the possible solutions to the problem. A description of the solution method is beyond the scope of this paper (see Lawson & Hanson 1974, Clark 1980), but the technique is one of finding

$$\hat{\mathbf{m}} = (\mathbf{H^t H} + \beta \mathbf{I}) \mathbf{H^t d}$$

but with the additional constraint that

$$\mathbf{P\hat{m}} \geqslant \mathbf{f}$$

where $\mathbf{P}$ is a matrix such that $\mathbf{P\hat{m}} = \mathbf{z}$. The vector $\mathbf{z}$ is the vector of total (not incremental) ice-sheet thickness variations through space and time. If $\mathbf{f}$ is the zero vector the desired non-negativity constraint is applied. Numerical algorithms exist for the solution of this 'quadratic-programming' problem and the one given by Lawson and Hanson (1974) is used here.

Model appraisal

The construction of this best-fitting constrained model of ice-sheet thickness answers only part of the problem, however. It is not certain that the best-fitting glacial history is necessarily the correct glacial history because of the ambiguities resulting from errors in the sea-level data. Since the predictions of the model need only to lie within the expected errors of the date, no single model is unambiguously 'correct.' For example, an ice-sheet history is constructed that best fits the data, subject to the imposed constraints. If one of those data points is actually incorrect because of, for example, measurement error, then the predicted ice-sheet history is not the correct history; rather, another history best fits the corrected data. Of course, we can never know what the correct data are but usually we know approximately the expected error about the data.

There are several approaches to the solution of this inverse problem. Perhaps the most common is a sensitivity analysis where one variable (for this study the ice-sheet thickness at a particular locality and time) is altered slightly in the forward calculation while all others remain fixed. The resulting degree of perturbation of the predictions gives an indication of how sensitive the model is to errors in the data. The point at which the prediction exceeds the error range is the expected range of the trial model parameter. The procedure is then repeated successively for each variable. My criticism of this approach is that there is no means to evaluate how compensating changes in the other 'fixed' variables would alter the results.

Another approach, first suggested by Backus and Gilbert (1967, 1968, 1970), is much more elegant and mathematically exact, but much more difficult to interpret in a physical sense. The method relies upon the fact that, although an infinite number of solutions can fit the data to within the same average error, only one of those solutions is the correct solution. The solution in hand, $\hat{m}$, is also one of these solutions and so the problem becomes one of finding something that the entire infinite set has in common. This requirement is satisfied because an averaging function exists called the 'resolving kernel.' The model in hand is a smoothed (averaged) function of the permissible models and Backus and Gilbert (1968) show that every permissible model, when smoothed by this resolving kernel, gives the model estimate already determined. Where the resolving kernel is broad, the model estimate, $\hat{m}$, is probably a poor estimate of the correct solution whereas a very concentrated kernel indicates the model estimate is probably very close to the actual solution. I used this method (Clark 1977) in a related problem where time dependence was not a factor in the calculation. However, for the present problem the resolving kernel averages not only over a spatial dimension, but also over a time dimension. Interpretation of such a function is difficult at best and I seriously doubted if I would gain any understanding from the exercise.

The method adopted here is similar to one proposed by Jackson (1976, 1979) termed 'most-squares analysis.' His approach is to find a range of models, all of which fit the data to within a prescribed accuracy and are physically plausible. His treatment did not include the quadratic programming method employed here so slight changes in implementation, but not philosophy, were necessary. The approach is analogous to sensitivity analysis except for one very significant change. Instead of holding all variables except one fixed, in the most-squares analysis implemented here only one variable is fixed and all others are free to change such that the error of fit is minimized. Should the fit be within the estimated error range, the value of the fixed variable is changed slightly and the process is repeated until a value is determined for the fixed variable such that no permissible change in the free variables can result in adequate fit to the data. This is the point at which the data constrain the fixed variable. The process is repeated for every variable. Although this is a tedious procedure, it is readily understood and interpreted. It results in a much more conservative estimate of uncertainty than that determined by sensitivity analysis because compensation among the variables in the most-squares method results in a greater variable range for a given error-of-fit.

The method is readily incorporated into the constrained least-squares method used here by augmenting slightly the constraints $P\hat{m} > f$ to include, in addition, an ice thickness of a prescribed amount and setting the appropriate element of f equal to this thickness. (See Clark 1980 for details.)

Inverse calculation: application & results

Quadratic programming of the most-squares method outlined above was applied to the problem of reconstructing the demise of the Laurentide ice sheet of North America from sea-level data. Although in principle all ice-sheets of the last ice age could be reconstructed using these methods in one calculation, the increased numerical difficulties do not warrant such an elaborate calculation for the present purposes. The contributions of the European, Greenland, and Antarctic ice sheets to changes in North American sea level were assumed to be as indicated in Figure 5.2. The sea-level data were adjusted for this component of eustatic sea-level rise before application of the procedure for the inverse calculation.

Ice sheet representation

To solve the inverse problem an additional simplification was used. The Laurentide ice sheet was represented by 8 large discs (Figure 5.3) and each disc was assumed to have a constant thickness of ice at any instant in time, although thicknesses among discs could vary. The locations and sizes of these discs were chosen to approximate areal extent of the Laurentide ice sheet, as indicated by Prest (1969), and the positions of different dynamic regions within the ice sheet. The locations of the 42 sea-level curves used in this study are also included in the figure. These curves are from compilations by Walcott (1972b) and Clark (1980) of previously published sea-level data. The sources of data

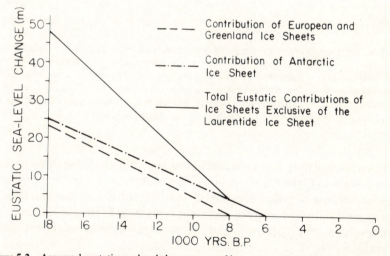

Figure 5.2 Assumed eustatic sea-level changes caused by retreat of ice sheets distant from North America. Because the inverse calculation only accounts for sea-level changes resulting from melting of North American (Laurentide) ice, all data indicating sea levels prior to 6000 yr BP must be corrected in elevation for the sea-level effects of distant ice sheets. The solid line represents this correction applied to all data. (From Clark 1980. Copyrighted by the American Geophysical Union. Used by permission).

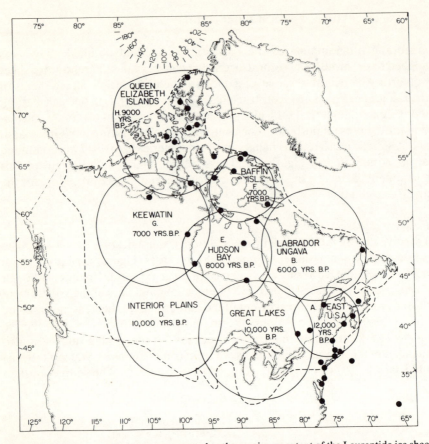

Figure 5.3 Eight ice grid elements compared to the maximum extent of the Laurentide ice sheet (dashed line, Prest (1969)) and the sea-level data. The ice elements are spherical caps and are slightly distorted in this figure because of the map projection. The time of assumed complete deglaciation of each ice element is indicated. The inverse calculation determines the amount of ice-thickness change over each of these ice elements as a function of time, that is consistent with the sea level data. (From Clark 1980. Copyrighted by the American Geophysical Union. Used by permission).

and the method of determining errors in the data are included in those papers. In previous work (Clark 1980) I assumed these discs were static. Here realism is enhanced because the discs are moved and reduced in size through time to better approximate the history of ice extent as given by Prest (1969). Figure 5.4 illustrates three 'snapshots' in time of this disc representation of the retreating Laurentide ice sheet.

The ice discs overlap in some regions and this will introduce error. It is possible to avoid such overlap and also to improve realism by employing rectangles instead of discs. This improvement is tractable but more difficult numerically and for the examples given here the simplified representation shown in Figure 5.4 is used. Because of these simplifications the ice-sheet

Figure 5.4 Retreat of the Laurentide ice sheet approximated in this study by movement and shrinking of the 8 ice discs shown in Figure 5.3. Only three of 11 time steps are illustrated here.

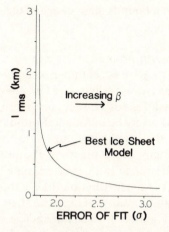

Figure 5.5 Trade-off curve for the quadratic programming inversion. The 'best' ice model that is everywhere nonnegative is indicated at the knee of the curve. The best model is one where 1000 year incremental ice thickness changes greater than ± 700 m are discouraged.

reconstruction results reported here are meant only as an illustration of the method—not as a definitive statement of ice-sheet history.

Trade-off curve
In applying the quadratic-programming method described earlier, the first step is to find an ice-sheet model that is a realistic compromise between its ability to fit the data and its average amplitude of ice-thickness fluctuation. This procedure, of increasing β until the 'right' model is found, is not as subjective as it might at first seem. Figure 5.5 shows the resulting trade-off curve between error-of-fit; $\sigma \equiv (\epsilon^2/n)^{\frac{1}{2}}$, and root-mean-square amplitude of fluctuation $I_{rms} \equiv \mathbf{m}^{\mathbf{t}}\mathbf{m}/M)^{\frac{1}{2}}$ n and M are the numbers of data points and ice elements, respectively. As β increases the amplitude fluctuation decreases dramatically from 2900 m to only 1050 m with virtually no loss in ability to fit the data. At this point additional reduction in the fluctuation of ice-sheet thickness causes progressively greater error-of-fit until an unacceptable increase in error-of-fit for a given decrease in ice-sheet fluctuation is attained. The best compromise in ice-sheet models is one near the 'knee' of the trade-off curve. The one selected has a root-mean-square thickness change from one 1000 year time step to the next of 700 m and an error-of-fit of 1.85.

Fit to data
Figure 5.6 illustrates the fit of predictions to sea-level observations at 5 of the 42 localities. The ice-sheet model selected in the previous section was used in the prediction. It is clear that the fits are good. Only in limited regions do predictions exceed the expected error of the observations. The coefficient of determination, R^2, is 0.91 ($R = 0.95$). The F statistic is 89.9 with 29 and 262 degrees of freedom. Therefore this fit is significant at the .005 level of probability. The next phase is to somewhat 'relax' these overly stringent fits to the data and to observe what bounds these data and their accompanying errors place upon the ice-sheet thickness.

Predicted ice sheet and its error bounds
The variation in predicted total ice-sheet thickness that resulted in the sea-level predictions is given for each ice disc as the solid line in Figure 5.7. It is tempting to make elaborate interpretations of ice-sheet history from this prediction, but such interpretations would surely be unfounded without considering the error bounds associated with the prediction. In fact, the range of ice-sheet thicknesses is of considerably more use than the predicted best-fitting ice history. This is because this range is a better estimate of the constraints that sea-level data place upon the reconstruction than is the reconstruction itself.

   An example of the method used to determine these error bounds for the reconstruction is given in Figure 5.8. Each line indicates how the overall fit to the data changes as the Hudson Bay ice disc varies in thickness at a given time. To calculate each curve, a thickness and a time were specified for the ice disc and the error for the resulting best-fitting model was determined. Changing the ice thickness allowed construction of the curve. The limit of acceptable ice

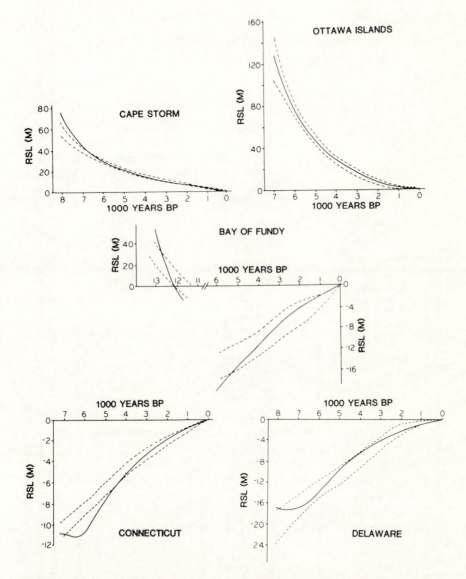

Figure 5.6 Five of the 42 observed sea-level curves compared to sea-level predictions of the 'best' nonnegative ice model for North America. The dashed lines indicate the range in data, and the solid line is the model sea-level prediction. All curves are relative to present sea level. Sources of the data: Cape Storm (Blake 1975), Ottawa Islands (Andrews & Falconer 1969), Bay of Fundy (Grant 1970), Connecticut (Bloom & Stuiver 1963), and Delaware (Belknap & Kraft 1977).

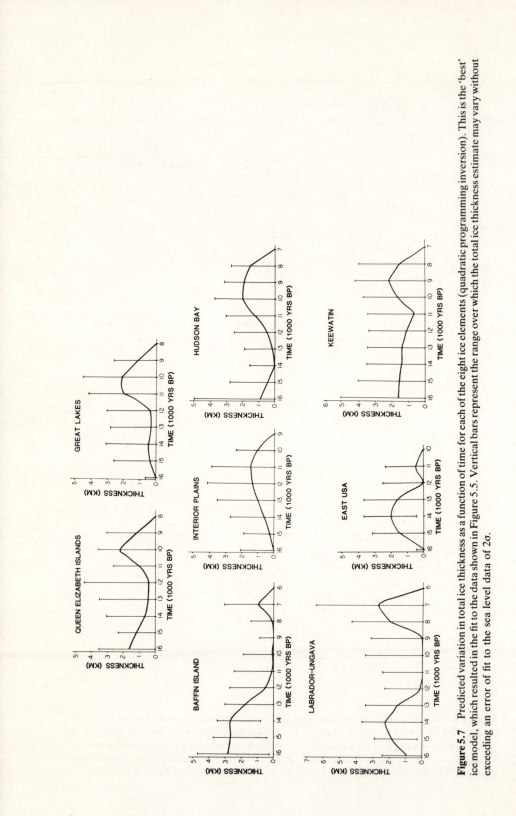

Figure 5.7 Predicted variation in total ice thickness as a function of time for each of the eight ice elements (quadratic programming inversion). This is the 'best' ice model, which resulted in the fit to the data shown in Figure 5.5. Vertical bars represent the range over which the total ice thickness estimate may vary without exceeding an error of fit to the sea level data of 2σ.

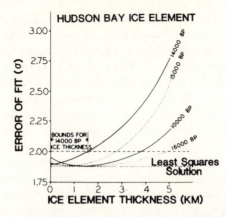

Figure 5.8 Relationship between error of fit, σ, and total ice-sheet thickness for the Hudson Bay ice element (*E*) at four different times. From relationships such as these, the range of permissible ice thicknesses can be determined from most-squares inversion. For example, if the greatest tolerable error of fit is 2σ, the range of permissible ice thicknesses 14 000 years ago on Baffin Island is 0–1550m.

thickness is arbitrarily chosen as the point where the predicted ice-sheet model no longer causes sea-level predictions to fit the data to within two standard deviations (2σ). This procedure was repeated for every ice disc and every time step.

The range determined from this method is indicated as vertical bars in Figure 5.7. Errors in total ice thickness vary from 0.5 km to 6.5 km although a 3 km error bar is typical. Clearly the data do not appreciably constrain the reconstruction. This conclusion, though pessimistic, is extremely important for the interpretation. Without following through with the inverse modeling exercise one might erroneously think, after fitting the data adequately by a forward calculation, that the ice-sheet history is determined or verified. Any scientist who uses models must exercise caution in order to avoid this type of error.

Despite the large errors associated with this reconstruction, some very general conclusions can be tentatively advanced. Over Hudson Bay a general thinning of the ice sheet between 15 000 BP and 13 000 BP is suggested. The eastern U.S. ice disc seems to thicken during the same time. The Keewatin ice element appears to have a rather stable thickness whereas the Labrador–Ungava ice disc is slightly thinner and more erratic in thickness changes. Ice-sheet predictions for other regions where apparent thickening occurs during the later stages of glaciation are more likely explained by the migration of the model ice disc toward the original center of the ice sheet and its decrease in radius.

Conclusion

The construction of a numerical representation of a geomorphic process can, without doubt, lead to greater understanding of the process and may help to resolve apparent conflicts in data. The forward calculation outlined here has offered an explanation for differences in Holocene sea levels observed around the world. There still remains, however, a number of regions where predictions do not match sea-level observations. Although one might suggest that the data are in error due to misinterpretations of the geology or incorrect dating of samples, it has been my experience that observations of sea-level change are remarkably reliable. The cause of any mismatch therefore resides in the numerical model and, in particular, in its assumptions or simplifications. The assumed Earth rheology (see Peltier 1976) or the assumed ice-sheet history are immediate suspects. If the forms of predicted and observed curves differ, as they do in a few regions such as the Aleutian Islands (Black 1980), then the Earth rheology is the most likely cause. On the other hand if the forms of the predicted sea-level curves are similar to observed curves, with amplitude or timing of the prediction the cause of mismatch, then the ice-sheet history, not the Earth rheology, is probably in error.

For regions in eastern North America it is the ice-sheet history that is the likely cause of incorrect predictions. Using a suitable inversion method a glacial history was found that efficiently fit the sea-level data. The good fit does not, in itself, indicate that the calculated model of ice sheet history is correct or even nearly correct. An appraisal of the ice sheet model, where the real constraints the sea-level data place upon the calculated glacial history are determined, resulted in a pessimistic conclusion. The errors about the predicted ice-sheet history were excessive indicating that the sea-level data did not significantly constrain the reconstruction of the ice-sheet. Every modeling study of the type illustrated here should perform an appraisal of the resulting model before acceptance of the model as correct. To end on a more optimistic note, it is likely that the extreme simplifications of the ice-sheet representation in the inverse calculation greatly affected the sensitivity to the data of the ice-sheet reconstruction. It is possible to eliminate these unrealistic assumptions and therefore greatly improve our understanding of the glacial history of North America.

Acknowledgments

Discussions with Arthur L. Bloom and Craig S. Lingle were of considerable benefit. W. Richard Peltier kindly supplied the necessary Green functions. Jan Woudenberg typed many drafts of the manuscript. My wife, Sue, offered encouragement and editorial criticism. Funds for this work were provided by the National Science Foundation under grants EAR77–13662 and EAR78–12977.

References

Andrews, J. T. and G. Falconer 1969. Late glacial and post-glacial history and emergence of the Ottawa Islands, Hudson Bay, N.W.T.: evidence on the glaciation of Hudson Bay. *Can. J. of Earth Sci.* **6**, 1263–76.

Backus, G., and F. Gilbert, 1967. Numerical applications of a formalism for geophysical inverse problems. *Geophys. J. Roy. Astron. Soc.* **13**, 247–76.

Backus, G., and F. Gilbert, 1968. The resolving power of gross Earth data. *Geophys. J. Roy. Astron. Soc.* **16**, 169–205.

Backus, G., and F. Gilbert, 1970. Uniqueness in the inversion of inaccurate gross Earth data. *Phil. Trans. Roy. Soc. London* **A266**, 123–92.

Belknap, D. F., and J. C. Kraft 1977. Holocene relative sea-level changes and coastal stratigraphic units on the northwest flank of the Baltimore Canyon trough geosyncline. *J. Sed. Petrol.* **47**, 610–29.

Black, R. F. 1980. Isostatic, tectonic, and eustatic movements of sea level in the Aleutian Islands. In *Earth rheology, isostasy, and eustasy*, N. A. Morner, (ed.), 231–48. New York: Wiley.

Blake, W., Jr. 1975. Radiocarbon age determinations and postglacial emergence at Cape Storm, southern Ellesmere Island, arctic Canada. *Geograf. Ann.* **57**, ser. A, 1–71.

Bloom, A. L. and M. Stuiver 1963. Submergence of the Connecticut coast. *Science* **139**, 332–34.

Cathles, L. M. 1975. *The viscosity of the Earth's mantle*. Princeton: Princeton University Press.

Chappell, J. 1974. Late Quaternary glacio- and hydro- isostasy on a layered Earth. *Quaternary Res.* **4**, 429–40.

Clark, J. A. 1977. An inverse problem in glacial geology: The reconstruction of glacier thinning in Glacier Bay, Alaska, between A.D. 1910 and 1960 from relative sea-level data. *J. Glaciol.* **18**, 481–503.

Clark, J. A. 1980. The reconstruction of the Laurentide ice sheet of North America from sea level data: method and preliminary results. *J. of Geophys. Res.* **85**, 4307–23.

Clark, J. A. and C. S. Lingle 1977. Future sea-level changes due to West Antarctic ice-sheet fluctuations. *Nature* **269**, 206–9.

Clark, J. A. and C. S. Lingle 1979. Predicted relative sea-level changes (18 000 yr B.P. to present) caused by late glacial retreat of the Antarctic Ice Sheet. *Quaternary Res.* **11**, 179–298.

Clark, J. A., W. E. Farrell, and W. R. Peltier 1978. Global changes in post-glacial sea level: A numerical calculation. *Quaternary Res.* **9**, 265–87.

Farrell, W. E., and J. A. Clark 1976. On postglacial sea level. *Geophys. J. Roy. Astron. Soc.* **46**, 647–67.

Grant, D. R. 1970. Recent coastal submergence of the Maritime Provinces, Canada. *Can. J. of Earth Sci.* **7**, 676–89.

Hughes, T., G. H. Denton, B. G. Anderson, D. H. Schilling, J. L. Fastook, and C. S. Lingle 1981. The last great ice sheets: A global view. In *The last great ice sheets*, G. H. Denton and T. Hughes (eds), 263–317. New York: Wiley.

Jackson, D. D. 1976. Most squares inversion. *J. Geophys. Res.* **81**, 1027–30.

Jackson, D. D. 1979. The use of a priori data to resolve nonuniqueness in linear inversion. *Geophys. J. Roy. Astron. Soc.* **57**, 137–57.

Jelgersma, S. 1966. Sea-level changes during the last 10 000 years, Proceedings of the International Symposium on World Climate from 8000 to 0 B.C. *Roy. Meteorol. Soc. London*, 54–71.

Jenssen, D. 1977. A three-dimensional polar ice-sheet model. *J. Glaciol.* **18**, 373–89.

Lawson, C. L., and R. J. Hanson 1974. *Solving least squares problems*. Englewood Cliffs, NJ: Prentice-Hall.

Lingle, C. S. and J. A. Clark 1979. Antarctic ice sheet volume at 18 000 years B.P. and Holocene sea-level changes at the West Antarctic margin. *J. Glaciol.* **24**, 213–30.

Paterson, W. S. B. 1974. Laurentide ice sheet: Estimated volumes during the late Wisconsin. *Rev. Geophys. Space Phys.* **10**, 885–917.

Peltier, W. R. 1972. The impulse response of a Maxwell Earth. *Rev. Geophys. Space Phys.* **12**, 649–69.

Peltier, W. R. 1976. Glacial isostatic adjustment II: the inverse problem. *Geophys. J. Roy. Astron. Soc.* **46**, 669–706.

Peltier, W. R. 1980. Ice sheets, oceans and the Earth's shape, In *Earth Rheology, Isostasy, and Eustasy*, Morner, N. A. (ed.), 45–63. New York: Wiley.

Peltier, W. R. and J. T. Andrews 1976. Glacial-isostatic adjustment-I. The forward problem. *Geophys. J. of the Royal Astron. Soc.* **46**, 605–46.

Peltier, W. R., Farrell, W. E., and J. A. Clark 1978. Glacial isostasy and relative sea-level: a global finite element model. *Tectonophysics* **50**, 81–110.

Prest, V. K. 1969. Retreat of Wisconsin and recent ice in North America. *Geol. Surv. Canada Map* 1257A.

Sugden, D. E. 1977. Reconstruction of the morphology, dynamics, and thermal characteristics of the Laurentide Ice sheet at its maximum. *Arctic & Alpine Res.* **9**, 21–47.

Walcott, R. I. 1970. Isostatic response to loading of the crust in Canada, *Can. J. Earth Sci.* **7**, 716–27.

Walcott, R. I. 1972a. Past sea levels, eustasy and deformation of the Earth. *Quaternary Res.* **2**, 1–14.

Walcott, R. I. 1972b. Late Quaternary vertical movements in Eastern North America: Quantitative evidence of glacio-isostatic rebound. *Rev. Geophys. Space Phys.* **10**, 849–84.

6

Coastal terraces generated by sea-level change and tectonic uplift

Arthur L. Bloom and Nobuyuki Yonekura

Introduction

Flights of marine terraces are classic features of emergent coasts. They have been the subject of serious attempts at global correlation by Quaternary stratigraphers because they were believed to represent stages in a progressive drop of eustatic sea level during late Cenozoic time, on which glacially controlled sea-level oscillations had been superimposed. Entire chronologies of the Pleistocene epoch were built on the model of oscillatory eustatic sea-level drop as the cause for the emerged terraces (Zeuner 1959). The terminology of the Mediterranean terrace sequence defined by Depéret (1918) is still common in the literature of the Quaternary Period. It is curious that only in the last decade an awareness has developed that plate-tectonic motions can cause uplift parallel to coasts over long distances. Even though the alleged uniform height of the Quaternary terraces in the Mediterranean and elsewhere was largely illusory, the idea that their heights must have been produced by eustatic, or world-wide, sea-level changes has died hard (Hey 1978).

A contrasting model for coastal terraces has been developed by structural geologists and tectonists working in obviously tectonic regions such as Japan, California, and New Zealand. Implicit or explicit in their tectonic models is the idea that tectonic uplift is irregular in amplitude and frequency. Extended intervals of stillstand during which terraces are built or eroded are punctuated by seismic pulses that isolate each surface and initiate a new erosional or depositional cycle at a lower level. The role of Quaternary sea-level oscillations in terrace evolution has generally been overlooked or minimized by these researchers. If the effect was considered at all, the reasonable assumption was made that interglacial sea levels returned to approximately the present level after each ice age. Because terrace formation is generally regarded as an interglacial phenomenon, glacially controlled sea-level oscillations could safely be ignored in tectonic terrace analysis. On the time scales of 10^5 to 10^6 years

Figure 6.1 Constructional coral reef terraces, Huon Peninsula, Papua New Guinea.

and with terraces hundreds of meters above sea level, the assumptions and conclusions of the tectonic evaluation of terraces heights remain largely valid, although they are generally untested because of a lack of suitable dating methods for terraces older than 200 000 years.

Coral-reef terraces (Fig. 6.1) offer exceptional opportunities to test hypotheses of terrace formation in general. Reef-building corals thrive only in warm, shallow water, and the internal structures of reef terraces prove that their constructional surfaces formed very close to low-tide level (Mesolella 1967, Verstappen 1960). Thus the upper surface of a reef terrace defines a datum plane at or within a few meters below the former sea surface. Except in areas of locally variable exposure to wind and surf, such as a coast with bold headlands and adjacent deep bays, the surface of a coral-reef terrace can be taken as an excellent record of former sea level. Furthermore, corals precipitate calcium carbonate as the mineral species aragonite, which has proved to be an excellent material for reliable radiocarbon and uranium-series dating. All diagenesis except for the most subtle changes is betrayed by the inversion of aragonite to calcite, so that spurious ages can be avoided by critical sample selection. Radiocarbon ages to about 30 000 years before present are reliable for well chosen coral samples, and uranium-series dates to about 200 000 years are reproduceable to within about 5 per cent. Because of their accurate record of former sea level and their ability to be dated directly, coral-reef terraces have become a major source of information about late Quaternary sea-level oscillations and about rates of tectonic uplift. The problem remains to separate and independently evaluate the two variables.

An extremely useful model for the origin of the Barbados reef terraces was offered by Mesolella *et al.* (1969, p. 259–60). They suggested that tectonic uplift be regarded as the analog of the paper drive in a strip-chart recorder, while glacial eustatic sea-level oscillations drive the recording pen. If the paper does not move in such a system, the pen moves back and forth over the same area, producing a thick but unreadable trace. If the paper moves forward slowly, large amplitude and low frequency pen motions can be distinguished, although high frequency oscillations remain superimposed. Progressively more rapid paper drive separates progressively finer-scale pen motions. This is the model that is tested in this paper. It is concluded that the assumption of a uniform rate of tectonic uplift gives good results on the time scale of 10^5 years, is correct to within a factor of 2 or 3 on the time scale of 10^4 years, and fails badly on the time scale of 10^3 years or less. With reference to late Quaternary time (10^4 or 10^5 years) glacial-eustatic sea-level fluctuations are the dominant factor in forming coastal terraces, and tectonic movements simply separate the terraces of successive interglacial and interstadial intervals. 'Jerky' tectonism dominates the development of coastal terraces on the 10^3 or 10^2 year time scale.

These principles are illustrated by an example (Fig. 6.2) in which moderately rapid uniform tectonic uplift of 2 mm per year (=2 m per 1000 years) is superimposed on an oxygen-isotope curve for benthic foraminifera (Streeter & Shackleton 1979). For simplicity, the oxygen-isotope fluctuation is assumed to represent fluctuation in glacier ice volume, so it is converted directly to a fluctuation of sea level. The position of each sample in the core was converted to time using the average sedimentation rate of 4.1 cm per 1000 years. The full-glacial to postglacial amplitude of the oxygen-isotope curve is assumed to represent 130 m of sea-level change. Terraces, especially coral reefs, are assumed to have formed at times of tangency between rising sea level and rising land. If sea level stood at +6 m during the last interglacial time, about 125 000 years ago, that terrace should be found at (125×2)m + 6 m = 256 m. At such a moderately rapid tectonic rate, most of the interstadials within the last glacial cycle should be recorded by emerged terraces even though they formed during sea-level maxima that were well below present sea level. Actual examples of such terraces are analyzed in subsequent paragraphs, but the hypothetical example of Figure 6.2 is useful to illustrate the analogy of the strip-chart recorder.

Terraces of Early and Middle Pleistocene age

Uplift at rates of a few meters per 1000 years is regarded as rapid (Bloom 1980, p. 513). If uplift is slow (for example, 0.1 m per 1000 years), the interstadial terraces predicted by Figure 6.2 will not be exposed above present sea level.

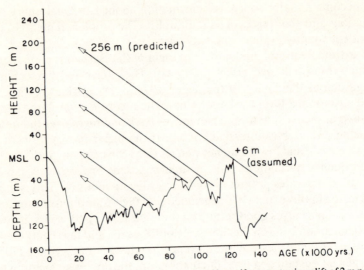

Figure 6.2 The hypothetical terrace sequence generated by uniform tectonic uplift of 2 m per 1000 years superimposed on sea-level oscillations modeled from the oxygen-isotope record of benthic foraminifera in core V29–179 (Streeter & Shackleton 1979). This core shows high amplitude interstadial oscillations, but the levels are different from those in Figures 6.4 and 6.5, and the figure is for illustrative purposes only. (After Streeter and Shackleton 1979; used by permission of the Am. Assn. for the Advancement of Science.)

However, terraces formed during the major interglacials of the last 2 million years may be uplifted and preserved. If major glacial cycles are about 100 000 years in length, an uplift rate of 0.1 m per 1000 years would separate successive interglacial terraces by only 10 m, barely in excess of the vertical range of storm waves and tsunamis. Considerable overlap of depositional units can be expected.

An excellent predictive model of the age of Middle Pleistocene and older terraces can be made if one of the lowest terraces in the sequence can be proved to be of last interglacial age, or about 125 000 years old. This situation is very common, because sea level was probably higher than present in last interglacial time (+6 m is a widely accepted estimate). On stable coasts or coasts with slow uplift, the last interglacial terrace is always prominent. Downtown Honolulu is built on it. The height of the last interglacial terrace above or below the assumed initial level of +6 m establishes a long-term (10^5 years) vertical tectonic rate, which can be cautiously extrapolated by an order of magnitude to predict the ages of higher and older terraces dating back to the early Pleistocene. In the few places where a tephrochronology is available, as in New Zealand (Pillans 1983) or Japan (Machida 1975); or where amino-acid racemization methods can be applied as in California (Muhs 1983) or reasonably inferred as in Baja California, Mexico (Ortlieb 1980), the predicted ages of older terraces are reasonably well supported. Correlations are often suggested with the odd-numbered oxygen-isotope stages of the deep-sea record, but those correlations should be used with caution, because the earlier oxygen-isotope stages are themselves dated only by interpolation to the Brunhes-Matuyama paleomagnetic epoch boundary that is about 700 000 years old.

Terraces of sea-level maxima during and since the last interglacial

The coral-reef terraces of the Huon Peninsula, Papua New Guinea (Fig. 6.1) preserve an exceptionally complete chronology of sea-level fluctuations for at least the last 140 000 years (Chappell 1974, 1983, Bloom *et al.* 1974). A great fault splinter on the northern coast of the Huon Peninsula has been rising and tilting in late Quaternary time, broken by numerous minor faults but maintaining its overall morphotectonic integrity. A succession of coral-reef terraces has been built on this block. Where the substrate surface was steep or the terrace-building interval was brief, the reefs were fringing. Where the preexisting slope was gentle or sea level stayed in the same position for a relatively long time (or repeatedly occupied the same level), the reef grew as a barrier seaward of a lagoon that was up to one km wide. The internal structure of the reefs show that they built upward and outward over their own fore-reef taluses. (Chappell 1974 Fig. 7, Chappell & Polach 1976). Their upper surfaces are usually level and composed of typical Indo-Pacific shallow-water corals and algae. Behind the reef crests, mollusc-rich back-reef and lagoon carbonate sand accumulated. Reef crests as old as 124 000 years to 140 000 years have no more than about 1 m of karst relief. Swallow holes and sinking streams mark the lagoon floors. Younger reef terraces show minor gully dissection and dripstone curtains down their fronts. Less than a meter of soil in weathered volcanic tephra has accumulated on the terrace treads.

Previous analyses of the terraces' ages and heights used an assumed constant uplift history and an assumed sea level of +6 m for the last interglacial stage (124 000 years) to derive the paleosea-level positions during the multiple interstadials of the latest glaciation. The resulting estimates of interstadial sea levels (Bloom *et al.* 1974) were consistent with similar estimates for terraces on Barbados, where the average rate of tectonic uplift was only about 10 per cent of the rate in Papua New Guinea. The converging estimates were regarded as reasonable estimates of the interstadial sea-level maxima, so that the height of a terrace of similar age elsewhere could be converted into an uplift rate by adding the present terrace height above sea level to its estimated original height (which is for all interstadial terraces at or below present sea level) and dividing the change in height by the age of the terrace.

The weak point of the argument was obviously the assumption that on the time scale of 10^4 to 10^5 years, tectonic uplift was at a uniform rate. A new graphic approach to this problem has been derived, based on previous Japanese work (e.g. Ota *et al.* 1968) but similar to the 'shoreline relation' diagnosis of Scandinavian workers (e.g. Donner 1965). The details of the method are to be presented in another paper (Yonekura & Bloom in prep.). For illustration, we can use data on the ages of terraces and their heights along six transects on the

Table 6.1 Measure of reef crest elevations (m) for six transects along the Huon Peninsula, Papua New Guinea. (After Tables 3 and 4 of Bloom *et al.* 1974). Data include arbitrary errors which were assigned to permit least squares evaluation. Errors assigned as follows: ± 5m for elevations $H \geqslant 60$ m; ± 2m for 60m $> H \geqslant 5$m and ± 1m for $H < 5$m.

Terrace	Age (ka)	Kanzarua	Blucher	Kwambu	Nama	Sambero	Kambin
VIIb	124	330	280	215	160	150	120
VI	105*	250	215	160	115	110	93
V	82	190	155	117	90	80	60
IV	60	125	****	70	48	****	28±2
IIIa	50–40**	90	65	42	****	****	****
IIIb	40	70	41	28	10	10	****
II	28	30	18	7	****	****	****
I	6***	15	10	6	5	5	2.5

* An age of 105 ka is assigned for the VI terrace, since Bloom *et al.* (1974) indicated 107 ka in text and 103 ka in Table 3.

** IIIa has not been dated directly.

*** After Chappell and Polach (1976).

**** Terrace crest uncertain or no satisfactory measurement.

Table 6.2 Regression equations for elevations (H_i) of terrace i against elevations of terrace VIIb. These equations are used for determination of paleosea levels (SL). See Table 6.1 and Figure 6.3. $H_i = a_i H_{VIIb} + b_i$; r is the correlation coefficient. SL is H_i when $H_{VIIb} = 6$ m.

H_i (age in ka)	a_i	b_i	r	SL (m)
H_{VIIb} (124)	1.00	0.00	1.000	+6 (assumed)
H_{VI} (105)	0.77	−4.55	0.999	+0.1
H_V (82)	0.60	−10.16	0.999	−6.6
H_{IV} (60)	0.46	−26.80	0.999	−24.0
H_{IIIa} (50–40)	0.41	−48.26	0.995	−45.8
H_{IIIb} (40)	0.32	−40.21	0.981	−38.3
H_{II} (28)	0.20	−36.25	0.995	−35.1
H_I (6)	0.05	−3.93	0.969	−3.6

Huon Peninsula, Papua New Guinea (Tables 6.1 and 6.2). These data are taken directly from Bloom *et al.* (1974, Tables 3 and 4). Arbitrary errors of ±5 m, ±2 m, and ±1 m were assigned to the reported terrace heights to permit least-squares error evaluation.

The last interglacial terrace, known as terrace VIIb, has an assumed age of 124 000 years and was formed when the sea was 6 m higher than at present. For each of the six transects, the heights of all lower and younger terraces are plotted against the height of terrace VIIb (Fig. 6.3). The data points on Figure 6.3 are drawn as closed circles large enough to include the probable errors of age and height. If we assume that a critical location terrace VIIb maintained a height of 6 m for the past 124 000 years (ignoring erosion) then at this location there would have been no uplift and differences in the elevation of the terraces would reflect changes in sea level only. Thus where the line of best fit for a given terrace intersects the vertical dashed line at $H_{VIIb} = 6$ m, it gives the height of sea level at the time of formation of that terrace.

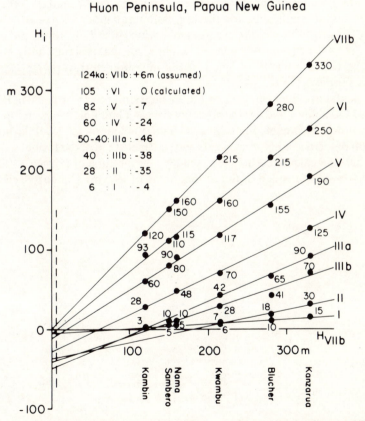

Figure 6.3 Regressions of height of terrace i (H_i) as a function of height of terrace VIIb (H_{VIIb}) based on six transects on the Huon Peninsula, Papua New Guinea (Yonekura and Bloom, in prep.). See text for explanation.

A regression may be performed to give a better estimate of the relationship. This yields the equation:

$$H_{i,t} = a_i H_{VIIb,t} + b_i \qquad\qquad (6.1)$$

where $H_{VIIb,t}$ is the height of terrace VIIb on transect t; $H_{i,t}$ is the height of an intermediate terrace i on the same intersect; a_i is the regression coefficient and b_i is the intercept. The intercept gives the elevation of terrace i assuming the elevation of terrace VIIb is zero. However, as we have seen, to calculate sea level at the time each terrace i was formed we must predict its value based on an elevation of 6 m for the terrace VIIb. This is analogous to predicting the heights of the terraces along a transect at the critical location where H_{VIIb} has remained at 6 m and uplift has therefore been zero.

Thus, by substituting the regression-line values of a_i and b_i in the equation with $H_{VIIb} = 6$ m, the paleosea-level estimates for the several Wisconsin-age interstadials can be calculated (Figs. 6.3 and 6.4, Table 6.2) without the further assumption of a constant uplift rate. The justification for the method is the very high correlation coefficients for the regression equations (Table 6.2). Only terrace IIIa (50 000 to 40 000 years ago), which has only three measured heights, has a least-squares predicted height error that is significantly greater than the errors arbitrarily assigned to the measured terrace heights on the transects (Table 6.1, Fig. 6.4).

The calculated values for terraces I to IV are similar to those listed in Bloom *et al.* (1974, Tables 3 and 4) where they were calculated on the assumption of constant uplift. However, the newly calculated height of sea level during the formation of terrace VI (105 000 years ago) is 0 m instead of the former value of −15 m, and the calculated height of sea level at the time of terrace V (82 000 years ago) is −7 m instead of the former value of −13 m. The purpose of this

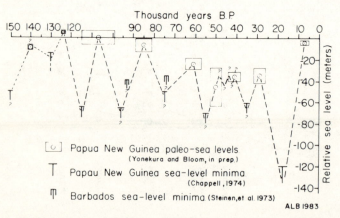

Figure 6.4 Revised estimate of the interstadial high sea levels of the last 105 000 years, based on the regression lines of Figure 6.3 and an assumed sea level of +6 m at 124 000 years ago. Boxes indicate probable errors of age and height.

paper is to review the method, not the results, so further discussion is deferred. However, the demonstration that a valid mathematical regression technique gives sea-level estimates that are quite similar to those estimated by assuming constant uplift rate is justification for the assumed constancy of uplift at the time scale of 10^5 years.

The last interglacial surface is widespread at depths of 6 to 10 m below a Holocene coral veneer on many atolls. If no more than 1 m of limestone has been lost from the reef surface during subaerial exposure, a lowering of the last interglacial reef surface from its assumed original height of +6 m to (for example) a present height of −6 m in about 120 000 years implies an atoll subsidence rate of about 0.1 m per 1000 years, a rate appropriate for subsidence of oceanic lithosphere during cooling (Sclater *et al.* 1971, Bloom 1980, p. 512).

Tectonic movements on the time scale of 20 000 years

The straight-line regression equations with high correlation coefficients demonstrated in previous paragraphs justify the assumption of constant uplift rate on the time scale of 10^5 years, but do not require it on shorter times. Careful analysis of Table 6.2 shows that there was variation in the Huon Peninsula uplift rate on the time scale of the 20 000 year sampling interval. For instance, terrace VI is 85 per cent the age of terrace VIIb ($105 \div 124$), but is only 77 per cent as high (the value of coefficient a_i, ignoring b_i). Uplift in the interval between 124 000 and 105 000 years ago was somewhat greater than the long-term average.

Another way of modeling uplift rates on the 20 000 year time scale is to accept sea-level estimates such as those in Table 6.2 and Figure 6.4 and from them calculate uplift rates for successive increments of dated uplift history. This method is appropriately called 'bootstrapping' in that each increment of uplift is used as the basis for calculating the next older increment. An example (Fig. 6.5) is drawn from work by J. Urmos (1985). The site is Araki Island, a small reef-terraced island 5 km south of the south coast of Santo Island, Vanuatu (for location, see inset, Fig. 6.6).

On Araki, a Holocene reef terrace about 5500 years old is 26 m above present sea level. Assuming, for simplification, that sea level in the region has not changed in the last 5500 years, the average late-Holocene uplift rate is 4.75 m per 1000 years. Above the large Holocene reef terrace on Araki is a succession of small stair-step terraces up to the flat reef-capped summit at 237 m. The next dated terrace on the hillside is about 38 000 years old and is now at a height of about 40 m. Since the rate of uplift for the last 5500 years is known, and assuming an original paleosea level of −41 m (Fig. 6.4), the increment of uplift between 5500 and 38 000 years is calculated to be about 1.75 m per 1000 years. The process is repeated for each step back in time, using the previously established estimates of sea level at the time of reef growth. The reef of the top of Araki Island is 105 000 years old and is now at 237 m. If sea

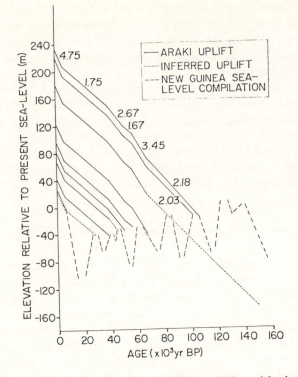

Figure 6.5   One model of incremental uplift rates (meters per 1000 years) for the last 105 000 years on Araki Island, Vanuatu. Note the unusually rapid late-Holocene rate (J. Urmos 1985). For location, see Figure 6.6. inset.

level at the time of origin was at present level (see previous paragraphs) then the average uplift rate for 105 000 years has been 2.26 m per 1000 years. However, successive increments of uplift rate range from 1.67 to as high as 4.75 m per 1000 years (Fig. 6.5), a factor of 2.8. In particular, the late-Holocene uplift rate has been faster than at any prior time in the last 105 000 years. We do not know if this is evidence of accelerated Holocene tectonic movement, or an artifact of the short sampling interval. We believe that the tectonic uplift of the region has accelerated in the Holocene, because nowhere in Vanuatu have we found a reef 28 000 years old, such as has been found in the Huon Peninsula of Papua New Guinea (Chappell & Veeh 1978). If the rapid uplift of the last 5500 years had continued for as long as the last 28 000 years, the interstadial terrace of that age would be far above the Holocene terrace, even though it started at a sea-level position 35 m below present sea level (Table 6.2, Fig. 6.4). However, the extreme size of the Holocene terrace on Araki Island could be caused by a relatively thin Holocene veneer over an older reef-terrace substrate, as suggested by the dashed trajectory of inferred uplift for a hypothetical 28 000 year old terrace on Figure 6.5. The presence of such a substrate under a Holocene veneer has been hypothesized from morphologic evidence in other parts of south Santo Island (Strecker *et al.* in press), but will only be verified by future drilling.

The accelerated uplift during the last 5500 years on Araki Island cannot be directly compared to either the long-term average rate or to older dated increments of uplift. It is possible that during any previous 20 000 year interval between interstadial high sea levels, much of the total movement was concentrated in brief intervals of 5000 years or less. There is no way that such 'jerkiness' could be detected with the 20 000 year sampling interval that is provided by the average spacing of interstadial sea-level oscillations. Therefore, we can only conclude that the last 5500 years of uplift, during which postglacial sea level has been at its approximate present level, has been unusual by comparison to the 20 000 and 100 000 year average rates, although we cannot disapprove that those average rates consisted of shorter intervals of alternately fast and slow vertical movements.

Uplift rates on the 1000-year time scale

The impressive late Holocene uplift rate of 4.75 m per 1000 years for Araki Island is comparable to Holocene uplift rates at two nearby places on the Santo Island coast (Jouannic *et al.* 1980). On the Tasmaloum Peninsula, a reef with a probable radiocarbon age of 6000 years is now almost 33 m above sea level; on nearby Tangoa Island, a reef crest 6700 years old is now 19 m above sea level (Fig. 6.6). At both places, numerous corals were collected at one-meter intervals between present sea level and the emerged Holocene reef crest and a selection of them were dated, to document the uplift history of the last 6000 to 6500 years. The results are subject to multiple interpretations, but offer

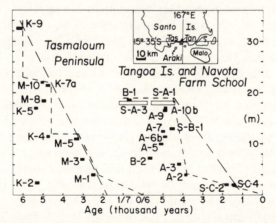

Figure 6.6 Models of possible late-Holocene uplift history of two coastal sites in southern Santo, Vanuatu. Symbol dimensions show probable errors of age and height. Closed symbols show radiocarbon-dated samples with 400 years added to their ages to compensate for probable metabolic fractionation effect; open symbols show samples dated by the uranium-series method.

interesting alternatives. All the dates were determined from shallow-water reef corals similar to those living on the shores today. For simplicity, no change of eustatic sea level is assumed. Heights were surveyed or hand leveled with one-half meter accuracy from the reference level of corals living on the modern reef, so the heights are realistic measures of the net relative change in level during the elapsed millennia. To correct for the probable metabolic fractionation effect of corals, all radiocarbon dates have had 400 years added to them. Two uranium-series dates (S-A-1 and S-A-3, Fig. 6.6) have inherently large counting errors of 600 and 700 years, and were not otherwise corrected.

Any relative sea-level curve or tectonic uplift curve must be drawn so that the dated corals were below sea level at the time they were alive. Corals thrive in water depths of up to 10 m, although any assumed initial water depths will give even faster inferred uplift rates. The long dashed lines on Fig. 6.6 are therefore minimal uplift trajectories.

For the Tasmaloum Peninsula, uplift of 33 m at a constant rate of 7.9 m per 1000 years between 6000 and 1800 years ago, followed by standstill for the last 1800 years, is a possible interpretation (long dashed line, Fig. 6.6). However, many of the corals that were selected for dating seem to have died simultaneously (within the probable error of the dating method) even though they are separated by vertical ranges of 5 to 10 meters. Most of the dated samples were from globose or 'brain' corals less than 30 cm in diameter, which were probably only 30 to 50 years old at their time of death, based on estimates of preserved annual growth increments. A more likely interpretation of the uplift of the Tasmaloum Peninsula is therefore that there was 10 m of uplift about 6000 years ago, followed by a pause of 1400 years, then another 10 m of uplift about 4600 years ago, then another quiet interval until 3200 years ago. In the last 3200 years, the coast has risen about 12 m, but only 4 m of that uplift has occurred in the last 2500 years (short dashed line, Fig. 6.6).

Comparable interpretations of Holocene uplift can be made from a suite of 14 dated samples collected from Tangoa Island and the adjacent part of south Santo Island in the Navota Farm School (Fig. 6.6). These samples were all collected from within 3 km of each other, in an area that seems to be on a single tectonic block, although probably not on the same tectonic block as either Araki Island or Tasmaloum Peninsula. As with the record of Tasmaloum Peninsula, the simplest interpretation (long dashed line, Fig. 6.6) is for 19 m of uplift between 4500 and 1000 years ago, at an average rate of 5.4 m per 1000 years, followed by 1000 years of stability. However, considering that the three lowest samples show only 4 m of uplift in the last 3800 years, a more realistic interpretation is for 15 m of uplift between 4500 and 3800 years ago followed by 3800 years of much slower uplift (short dashed line, Fig. 6.6). The maximum mapped height of Tangoa Island is 22 m, but the map is of uncertain accuracy. The hand leveled height of 18.6 m for sample B-1 (Fig. 6.6) on Tangoa Island and the surveyed height of 18 m for samples S-A-1 and S-A-3 on Navota Farm School, only 3 km distant on the main island, suggest that for several thousand years prior to 4500 years ago, the entire area was submerged to about the present 20 m contour and was the site of vigorous growth of patch reefs. Then quite suddenly about 4500 years ago the region emerged and the relict lagoon-floor relief became the present landscape (Jouannic *et al.* 1980, Strecker *et al.* 1984). Tasmaloum Peninsula and the Tangoa Island-Navota Farm School area share with Araki Island evidence of very rapid late-Holocene uplift in the last 5500 to 6500 years. However, the multiple dates from Tasmaloum and Tangoa show that the rapid late-Holocene uplift probably had a fine structure of 'jerkiness', in which brief intervals of as much as 10 m of vertical movement were followed by 1500 years or more of relative stability. The intervals of uplift are not resolvable within the hundred-year standard errors of the radiocarbon dates.

Earthquakes of magnitude 7.0 or greater are frequent in Vanuatu, and many have been associated with observed vertical uplift. About 50 km of coastline on northeast Malekula Island experience uplift ranging up to 1.2 m during an August, 1965 series of large earthquakes (Taylor *et al.* 1980). The 1500-year or longer intervals of relative quiet shown by the interpretations of Figure 6.6 are inconsistent with the known frequency of large earthquakes in the area. Perhaps the brief intervals of as much as 10 m of vertical movement resulted from a series of 10 or more major seismic shocks within a few centuries, although such clustering has never been reported from seismic regions.

The suggested uplift curves of Figure 6.6 are similar to the 'time-predictable' recurrence model for large earthquakes proposed by Shimazaki & Nakata (1980). Figure 6.7b summarizes their hypothesis, in which the time interval between two successive large earthquakes is proportional to the amount of seismic displacement of the *preceding* earthquake. While the intervals of strong uplift in Vanuatu are not likely to have been caused by single earthquakes, they could have been caused by clusters of closely spaced earthquakes that relieved some proportion of the elastic strain built up by continuing subduction, and

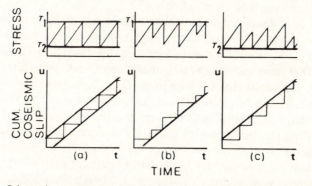

Figure 6.7 Schematic recurrence models for large earthquakes (Shimazaki and Nakata 1980, Fig. 1). Model (b), time-predictable, seems to apply to the uplift histories of Figure 6.6. The time interval between two successive large earthquakes is proportional to the amount of seismic displacement of the preceding earthquake. (From Shimazaki & Nakata 1980, Fig. 1. Used by permission of the American Geophysical Union and Dr. Shimazaki).

allowed following intervals of relative stability whose length was proportioned to the previous strain release. Such a model would explain, for example, the tendency for the Tasmaloum uplift curve (Fig. 6.6) to return to or follow along the mean trend line of long-term uplift until about 2000 years ago.

The lack of major uplift for the last 2000 to 4000 years at both Tasmaloum and Tangoa is inconsistent with the 'time-predictable' model of Shimazaki and Nakata. By that model, both places are long overdue for major uplift events or clusters of events. Since large earthquakes occur in the area every year or two, the reason for insignificant vertical movements in the last few thousand years remains to be resolved.

Summary

This review of emerged coastal terraces has focused on different processes, acting on different time scales. On the scale of 10^5 years, glacially controlled eustatic fluctuations of sea level are the dominant factor in generating coastal terraces. If each interglacial sea level returned to within a few meters of present level, flights of terraces represent the cumulative tectonic uplift during the oscillatory glacial cycles. On this time scale, a uniform average uplift rate gives good predictions of terrace ages.

Within the 10^5 years of a typical full-glacial cycle, brief interstadial intervals of relatively high sea levels occur at about 20 000 year intervals. None of the interstadials of the last glacial age, except the first one about 105 000 years ago, had sea levels as high as the present or during the last interglacial. Successive uplift increments based on 20 000 year intervals between dated coral-reef terraces lend support to the hypothesis that tectonic uplift on this time scale is also relatively constant, with the uplift rates of successive 20 000 year intervals varying by a factor of only 2 or 3.

Within the 6000 years of late Holocene time, when glacially controlled sea level has been relatively stable, tectonic uplift has been measurably erratic in space and time. At several Vanuatu localities, the overall uplift rate for the last 5000 to 6000 years exceeds the longer-term uplift rate by a factor of 3 to 5. Furthermore, this rapid Holocene uplift seems to have been subdivided into intervals of perhaps a few centuries during which as much as 10 meters of uplift accumulated, followed by 1500 years or more of almost no net uplift.

Finally, within the time scale of centuries or decades, tectonic uplift can produce 1 meter or more of emergence on 50 to 100 km of coast with no movement on adjacent segments. Cumulative effects of such uplift, which are usually associated with major earthquakes, are generally positive, although interseismic reversals are known. It seems that every order-of-magnitude time interval has its own distinctive set of processes that operate to create terraces on rising coasts. The processes that control one time interval may be negligible in their impact on shorter or longer time scales.

Acknowledgment

This is Cornell University Dept. of Geological Sciences contribution No. 804.

References

Bloom, A. L. 1980. Late Quaternary sea level change on South Pacific coasts: a study in tectonic diversity. In *Earth rheology, isostasy and eustasy*. N.-A. Mörner (ed.), 505–16. Chichester: John Wiley.

Bloom, A. L., W. S. Broecker, J. M. A. Chappell, R. K. Matthews, and K. J. Mesolella 1974. Quaternary sea level fluctuations on a tectonic coast: New ^{230}Th/^{234}U dates from the Huon Peninsula, New Quinea. *Quatern. Res.* **4**, 185–205.

Chappell, J. 1974. Geology of coral terraces, Huon Peninsula, New Guinea: A study of Quaternary tectonic movements and sea-level changes. *Geol. Soc. Am. Bull.* **85**, 553–70.

Chappell, J. 1983. A revised sea-level record for the last 300,000 years from Papua New Guinea. *Search* **14**, 99–101.

Chappell, J. and H. A. Polach 1976. Relationship between Holocene sea level change and coral reef growth at Huon Peninsula, New Guinea. *Geol. Soc. Am. Bull.* **87**, 235–40.

Chappell, J. and H. H. Veeh 1978. ^{230}Th/^{234}U age support of an interstadial sea level of −40 m at 30 000 yr BP. *Nature* **276**, 602–3.

Depéret, C. 1918. Essai de coordination chronologique générale des temps Quaternaires. *Comptes rendus de l'Académie des Sciences Paris* **167**, 418–22.

Donner, J. J. 1965. Shore-line diagrams in Finnish Quaternary research. *Baltica* **2**, 11–20.

Hey, R. W. 1978. Horizontal Quaternary shorelines of the Mediterranean. *Quatern. Res.* **10**, 197–203.

Jouannic, C., F. W. Taylor, A. L. Bloom and M. Bernat 1980. Late Quaternary uplift history from emerged reef terraces on Santo and Malekula Islands, central New Hebrides island arc. *UN ESCAP, CCOP/SOPAC Tech Bull.* **3**, 91–108.

Machida, H. 1975. Pleistocene sea level of south Kanto analysed by tephrochronology. *Roy. Soc. New Zeal. Bull.* **13**, 215–22.

Mesolella, K. J. 1967. Zonation of uplifted Pleistocene coral reefs on Barbados, West Indies. *Science* **156**, 638–40.

Mesolella, K. J., R. K. Matthews, W. S. Broecker and D. L. Thurber 1969. Astronomical theory of climatic change: Barbados data. *Jour. Geol.* **77**, 250–74.

Muhs, D. R. 1983. Quaternary sea-level events on northern San Clemente Island, California. *Quat. Res.* **20**, 322–41.

Ortlieb, L. 1980. Neotectonics from marine terraces along the Gulf of California. In *Earth rheology, isostasy and eustasy*, N.-A. Mörner (ed.), 497–504. Chichester: John Wiley.

Ota, Y., S. Kaizuka, T. Kikuchi and H. Naito 1968. Correlation between heights of younger and older shorelines for estimating rates and regional differences of crustal movements. *The Quatern. Res.* (Japanese) **7**, 171–81.

Pillans, B. 1983. Upper Quaternary marine terrace chronology and deformation, South Taranaki, New Zealand. *Geology* **11**, 292–7.

Sclater, J. G., R. N. Anderson and M. L. Bell 1971. Elevation of ridges and evolution of the central eastern Pacific. *Jour. Geophys. Res.* **76**, 7888–915.

Shimazaki, K. and T. Nakata 1980. Time-predictable recurrence model for large earthquakes. *Geophy. Res. Lett.* **7**, 279–82.

Strecker, M. R., A. L. Bloom and J. Lecolle 1984. Time span for karst development on Quaternary coral limestones: Santo Island, Vanuatu. *25th Internat. Geogrl Cong. Proc.* Paris (In press).

Streeter, S. S. and N. J. Shackleton 1979. Paleocirculation of the deep North Atlantic: 150 000-year record of benthic foraminifera and oxygen-18. *Science* **203**, 168–71.

Taylor, F. W., B. L. Isacks, C. Jouannic, A. L. Bloom and J. Dubois 1980. Coseismic and Quaternary vertical tectonic movements, Santo and Malekula Islands, New Hebrides island arc. *Jour. Geophys. Res.* **85**, 5367–81.

Urmos, J. P. 1985. Oxygen isotopes, sea level and uplift of reef terraces, Araki Island, Vanuatu. M.S. thesis, Dept. of Geology, Cornell Univ.

Verstappen, H. Th. 1960. On the geomorphology of raised coral reefs and its tectonic significance. *Zeits. für Geomorph.* **4**, 1–28.

Yonekura, N. and A. L. Bloom, in preparation. *Graphic analysis of paleosea-levels and tectonic uplift based on emerged reef terrace data of Huon Peninsula, Papua New Guinea and Barbados, West Indies.*

Zeuner, F. E. 1959. *The Pleistocene Period*, 2nd edn London: Hutchinson.

7

Computer models of shoreline configuration: headland erosion and the graded beach revisited

Paul D. Komar

Introduction

Numerical computer models have been successfully employed to simulate a variety of physical processes. Perhaps most familiar are those utilized in fluid dynamics or by physical oceanographers and engineers to study water movements, whether it be the circulation of ocean currents or the flow in an estuary. Rarer are numerical models that simulate sedimentological processes, the movement of sand in a river, estuary or in the sea.

Some success has been achieved in the development of computer models to examine shoreline changes resulting from the transport of sand along beaches by waves and currents. The first models of this type were those of Price *et al.* (1973) that examined the equilibrium shoreline shape in a groin field, and those of Komar (1973) that analyzed the shoreline configuration of river deltas. Subsequent studies have modeled the 'hooked' or 'logarithmic-spiral' beach (Rea & Komar 1975, LeBlond 1979, Dean 1979), the effects on the shoreline of offshore dredging (Motyka & Willis 1975, Horikawa *et al.* 1977), and shoreline changes due to jetty construction (Komar *et al.* 1976, Perlin & Dean 1979). Reviews of computer modeling of shorelines can be found in Dean (1979) and Komar (1977, in press).

In the present study the techiques of numerical models of shoreline changes will be illustrated by a model of headland erosion in which the sand is transported into an adjacent embayment. This leads to a consideration of the graded shoreline, one whose curvature in plan view is adjusted in such a way that the waves impinging on the shore provide precisely the amount of energy required to transport the load of sediment supplied to the beach. In this analysis of the graded shoreline, models will be employed to examine the shapes of equilibrium river deltas and of pocket beaches with and without significant sand sources.

Computer models are ideal for quantitatively analyzing such geomorphic concepts because they permit systematic changes in the parameters of interest while eliminating other complicating effects which would obscure the results. At the same time, many of the models prove to be of practical relevance in that they predict or explain patterns of coastal erosion and deposition.

Techniques of modeling

The techniques of developing computer models of shoreline changes parallel the formulation of fluid-flow models. The relationships that govern the flow of a fluid, the Navier–Stokes equations, are replaced by equations that relate the sand movement along the beach to the nearshore waves and currents. A continuity equation maintains an account of the total volume or mass of sand, analogous to the fluid continuity relationship.

Such models divide the environment into cells as depicted in Figure 7.1A where the smooth shoreline is approximated by a series of cells of width Δx. The smaller their Δx widths, the more closely they represent the smooth shoreline. The end of each cell terminates in a schematic beach profile, as is shown in Figure 7.1B. The basis of any shoreline model is the shifting of sand from one cell to another; this causes changes in the shoreline position. The computer is used to evaluate these shifts as produced by the physical processes and to keep track of the quantities of sand in each cell.

The resulting change in the shoreline position in any particular cell is produced by the balance of sand input versus exit, a mini-budget of sediments. Focusing on the cell of Figure 7.1B, if Q_i is the rate of littoral drift from cell i into cell $i+1$ and Q_{i-1} is the drift between cells i and $i-1$, then the net gain or loss of sand volume for cell i is simply

$$\Delta V_i = (Q_{i-1} - Q_i)\Delta t \qquad (7.1)$$

where Δt is an increment of elapsed time. The sand movement, Q, could have units of m³/day such that $Q\Delta t$ would be the sand volume. Note that ΔV_i can be positive, negative, or zero depending on the balance between Q_{i-1} and Q_i. If more sand leaves the cell than enters during the Δt time interval, then $\Delta V_i = -$, signifying erosion. Similarly, $\Delta V_i = +$ represents a net sand accumulation. A balance between losses and gains yields $\Delta V_i = 0$.

To examine shoreline variations we must convert this volume change, ΔV_i, into a change in the on-offshore length, y_i, of the cell (Fig. 7.1A). If Δy_i is the corresponding change in y_i in the time interval Δt, then from the geometry of the cell shown in Figure 7.1B

$$\Delta V_i = d\Delta y_i \Delta x \qquad (7.2)$$

The linear dimension d must be chosen to yield the correct correspondence between the volume change and the shoreline change; its value depends on the

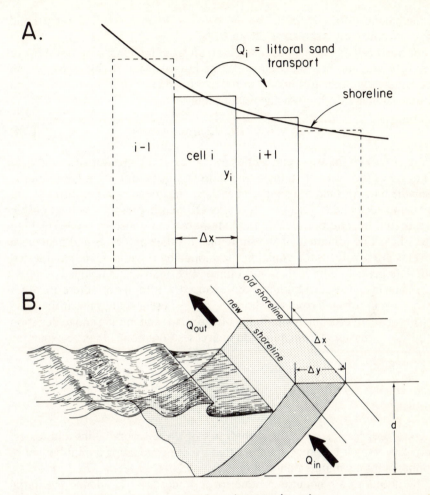

Figure 7.1 A. The approximation of a smooth shoreline with a series of cells in width Δx and variable y_i lengths. B. Each cell ends in a schematic beach profile with the change in the sand volume ΔV in the cell producing a change in shoreline position, Δy. (After Komar 1973. Used by permission of the Geological Society of America).

nature of the beach profile and water depth to which deposition or erosion occurs.

Combining Equations 7.1 and 7.2 to eliminate ΔV_i yields

$$\Delta V_i = (Q_{i-1} - Q_i)\,\frac{\Delta t}{d\Delta x} \tag{7.3}$$

for the change in shoreline position of cell i. Since in the models the parameters Δt, d, and Δx generally are fixed, the Δy shoreline changes are produced simply by the balance of the Q terms. Again, $\Delta y_i = -$ signifies shoreline retreat and $\Delta y_i = +$ represents a shoreline advance.

For most cells the principal Q terms result from the longshore transport of sand by waves and nearshore currents (the littoral drift). In this case they can be evaluated with the formulae that relate the sand transport to the waves and currents. I generally employ the relationship

$$Q_s = 6.85(ECn)_b \sin \alpha_b \cos \alpha_b \qquad (7.4)$$

where $(ECn)_b$ is the wave energy flux evaluated at the breaker zone and α_b is the angle of the wave breaking at the shoreline. Equation 7.4 is based on the available data for sand transport on beaches, and in this empirical form Q_s has units of m^3/day and $(ECn)_b$ is watts/meter (Komar in press). This relationship is restricted to conditions where the longshore sand transport is due solely to waves breaking obliquely at the shoreline. If other processes and nearshore currents are involved, the equations developed by Bagnold (1963) and tested with data presented by Komar and Inman (1970) must be employed.

In some instances, in addition to those caused by littoral drift, there are other losses or gains of sand from the shoreline cell. These include gains from rivers or from sea-cliff erosion, losses to the offshore, to sand mining and to deflation by onshore winds that form dunes. In such cases Equation 7.3 can be expanded to

$$\Delta y_i = (Q_{i-1} - Q_i \pm Q_r) \frac{\Delta t}{d \Delta x} \qquad (7.5)$$

where Q_r represents collectively the rates of these several gains and losses. Here especially it becomes apparent that we are conducting a mini-budget of sediments for each individual cell of the model.

The inputs to a computer routine for a model are : (1) define the initial shoreline configuration; (2) establish any sources and losses of sand (the Q_r terms); (3) provide the model with the offshore wave conditions; and (4) indicate how the littoral transport of sand will be governed by these waves (Equation 7.4). The model itself runs through increments of time, Δt, for some total span of time, determining how the shoreline evolves from its initial configuration. In a forthcoming paper (Komar in press) I present more details on the development of models, including the portion of a Fortran program which is the heart of most of the shoreline models I have developed.

Headland erosion

A simple illustration of numerical shoreline models is provided by the analysis of headland erosion and the transport of the eroded sediment into an adjacent embayment. In part this was selected as an example so that the present analysis can be compared with the a–b–c model of headland erosion presented by May and Tanner (1973) as part of this series.

The initial shoreline of the model (Fig. 7.2) was taken as sinusoidal with a total headland to embayment relief of 2 km and a longshore wavelength of 10 km (headland to embayment distance of 5 km). Thus the feature analyzed is large. It is on the scale of a medium-size embayment or roughly equivalent in size to the 'secondary capes' on the North Carolina coast studied by Dolan *et al.* (1974). In the model of Figure 7.2 the wave conditions were held constant with $(ECn)_b = 10^4$ watts/m and with the waves approaching straight on to the apex of the headland. At this stage wave refraction is not included, but its effects will be discussed later in this section. The only Q term is that due to littoral drift and evaluated with Equation 7.4. In that $(ECn)_b$ is taken as constant, the longshore variations in Q result directly from longshore changes in the breaker angle α_b. The cell widths were set at $\Delta x = 100$ m with $d = 4$ meters. The time increment was $\Delta t = 0.1$ day, the total run covered 360 days, and the shoreline was printed at 10-day intervals.

This is a comparatively straightforward model and the results are largely what one would anticipate: the headland erodes back and the embayment progressively fills with the eroded material. Figure 7.2 diagrams the shoreline at monthly (30 day) intervals for the total 360-day 'year'. There is a perfect symmetry between the eroding headland and the filling bay with the node remaining at the midpoint between the headland and embayment. The headland erosion rate (Fig. 7.2B) averaged over 10-day intervals is initially very high (6.2 m/day). Subsequently the rate decreases with time, at first rapidly and then more slowly. At the end of 360 days the headland erosion rate had decreased to 0.7 m/day. Being a mirror reflection, the deposition rates at the embayment minimum have the same values as the headland erosion. The ultimate shoreline would be straight and parallel to the incoming waves, as represented by the dashed line in Figure 7.2A. In theory the model will approach but never achieve this ultimate orientation because all breaker angles become very small as does the Q littoral drift required to bring about the shoreline changes. The erosion rate curve of Figure 7.2B similarly asymptotically approaches a zero value.

Although the model results, as diagramed in Figure 7.2, conform with what one would anticipate, they do not agree with the a–b–c model of May and Tanner (1973). According to their analysis the maximum erosion would not occur at the apex of the headland (their point a), but instead on the flank (point b) at some unspecified distance between the headland apex and the nodal point c where embayment deposition begins. Similarly, according to the a–b–c model the maximum deposition would not be at the embayment minimum (point e) as found in the numerical model of the present study, occurring instead on the flank of the embayment (their point d). May and Tanner correctly deduce that the breaker angle and hence the littoral drift Q_s decrease as the headland is approached, both becoming zero at the headland apex itself. However, the local rate of erosion depends on the longshore variation or gradient of the

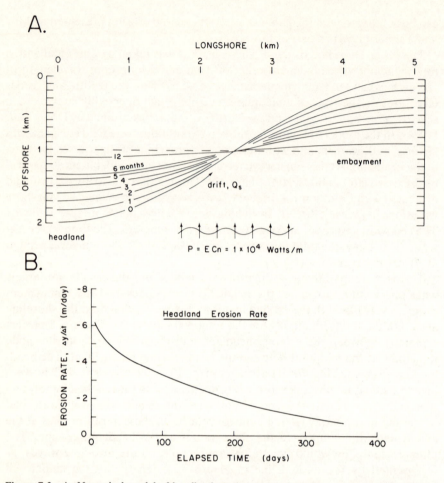

Figure 7.2 A. Numerical model of headland erosion and sediment transport into an adjacent embayment. B. The erosion rate at the headland apex during the 360 days of the model run.

littoral drift, dQ_s/dx, and according to May and Tanner this gradient also approaches zero at the headland, reaching a maximum at their point b on the flank. It is here that their analysis is incorrect. Although Q_s decreases to zero at the headland apex, dQ_s/dx does not, for it actually reaches a maximum there. It is apparent that their model is also intuitively implausible; having dQ_s/dx become zero at the headland apex implies that it will not even erode! Similarly there would be no deposition in the deepest part of the embayment, only on its flanks.

The relationship between the gradient dQ_s/dx and the local shoreline erosion or deposition is apparent in the continuity relationship (Eq. 7.3), especially if we modify it somewhat. Setting $\Delta Q = (Q_i - Q_{i-1})$, Equation 7.3 becomes

$$\frac{\Delta y_i}{\Delta t} = -\frac{1}{d}\frac{\Delta Q}{\Delta x} \tag{7.6}$$

and then shrinking the finite elements to their limits yields

$$\frac{dy}{dt} = -\frac{1}{d}\frac{dQ_s}{dx} \tag{7.7}$$

The time derivative of the shoreline position, dy/dt, is the rate of shoreline erosion ($dy/dt = -$) or deposition ($dy/dt = +$). It can be seen that an increasing quantity of littoral drift ($dQ_s/dx = +$) produces shoreline erosion, and the larger the magnitude of dQ_s/dx the greater the resulting erosion. Similarly, $dQ_s/dx = -$ yields $dy/dt = +$; shoreline deposition. Again, according to the a–b–c model, the mistaken conclusion that $dQ_s/dx = 0$ at the headland apex and embayment center yields $dy/dt = 0$; there are no shoreline changes.

One advantage of having assumed a sinusoidal initial shoreline in the present analysis is that it becomes a simple matter to determine analytically the longshore variations in Q_s and dQ_s/dx, and hence to evaluate quantitatively the patterns of the initial erosion and deposition. Since we assumed that $(ECn)_b$ is constant alongshore, from the littoral drift relationship of Equation 4 we obtain

$$\frac{dQ_s}{dx} = 6.85(ECn)_b(\cos^2\alpha_b - \sin^2\alpha_b)\frac{d\alpha_b}{dx} \tag{7.8}$$

where it is seen that the local value of dQ_s/dx depends both on the magnitude of the breaker angle and on its longshore gradient, $d\alpha_b/dx$. The values of α_b and $d\alpha_b/dx$ can be determined from the equation for the sinusoidal shoreline,

$$y = A\cos(kx) \tag{7.9}$$

where A is the amplitude (1000 m in Fig. 7.2) and $k = 2\pi/L$ ($L = 10\,000$ m). Note that the headland occurs at $x = 0$ and the embayment minimum at $x = L/2$. The breaker angle then becomes

$$\alpha_b = \arctan\left(\frac{dy}{dx}\right) = \arctan(-kA\sin(kx)) \tag{7.10}$$

and

$$\frac{d\alpha_b}{dx} = -\frac{1}{1+[kA\sin(kx)]^2}\,k^2A\cos(kx) \tag{7.11}$$

Note that at the headland apex ($x = 0$)

$$\frac{d\alpha_b}{dx} = -k^2A \tag{7.12}$$

which is the maximum magnitude of $d\alpha_b/dx$ even though α_b becomes zero.

The evaluated distributions of Q_s and dQ_s/dx are plotted in Figure 7.3 together with $(\cos^2\alpha_b - \sin^2\alpha_b)$ and $d\alpha_b/dx$ factors that determine dQ_s/dx according to Equation 7.8. Both factors are seen to have maxima at the headland apex, producing the maximum in dQ_s/dx at this position, not zero as given by the a–b–c model (May & Tanner 1973). Similarly, dQ_s/dx achieves its largest negative value at the embayment center implying that the deposition rate will be a maximum. It is interesting that the dQ_s/dx longshore variation is not simply sinusoidal as might at first be assumed. Also due to the direct

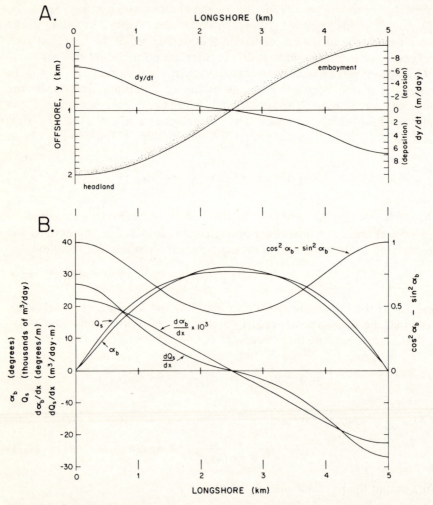

Figure 7.3 Patterns of longshore variations in the breaker angles, α_b, the littoral drift rate, Q_s, their gradients $d\alpha_b/dx$ and dQ_s/dx, and the resulting erosion–deposition, dy/dt, for the sinusoidal headland–embayment shoreline.

proportionality between the rate of shoreline change dy/dt and dQ_s/dx, Equation 7.7, the variation in dy/dt (Fig. 7.3A) shows the exact same pattern. Note that the erosion is accentuated around the headland apex and the deposition at the embayment minimum even without the effects of wave refraction (which would further accentuate these patterns) having been included. The erosion rate, dy/dt, evaluated with Equation 7.7, basically agrees with the initial rate determined in the numerical model (Fig. 7.2B); the small difference results from the value in Figure 7.2B representing an average for the first 10 days rather than the actual initiation of shoreline changes as given in Figure 7.3. Although we can obtain analytical relationships, Equations 7.8, 7.10 and 7.11, for the initial stage of shoreline changes in the headland erosion, the subsequent patterns of shoreline positions, breaker angles, dQ_s/dx values, and the dy/dt erosion or deposition rates can by evaluated only with the numerical model. The above analytical results apply only to a sinusoidal shoreline, and the evolving shoreline quickly departs from the initial sinusoidal form.

The above numerical model (Fig. 7.2) and the analytical analysis neglected any consideration of wave refraction effects that would produce wave convergence and an increase in $(ECn)_b$ at the headland, and wave divergence and a decrease in $(ECn)_b$ within the embayment. The magnitudes of the breaker angles also would decrease, affecting the Q_s drift rates. A full analysis including wave refraction requires coupling the shoreline model with a computer evaluation of wave refraction. This of course greatly increases the complexity of the model as the offshore depths must be specified as well as the shoreline configuration. Models of this type have been developed, for example by Motyka and Willis (1975) in their analysis of the effects of offshore dredging on shoreline changes. In the present study only a simple analysis was carried out to indicate the effects of wave convergence and divergence on the patterns of headland erosion. This involved the inclusion of a wave-refraction coefficient that increased $(ECn)_b$ at the headland apex by a 1.25 factor, and a 0.75 factor decrease at the embayment center. The refraction factor changed linearly alongshore between these two values, being 1.00 at the node between the headland and embayment. The effects of this did not produce large changes from the above results that neglected refraction considerations. As expected, the exact symmetry of the erosion and deposition patterns was lost. The erosion rate at the headland apex increased while the deposition rate in the embayment center decreased from the values obtained in the analysis without refraction. There was a delay in the transport of the eroded material from the headland into the embayment due to the lower $(ECn)_b$ values within the embayment. This produced some bulging on the headland and embayment flanks, with the nodal point no longer being midway between the headland and embayment but instead shifting position with time. The maximum deposition rate still occurred at the embayment center, and will apparently not shift to the flanks unless $(ECn)_b$ is greatly reduced within the embayment center. Even in this case the ultimate shoreline again would be straight, so that the maximum amount of total deposition would be at the embayment center.

The graded shoreline

The concept of the graded river is well established and an extensive literature exists dealing with its analysis and applications. Analogous to the graded river, one can also consider a graded shoreline. This concept may have been recognized by G. K. Gilbert (Pyne 1980, p. 140), but it is best developed by Tanner (1958) who used the term 'equilibrium beach' for this same concept. Tanner defined the equilibrium beach as one having '. . . curvature and sand prism characteristics adjusted to each other so delicately that the potential littoral motion provides precisely the energy needed to transport the detritus supplied at the up-current end. The time element in this balance is long-term rather than instantaneous.' (Tanner 1958, p. 889). The conceptual similarity with the graded river is apparent.

From this definition it is clear that the developing shoreline within the embayment of the above model (Fig. 7.2) is in effect graded. The shoreline curvature that is quickly achieved is precisely that required to transport the material eroded from the headland and deposit it in the embayment. The quantity of this material decreases with time as the headland erodes, so that the shoreline curvature similarly changes with the breaker angles becoming smaller in response to the reduced littoral drift. At any given instant the curvature of the shoreline is such that the breaker angles decrease toward the embayment center. This occurs because of the decreasing quantity of sand to be transported with some being left behind in the form of deposition along the entire length of the embayment shoreline.

A similar graded shoreline, but one easier to analyze, forms on a river delta and fits very closely into the above definition provided by Tanner (1958). The delta shoreline achieves a curvature that permits the ocean waves to distribute the river sand along the coast. Computer models of delta growth have been developed by Komar (1973) utilizing the techniques discussed earlier. Figure 7.4A reproduces the results of one model with a river supplying sand at the rate $2 \times 10^4 \, \text{m}^3/\text{day}$ with waves of energy flux $(ECn)_b = 3 \times 10^3$ watts/m operating to move the sand alongshore as littoral drift. It can be seen that the delta builds out rapidly and that its form quickly achieves a constant shoreline curvature that must be graded or in equilibrium. Figure 7.4B shows the effects of different levels of the wave energy flux; as expected, the higher the wave conditions the flatter the equilibrium delta shape. This result is due to smaller values of the breaker angles being required to redistribute the river sand when $(ECn)_b$ is greater. A similar result is obtained when the waves are held constant but the quantity of sand supplied by the river is changed.

This particular graded shoreline can be analyzed to obtain analytical relationships, much as was done for the eroding headland model. To maintain a constant shoreline shape through time the deposition rate must be the same everywhere along the delta, that is, $dy/dt = $ constant. From Equation 7.7 we have in turn that $dQ_s/dx = $ constant, and direct integration of this yields

$$Q_s = ax + b \tag{7.13}$$

where a and b are constant coefficients. At the river mouth ($x = 0$) the littoral drift on each delta flank must account for half of the sand supplied by the river so that $Q_s = Q_r/2$ at $x = 0$, and this condition yields $b = Q_r/2$ in the above relationship. The coefficient a is determined by equating the total sand accreting on the delta to that supplied by the river. This is more easily done if we limit the growth of the delta flank to a longshore length s, determined by

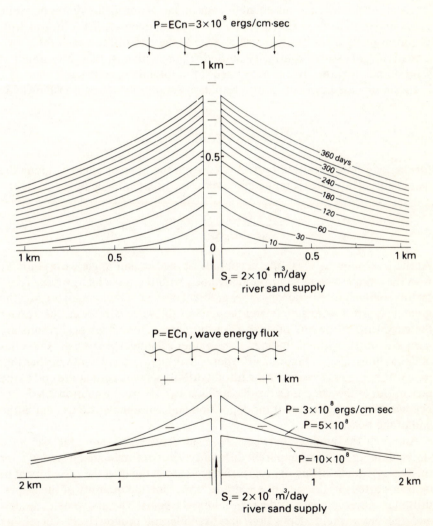

Figure 7.4 Numerical models of delta growth as an example of the graded shoreline. A. Patterns of shoreline changes during one year of delta growth. B. Changes in the equilibrium graded delta as a function of the wave energy flux, $P = (ECn)_b$. (From Komar 1973. Used by permission of the Geological Society of America).

some geomorphic or engineering obstacle. The rate of increase of sand volume on the beach is then $(dy/dt)sd$ where d is again the thickness of this accretion (Fig. 7.1B). Since this must equal the sand supplied by the river to this delta flank, $(dy/dt)sd = Q_r/2$. Combining Equations 7.7 and 7.13, we also have $dy/dt = -a/d$ so that $a = -Q_r/2s$ and Equation 7.13 becomes

$$Q_s = \frac{Q_r}{2}\left(1 - \frac{x}{s}\right) \tag{7.14}$$

Thus, in order to have an equal advancement of the delta shoreline while maintaining the same shape, there must be a linear decrease in the littoral drift, Q_s, from $Qr/2$ at the river mouth to zero at the end of the beach (at $x = s$). Equation 7.14 assures continuity of sand volume—that is, that the amount of sand supplied by the river is accounted for in total delta accretion.

Evaluating the littoral drift with Equation 7.4, the above relationship yields

$$\sin(2\alpha_b) = \frac{Q_r}{6.85(ECn)_b}\left(1 - \frac{x}{s}\right) \tag{7.15}$$

for the longshore variation (x-dependence) in the breaker angle α_b along the delta flank. The corresponding gradient of the breaker angle is

$$\frac{d\alpha_b}{dx} = -\frac{Q_r}{6.85(ECn)_b(2s)}(\cos^2\alpha_b - \sin^2\alpha_b)^{-1} \tag{7.16}$$

At the river mouth, $x = 0$, the magnitude of the breaker angle varies directly with the river sediment supply and inversely with the wave energy flux. These factors similarly affect the longshore gradient of the breaker angle, $d\alpha_b/dx$; the greater the river sand supply and the smaller the wave conditions the greater the curvature of the equilibrium delta shoreline. These analytical results are seen to conform with the findings of the numerical models, especially the series of deltas illustrated in Figure 7.4B where $(ECn)_b$ was varied. Note further that $\alpha_b = 0$ at $x = s$, a necessity for the littoral drift to become zero at the end of the beach. But at the same time $d\alpha_b/dx$ and thus dQ_s/dx from Equation 7.8 do not become zero, a requirement for deposition and shoreline advancement at the end of the beach.

Although the analogy between the graded river and shoreline has been stressed, there are some apparent differences. For example, in the graded river there is essentially always a progressive increase in the quantity of sediment to be transported. In the case of the graded shoreline the quantity of the littoral drift may either increase or decrease along the coast. The above delta models and the embayment deposition are examples of progressively decreasing transport. However, one could equally consider segments of coastline with many sand sources that result in an increasing littoral drift alongshore. The configuration of the headland as it is being eroded provides an example.

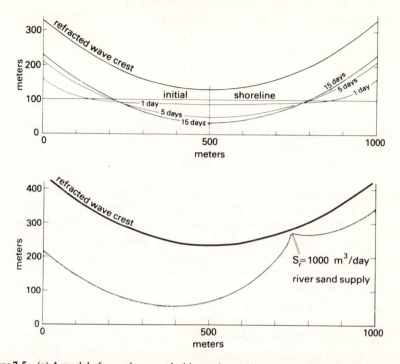

Figure 7.5 (a) A model of an embayment lacking major sand sources or sinks, demonstrating that the shoreline quickly attains the same curvature as the wave crests such that $Q_s = 0$ everywhere, the zero-drift configuration. (b) The same model but with a river source. The delta achieves a graded shoreline that systematically departs from the zero-drift configuration that acts as a 'base level'. (From Komar 1976, pp. 258 & 262. Used by permission of Prentice Hall).

Tanner (1958) also argued that the equilibrium shoreline lacks a counterpart to the base level found in rivers. However, a reasonable analogy is provided by the shoreline where the net littoral drift is everywhere zero. Such a zero-drift shoreline is best achieved within a pocket beach having negligible sand sources and losses. The numerical model of Figure 7.5A illustrates that the shoreline curvature quickly takes on the shape of the incoming wave crests where $\alpha_b = 0$ and hence $Q_s = 0$ everywhere along the shore. If a sand source such as a river is introduced into the pocket beach a new equilibrium shoreline is achieved as shown in the model of Figure 7.5B. The delta formed is superimposed on the former zero-drift shoreline curvature, the zero-drift configuration being approached with increasing distance from the river mouth. The zero-drift shoreline thus serves as a base or reference shoreline in the analysis of the river-source effects and delta growth, and this role is similar to the base level of rivers.

One fascinating result in the delta model of Figure 7.5B is the asymmetry of its equilibrium form. To maintain a constant shoreline equilibrium, more river sand is required for the longer stretch of beach to the left of the river than for the short stretch to the right. To account for this difference, the delta attained an asymmetrical form where the breaker angles and thus the littoral drift rates are greater on the left flank than on the right, thereby moving more sand into that segment of the bay.

Discussion

The above analyses illustrate the usefulness of numerical computer models for investigating geomorphic concepts of shoreline configuration. The models are able to examine how the shoreline changes through time and in many cases approaches an equilibrium or graded form. The analysis is obviously more powerful than conceptual models such as the *a–b–c* model of headland erosion presented by May and Tanner (1973), and is less likely to result in erroneous conclusions. With the computer-simulation models we can sufficiently simplify the problem so that we are examining only one or two parameters. For example, in the delta models we considered the simple case of a river supplying sand at a constant rate with waves of fixed energy flux and direction. We were able thereby to examine how these two opposing parameters control the overall equilibrium delta shape. In this very simple case we were also able to derive analytical equations from the delta form, but this is not normally possible in more complicated models. For example, the next step in the analysis of delta shapes could involve seasonally varying the river sand supply and considering a complete wave climate of changing wave energies and directions. One problem to examine would be an assessment of the degree to which these variations alter the graded delta curvature from the above simple models. I developed a model of a complex delta (Komar 1977), reproduced in Figure 7.6, in which after an initial stage of delta growth the river mouth suddenly shifted position to the flank. While a new delta began to form there, the area of the former river mouth eroded back under the continuing waves. The result is the

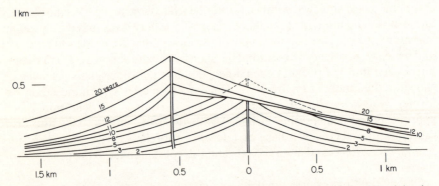

Figure 7.6 Computer model of the growth of a complex delta formed by the shift of the river mouth to the delta flank after 10 years of the initial growth. The dashed shorelines gives the extent of the original delta before it began to erode back when the river mouth shifted. (From Komar 1977, p. 510. Copyright by John Wiley and Sons, Inc. Used by permission).

complicated pattern of growth lines seen in Figure 7.6, which can be taken as being equivalent to a sequence of beach ridges representing the delta growth. Such a model therefore could be employed in a study of patterns of beach ridges. As I pointed out in that study (Komar 1977), it is also possible to run the model 'backwards.' For example, the growth pattern of the complex delta in Figure 7.6 or of a real delta, could be the beginning point of the analysis, with the model evaluating the littoral drift but then transporting the sand back up the coast and moving it up the river. The model thereby 'peals' off the series of accreted beach ridges, working back through time.

Acknowledgment

This research was completed under the support of the NOAA Office of Sea Grant, Department of Commerce, grant NA81AA-D-00086 with Oregon State University.

References

Bagnold, R. A. 1963. Beach and nearshore processes. In *The Sea*: the earth beneath the sea. V. 3, M. N. Hill (ed.), 507–528. New York: Wiley-Interscience.

Dean, R. G. 1979. Beach erosion: causes, processes, and remedial measures. *CRC critical reviews in environmental control.* **6**, 259–96. Boca Raton, Fla.: CRC Press.

Dolan, R., L. Vincent and B. Hayden 1974. Crescentic coastal landforms. *Zeit. Geomorph.* **18**, 1–12.

Horikawa, K., T. Sasaki and H. Sakuramoto 1977. Mathematical and laboratory models of shoreline changes due to dredged holes. *J. Faculty Engr., Univ. Tokyo* **34**, 49–57.

Komar, P. D. 1973. Computer models of delta growth due to sediment input from rivers and longshore transport. *Geol. Soc. Am. Bull.* **84**, 2217–26.

Komar, P. D. 1976. *Beach processes and sedimentation.* Englewood Cliffs. N.J.: Prentice-Hall.

Komar, P. D. 1977. Modeling of sand transport on beaches and the resulting shoreline evolution. In *The Sea: marine modeling.* V. 6, E. D. Goldberg, I. M. McCave, J. J. O'Brien, and J. H. Steele (eds), 499–513. New York: Wiley-Interscience.

Komar, P. D. in press. Nearshore currents and sand transport on beaches. In *Physical oceanography of coastal and shelf seas*, B. Johns (ed.). Amsterdam: Elsevier.

Komar, P. D. in press. Computer models of shoreline changes. In *Coastal processes and erosion*, P. D. Komar (ed.). Boca Raton, Fla.: CRC Press.

Komar, P. D., J.R. Lizarraga-Arciniega and T. A. Terich 1976. Oregon coast shoreline changes due to jetties. *J. Waterways, Harbors and Coastal Engng Div., Am. Soc. Civ. Engrs* **102**, 13–30.

LeBlond, P. H. 1979. An explanation of the logarithmic spiral plan shape of headland-bay beaches. *J. Sed. Petrol.* **49**, 1093–1100.

May, J. P. and W. F. Tanner 1973. The littoral power gradient and shoreline changes. In *Coastal geomorphology*, D. R. Coates (ed.), 43–60. London: George Allen & Unwin.

Motyka, J. M. and D. H. Willis 1975. The effect of refraction over dredged holes. *Proc. 14th Conf. Coastal Engng*, 615–25.

Pyne, S. J. 1980. *Grove Karl Gilbert, a great engine of research.* Austin, Texas: Univ. of Texas Press.

Perlin, M. and R. G. Dean 1979. Prediction of beach planforms with littoral controls. *Proc. 16th Conf. Coastal Engng*, 1818–38.

Price, W. A., K. W. Tomlinson and D. H. Willis 1973. Predicting the changes in the plan shape of beaches. *Proc. 16th Conf. Coastal Engng*, 1321–29.

Rea, C. C. and P. D. Komar 1975. Computer simulation models of a hooked beach shoreline configuration. *J. Sed. Petrol.* **45**, 866–72.

Tanner, W. F. 1958. The equilibrium beach. *Trans. Am. Geophys. Union* **39**, 889–91.

8

Sediment transport in relation to a developing river delta

Charles E. Adams, Jr., John T. Wells, and James M. Coleman

Introduction

The deltaic plain of the Mississippi River system spans a distance of more than 350 km along the Louisiana shoreline of the northern Gulf of Mexico. This environmentally interesting and socio-economically important sedimentary depositional province is comprised of a succession of juxtaposed and imbricated sedimentary lobes (Fig. 8.1). The distal margin of each lobe represents the seaward terminus of a former course of the Mississippi River. The bird-foot or Balize Delta of the modern Mississippi River represents the most recent of these lobes.

The Balize Delta is the most prominent feature of the northern Gulf coastline (Fig. 8.2, inset). Extending to within approximately 25 km of the continental shelf break, it can both influence and respond to a variety of marine processes. For example, the orientation of the delta (i.e. normal to the local shoreline) prevents significant shore-parallel transport of water and sediment in this part of the Gulf over all but the outer one-quarter of the shelf. The results of process modifications by the delta should be reflected in the physical and biological characteristics of adjacent coastal areas.

Major course changes of the river (i.e. the abandonment of a channel for a hydraulically more favorable one), which give rise to the deltaic lobes of the Mississippi system, occur on the order of once each 1000 years. The modern Mississippi River has occupied its present channel for approximately 800–900 years (Fisk 1944, Table 3) and presently is at a hydraulic disadvantage relative to the Atchafalaya River, a major distributary channel that enters the Gulf approximately 200 km to the west (Fig. 8.2). The latter now carries approximately 30 percent of the total discharge of water and sediment of the combined river system. Increasingly larger percentages would be carried by the Atchafalaya if not for regulatory works, constructed in 1963, that limit it to this amount.

As a Mississippi distributary, the Atchafalaya River has carried large volumes of water and sediment for over 400 years (Fisk 1952). During much of

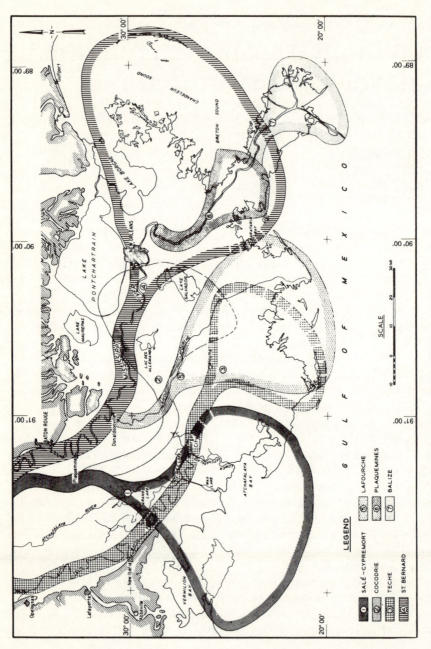

Figure 8.1. Recent deltas of the Mississippi River (from Coleman 1966).

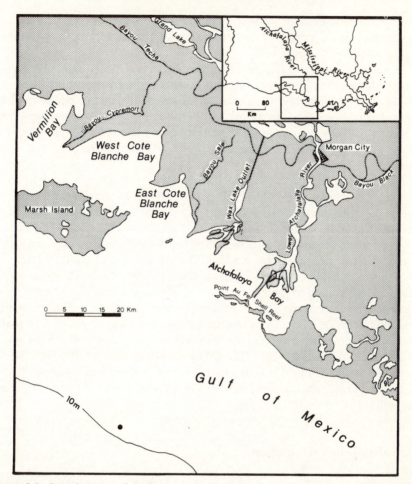

Figure 8.2 Location map of the lower Atchafalaya River and Bay and the continental shelf current monitoring site (solid circle).

this time the sediment load carried by the river was deposited within the many lakes and other naural catchment basins in the upper reaches of the river. In the 1950s clay-sized sediments began appearing in Atchafalaya Bay as a result of the filling of the catchment basins upriver. In the following decade (1962–72) the river sediment load coarsened; the dominance of silt- and sand-sized material carried by the river during this period resulted in the development of distal and distributary-mouth bars associated with the now rapidly prograding delta in Atchafalaya Bay (Roberts *et al*. 1980). During the past decade the delta has been growing subaerially. As the bay continues to fill with sediment, more coarse material will be available for delivery to the exposed coast and beyond to the continental shelf. Continued sediment supply should eventually lead to extension of the sedimentary lobe and associated delta onto the continental shelf. The time scale of this activity is presented by Wells *et al.* (1982, Fig. 37), who project significant progradation of the Atchafalaya Delta onto the continental shelf by the year 2030.

Changes in coastal configuration in the vicinity of the developing Atchafalaya delta will be, in part, a function of shelf sediment transport form and pattern. The changes may be expected to further modify local continental shelf physical processes. The proximity of the new delta to extensive and biologically productive marsh areas at or very near present sea level makes the nature and magnitude of those changes of great practical and economic interest.

Frequently, near-bottom current measurements alone are used to infer patterns of sediment transport and dispersion. It will be shown that the failure to consider the nature of the bottom sediments and their effects on flow dynamics can yield misleading results.

The purpose of this paper is to report on the implications of some near-bottom current measurements made on the continental shelf offshore of the developing Atchafalaya Delta. The objective of the data analysis was to gain a better understanding of the near-bottom flow in the region and to determine relationships between the flow and the distribution of coarse sediments, which form the backbone of the delta structure. Thus it should be possible to make inferences concerning processes and from those inferences to develop models of future delta growth patterns. The data encompass a $4\frac{1}{2}$-month period during the winter, when wind-driven shelf currents, which are principally responsible for the sediment transport process on this part of the Gulf shelf, are most energetic.

Continental shelf circulation

Winds near the northern shore of the Gulf of Mexico are predominantly from an easterly direction throughout the year. Tidally driven flows are unimportant for the most part (see below for an exception), and thus the net motion of shelf waters along the central Louisiana coast west of the Mississippi delta is to the west (Murray 1976), following the wind. During winter, however, migrating continental low-pressure centers frequently intrude into the northern Gulf of Mexico, bringing strong winds from the north (Brower *et al.* 1972). These northerly winds are associated with the passage of intense frontal systems. Depending on the angle between the front and the coastline, the wind may have either a westerly or an easterly directed longshore component. In the latter case the wind frequently is of sufficient magnitude to generate a vigorous easterly directed flow at all levels in the water column; in the former the local wind augments the net westward flow. The data to be presented indicate that westerly directed water motions associated with a frontal passage are relatively unimportant. The passage of 15 to 20 frontal systems through the Louisiana coastal region during a normal winter is common (Brower *et al.* 1972) and is believed to be a significant factor in the distribution of sedimentary materials on the open shelf.

Although there are definite and strong relationships between the current system along the Louisiana coast and the patterns of regional and local wind systems, other driving factors may also be important during certain times of the year. For example, during summer, when winds are generally weak and wind-associated surface stresses are small, tidal forces become dominant (Murray 1976) and transshelf motions associated with the rotary diurnal tide may mask the slow westward drift. During spring floods the buoyant plume of the Mississippi and Atchafalaya Rivers may create pressure gradients that strongly reinforce net westward motion and thus further inhibit eastward excursions of water and sediment.

Sediment transport theory

In the relatively shallow waters of inner continental shelf regions the movement of particulate materials, commonly referred to as sediment transport, is a conspicuous natural process. The movement or transport of shelf sediments results, in large part, from interactions between the sediments themselves and water motions associated with currents (unidirectional and tidal) and surface waves, which extend to the bottom in these depths. When the water motions are sufficiently energetic, large quantities of sediment may be moved as either bedload or suspended load. In the coastal waters of central Louisiana outside the surf zone intense transport of noncohesive sediments (coarse silt, sand, and fine gravel sizes) is associated primarily with storm-related waves and currents (Adams *et al.* 1982). In this case suspension transport frequently is the dominant mode of sediment motion because the mechanism of bedload transport is relatively inefficient (Smith, J. D. 1977) and the sediments are predominantly fine grained.

In sediment transport problems the principal index of incipient motion is the bottom stress, $\tau_0 (= \rho u_*^2;$ $\rho = $ *fluid density*, $u_* = $ friction velocity), and knowledge of its magnitude is crucial in determining the mode of transport. When τ_0 is less than some critical value, τ_c, the bottom and the flow do not interact; the flow remains clear. However, when $\tau_0 > \tau_c$, erosion takes place and sediment is transported. For a knowledge of the sediment transport field, then, it is sufficient to know the spatial distribution of τ_0 and the value of τ_c, which is a function of the physical nature of the bottom sediment (grain size, density, etc.), the fluid (density, viscosity), and the flow (velocity, turbulence intensity). Shields's criterion (e.g. Komar 1976a), which has been shown by Madsen and Grant (1975) to be applicable to oscillatory as well as unidirectional flow, normally is used for obtaining estimates of τ_c when the sediments are cohesionless.

The stress at the seabed is due to water motions associated with both currents and surface waves. When both are present, the latter phenomenon normally contributes more to the stress, even when the associated motions are of comparable magnitude. As justification for this, consider a boundary layer in which the gradient form of the law of the wall (Schlichting 1955) may be written

$$\frac{\delta u}{\delta z} = \frac{u_*}{\kappa z} \tag{8.1}$$

where $\kappa = 0.4$ is von Karman's constant, z is the vertical coordinate (positive upward), and u is the fluid velocity at height z above the boundary. The velocity change across a boundary layer is proportional to the magnitude of the associated fluid motions, and for the case in consideration is approximately the same for both the wave and the current boundary layers. The latter feature, however, has a much larger length scale (i.e. the layer is thicker) than the former, and thus the velocity gradient is smaller. According to Equation 8.1, therefore, the stress ($u_* \sim \tau_0^{\frac{1}{2}}$) associated with the wave is greatest. This argument suggests that waves may initiate sediment movement when a current of comparable magnitude is unable to do so. To a first approximation, over a wave period, net transport associated with wave motion is small or entirely absent. Significant transport (redistribution of sedimentary material) implies that other, unidirectional currents are present simultaneously. However, it should be noted from studies by J. D. Smith (1977) and Grant and Madsen (1979) that shear stress associated with a combination of waves and currents is different from the stresses that would be expected in the cases of pure waves or currents.

To obtain estimates of the threshold index τ_0, marine geologists frequently measure near-bottom current speeds and infer values of u_* from speed profiles assuming that the logarithmic velocity profile

$$\bar{u} = (u_*/\kappa) \ln z/z_0 \tag{8.2}$$

is valid near the boundary. Here $\bar{u}$ is the average current speed at some height z above the bottom and z_0 is the so-called roughness length. Equation 8.2 applies in steady, uniform flows; it is the integrated form of Equation 8.1. The term u_*/κ in Equation 8.2 is the reciprocal of the slope of the velocity profile when depth versus speed is plotted on log-linear coordinates with speed as the abscissa. Thus, when velocity profile data are available, Equation 8.2 may be invoked to obtain estimates of u_* (and thus τ_0). Use of Equation 8.2 to obtain estimates of u_* tacitly assumes that either suspended sediment is not present in the flow or, if present, it has no influence on flow dynamics.

There is increasing evidence that particle concentrations at levels found in many natural fluid flows (i.e. when sediment is being transported) can alter the dynamical characteristics of the entraining fluid. One significant effect of the two-phase behavior of a sediment-laden flow is an apparent reduction in the turbulent drag force—that is a reduction of τ_0 (u_*). Evidence for this change comes from laboratory studies (Vanoni 1953, Gust 1976) and from continental shelf field studies (McCave 1973, Nihoul 1977, Cacchione & Drake 1982).

Results of a model by Adams and Weatherly (1981a & b) indicated a significant reduction in u_* and an increase in the turning of the current vector with depth (Ekman veering) in a sediment stratified flow. They found an equation of the form

$$\bar{u} = (u_*/\kappa') \ln z/z_0 \tag{8.3}$$

to be appropriate in a sediment-laden flow. In Equation, 8.3 $\kappa' = \kappa/(1+BR_f)$, where B = constant $\simeq 5$ and R_f = flux Richardson number, a stability index (e.g. Turner 1973). These results agree qualitatively with those of Smith and McLean (1977) and with a model of an atmospheric boundary layer stably stratified by a temperature gradient (Businger & Arya 1974).

Another method of estimating bottom shear stress is with the quadratic relationship (e.g. Komar 1976b), which may be written as

$$\tau_0 = \rho C_{100} U^2_{100} \tag{8.4}$$

where ρ is fluid density, C_{100} is a drag coefficient, and U_{100} is the mean current speed, both evaluated at 100 cm above the bottom $(u_{100} = U^2_{100}+v^2_{100})^{\frac{1}{2}}$. The drag coefficient is normally taken to be a constant and equal to 3×10^{-3} (Sternberg 1968). The quadratic law, although of great practical value, embodies less flow physics than does the law of the wall.

Oceanographic observations

The oceanographic data presented here are from two bottom-mounted platforms located within 15 m of each other approximately 50 km offshore (Fig. 8.2). Detailed descriptions of the instrumentation, calibration procedures, and platform configurations are given by Frey and Appell (1981). The mooring site is relatively flat, with a regional slope of about 0.0006 (0.03°). An elongate shoal with a relief of about 2 m lies approximately 3 km to the northeast. Bottom sediments near the current meter platform consist primarily of sand (>70%) with lesser amounts of silt- and clay-sized particles present (Hausknecht 1980). The sand-silt ratio was seasonably variable, ranging from a vernal high of about 5 to an autumnal low of about 3.5. The sand fraction was always at least moderately well sorted, with an average predominant mode of 3.5 ϕ (0.09 mm). Unpublished data indicate that clays and silts dominate the inner shelf in this area to water depths of 10–20 m, while the less common sands are believed to be remnants of bar-mouth deposits of former delta lobes.

The two rigid platforms, set on the bottom in 9 m of water, each supported two Grundy current meters with rotors that turned in either direction. The meters were located at heights of 1.0 m and 3.0 m above the footpads on one platform and at heights of 2.0 m and 4.0 m above the footpads on the other. The meter at 4.0 m failed to provide validated data (H. Frey, personal communication). Current scour about the base of the platforms during periods of intense flow resulted in some uncertainty in both the height above the bottom and the relative separations between the meters. To avoid confusion the current meters are hereafter referred to as the 1-m, 2-m, and 3-m current meters, the values representing approximate heights above the bottom.

Because the sampling interval is much longer (average greater than 10 minutes) than the period of ordinary surface gravity waves, and because the meters respond to backflow as well as forward flow (although not necessarily with equal efficiency), the contribution of wave orbital motions to the observed mean current speed is thought to be small. Continuous current-meter records at the three levels were available for the $4\frac{1}{2}$-month period of interest except for the 5 days (January 23–28, 1979) during which no current meter data were taken. For analysis purposes, current velocity was resolved into long-isobath (u, positive to the east) and cross-isobath (v, positive onshore) components. Significant wave heights based on spectral estimates were available only for the period November 1, 1978, to January 22, 1979.

Results

Quadratic shear stress

The general character of the flow field during the $4\frac{1}{2}$-month period (November 1, 1978 – March 12, 1979), as represented by the current meter at the 1-m level, is shown in a progressive vector plot of water motion (Fig. 8.3), in which the x-axis is long-isobath and the y-axis is onshore. The data for Figure 8.3 have been smoothed over 12 hours, thus concealing the rotary nature of the flow, which Daddio et al. (1978) attributed to wind-forced inertial currents. The net flow during the study period is in a westerly direction with a substantial onshore component. From November through the end of January, flow is mostly longshore; whereas from early February to the end of the data record, net motion is dominantly cross-isobath. Caution must be used in making a Lagrangian interpretation of Eulerian data, particularly in proximity to a boundary. For example Figure 8.3 indicates a shoreward movement of water of about 150 km during the observation period, whereas the current meter station is only about 50 km offshore. The data of N. P. Smith (1977) showed similar results. This shortcoming of the presentation method notwithstanding, it is clear from Figure 8.3 that in this area transshelf motions during the winter may be as important to the dispersion of particulate materials as are longshelf flows. Even though the observation period represents the time of the year when water motions are most energetic, the net speed of the water during the entire period was less than 5 cm/s, and to the west.

In contrast to the net westerly drift are two conspicuous periods of about 1 day each in which the water moved rapidly in an easterly and off-shelf direction (Fig. 8.3). Both of these easterly flow events coincided with frontal passages, which are characterized by high winds from a northerly direction and by

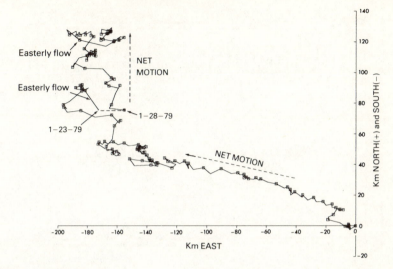

Figure 8.3 Progressive vector plot of water displacement at the 1-m level for the period November 1, 1978, to March 12, 1979. Open squares are drawn after each 24-hour interval. The data have been smoothed over 12 hours. Dashed line signifies no record. Note the two periods of prolonged eastward movement (from Adams *et al.* 1982. Used by permission of the Port Aransas Marine Laboratory).

relatively high surface waves. Current speed at the 1-m level during these two easterly flow events exceeded 50 cm/s.

The u and v components of velocity at the 1-m level are shown in Figure 8.4A and 8.4B, respectively. The v component clearly exhibits the diurnal tidal signal almost continuously throughout the record length. Subtidal events are more evident in the u component than in the v component record, which suggests that subtidal motions (wind-forced events) are principally directed parallel to the isobaths. The two easterly flow events mentioned above are clearly evident in the velocity component records, with peak velocities occurring at $t \simeq 1945$, 2795 hr.

The critical suspension curve of Bagnold (1966), as revised by McCave (1971), indicates that sediment finer than about 2ϕ (0.25 mm) will be transported almost entirely in suspension. The critical suspension curve also yields a critical bottom stress for resuspension of the 3.5ϕ (0.09 mm) sediments as $\tau_c = 1.7$ dynes/cm^2. As may be seen from the time series plot of shear stress (calculated from Eq. 8.4 in Figure 8.4C), the shear stress required to resuspend the 3.5ϕ sand was exceeded briefly on five separate occasions between November 1, 1978, and January 21, 1979. During four of these periods of elevated bottom stress the flow near the bottom was directed toward the east ($+u$) (Fig. 8.4A). A prolonged period of supercritical bottom stress, during which resuspension of significant quantities of sand-sized sediments might be expected, did not occur until January 20, 1979, when τ_0 reached a maximum of about 7 dynes/cm^2 and $\tau_0 > \tau_c$ for approximately 15 hours. From January 29 to March

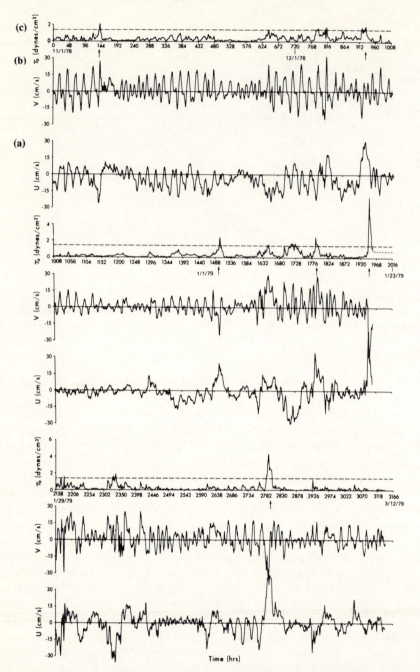

Figure 8.4 Plots of *u* (a) and *v* (b) components of velocity at the 1-m level and time series of bottom stress (τ_0) (c) as calculated from Equation 8.4 for the period November 1, 1978, to March 12, 1979. The straight line represents the critical shear stress for a 3.50ϕ size component ($\tau_c \simeq 1.7$ dynes/cm^2). Logarithmic layer analysis was made for the data between 2712 hr $< t \leqslant$ 2832 hr (from Adams *et al.* 1982. Used by permission of the Port Aransas Marine Laboratory).

12, 1979, the maximum $\tau_0 \simeq 4.5$ dynes/cm^2. The dynamics of the latter event will be presented in a later paper. In all cases of high bottom stress the flow event resulted from the passage of a frontal system.

Boundary layer flow

In a steady, neutrally stable boundary shear flow, a logarithmic layer is present immediately above the boundary. The velocity in this logarithmic layer is given by Equation 8.2. In the absence of simultaneous current velocity and suspended sediment concentration profiles, it is possible to exploit the velocity shear variation in the flows of sediment-laden and clear fluids to gain additional insight into the nature and the intensity of two-phase effects and in particular to make inferences concerning the stability of sediment-laden flows. This type of approach was used by McCave (1973) in a study of suspension transport of sands by tidal currents. The same methods of analysis can be applied to the suspension transport of coarse sediments by unidirectional currents. As an example, during periods of relatively low flow, when accelerative effects are small, bottom shear stresses are low and little or no sediment erosion occurs; the velocity profile is that of an essentially clear flow and is neutrally stable. As the velocity field develops in response to strong winds, turbulent shear stresses at the bottom increase and sedimentary particles stripped from the bottom are carried along with the flow. In the absence of thermal stratification and acceleration effects, any observations of increased velocity shear and thus enhanced stability can be attributed directly to suspended sediment effects. Comparison of velocity profiles in these two well-defined situations allows estimates to be made of relative suspended sediment concentrations associated with the respective flows. Because the period 2712–2832 hours (Fig. 8.4A, B) offered two distinctly contrasting flow regimes, a dynamical analysis of the current meter data was made, the results of which are summarized below.

Using the method of least squares, hourly averaged velocities at the three sampling levels were fit to Equation 8.2. A velocity profile was defined as being logarithmic when the coefficient of determination $r^2 \geq 0.9755$. This is the t test criterion for a 90% confidence level in a three-point fit.

Overall, mean velocity profiles satisfying the logarithmic criterion occurred about 50% of the time. An examination of the profile data showed a marked difference in the frequency of occurrence of logarithmic profiles during the early period (2712 hr $< t \leq 2783$ hr), when periodic influences were strong, and during the wind-forced unidirectional flow episode ($t \geq 2784$ hr). During the former, logarithmic profiles were found in only about 40% of the data, whereas during the latter they were found in more than 65% of the data. The wind-forced flow episode can be partitioned further into three distinct events. The first is the 10-hour period when the water column was accelerating; none of the velocity profiles for this period were logarithmic. The second is the interval 2794 hr $\leq t \leq 2805$ hr, during which about 85 per cent of the mean profiles were logarithmic. The interval 2806 hr $\leq t \leq$ 2831 hr had a percentage of logarithmic mean profiles similar to the preceding one; however, the slopes of the profiles were significantly larger and more variable. Thus on the basis of the occurrence of logarithmic profiles and on the slopes of these profiles, two distinct flow events were identified for analysis. These are the intervals 2712 hr $< t \leq$ 2783 hr and 2794 $\leq t \leq$ 2805. The early period was assumed to represent an episode of predominantly clear flow, and the later one was thought to be a period during which substantial qualities of sediment were being transported in suspension.

The speed data for the logarithmic velocity profiles within each of the two time intervals were averaged and those values again fitted to Equation 8.2 by the least-squares method. The resulting profiles are shown in Figure 8.5. Of interest in these profiles are the slopes from which values of u_* (τ_0) presumably may be calculated, and the relative differences of this parameter among the various profiles.

Neutral velocity profile

The composite velocity profile for 2712 hr $\leq t \leq$ 2783 hr (profile 1) is characterized by a relatively large slope, which yields $u_* = 1.0$ cm/s. This value of u_* should be representative of the shear velocity of the mean flow, assuming that wave orbital motions have been excluded from the current meter data. Thus it is the value that may be associated with the sediment transport process in the problem being considered. Wave oscillatory currents, although not contributing significantly to the mean flow, create stresses that influence

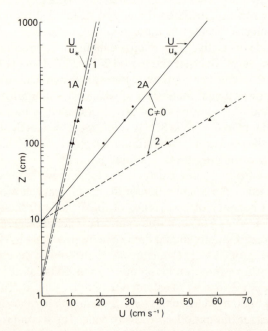

Figure 8.5 Plots of vertical current shear for sediment-laden and sediment-free flows.

resuspension rates and must be accounted for. To a first approximation, mean and wave stresses may be added vectorially to give an estimate of the total bottom stress acting to resuspend sediment. It is hoped that the wave-current interactions are highly nonlinear (J. D. Smith 1977; see also Grant & Madsen 1979), and estimates based on a linear superposition of components will likely result in underestimating the total stress. Inclusion of non-linear effects, however, seems unwarranted, given the nature of the wave data.

For monochromatic waves with a period of 4 s, the technique presented by J. D. Smith (1977) yields $u_{*w} \simeq 0.2$ cm/s. The subscript w refers to the waves. The total friction velocity at the bottom in the presence of waves and mean current for the period 2712 hr $\leqslant t \leqslant$ 2783 hr thus is $u_{*T} \simeq 1.2$ cm/s ($\tau_0 \simeq 1.4$ dynes/cm^2), which is less than the critical friction velocity required for resuspension of the 3.5ϕ sediments, according to McCave (1971). Here $u_{*T} = u_* + u_{*w}$. At the lower limit of validity of the McCave curve (coarse silt, about 4.5ϕ (0.04 mm)), the critical shear stress needed for resuspension is only slightly greater than the ambient value u_{*Total}. For silt-sized and finer materials, the combined wave-current stress should be sufficient for resuspension. The paucity of fine-grained sediments (silts and clays) at the Louisiana shelf site during the period of study, and the presence of these materials during other seasons when combined bottom stress might be expected to be lower, supports this interpretation. Thus, insofar as the initial 71-hour period is representative of the local mean flow, the combined tidal-inertial and oscillatory flows are only competent to resuspend the finer materials present.

Stable velocity profile
In the interval 2794 hr $\leqslant t \leqslant$ 2805 hr the slope of the velocity profile (Fig. 8.5, profile 2) is significantly smaller than that of profile 1. The marked decrease in slope (increase in shear) is strongly suggestive of a change from a clear flow to one that is sediment laden (or at least a change from a neutral to a stably stratified flow). However, implicit in Equation 8.2 is the assumption of neutral stability; thus buoyancy effects, which promote increased stability and which may have an influence on the flow dynamics, have been excluded.

From a practical standpoint, the use of Equation 8.2 in a stratified flow may result in spurious estimates of u_*. For example, as estimated from profile 2, $u_* = 7.6$ cm/s, which appears to be an abnormally high value for the observed current velocities. Another estimate of bottom stress may be obtained by assuming that the downstream flow is frictional, i.e. that surface wind stress is balanced by bottom stress ($\tau_w - \tau_0 = 0$). Longshore coastal currents satisfying this simple frictional balance have been observed by Huyer *et al.* (1978) and by Winant and Beardsley (1979). Data from Figure 8.4A and B indicate that accelerations during the period 2794 hr $\leqslant t \leqslant$ 2795 hr are (10^{-3}–10^{-4} cm/s^2). With the vertical eddy viscosity, K_v, having an order of magnitude of (10^2 cm^2/s), which is a reasonable estimate in a turbulent shear layer, the diffusion term in the long-isobath momentum equation is one or two orders of magnitude larger than the temporal acceleration term. If the long-isobath pressure gradient may be assumed small (no larger than the acceleration term), then the frictional assumption is a reasonable first approximation for the flow during this period. The average value of the long-isobath wind stress component from Figure 8.1 is $\tau_w = \tau_0 = 3.5$ dynes of $u_* = 1.9$ cm/s. The data of McCave (1971, Fig. 13) indicate that 3.5ϕ particles move as suspended load when $u_* \geqslant 1.3$ cm/s.

Thus during this period current effects alone are adequate to resuspend the coarser sand-sized sediments found at the site. It should be noted that τ_0 calculated from the assumption of a frictional boundary layer yields a value that is about 20% smaller than the value calculated from Equation 8.4.

The velocity data of profile 2, nondimensionalized by the reduced u_*, are plotted as profile 2A (Fig. 8.5). The dimensionless profiles (1A, 2A) are similar to those found by others (Vanoni & Nomicos 1959, McCave 1973, Adams & Weatherly 1981a) who have sought to contrast the velocity structures of sediment-laden and sediment-free flows. The combination of high bottom stresses and eastward flow during the period 2794 hr $\leqslant t \leqslant$ 2805 hr suggests that sand-sized and finer sediments are being resuspended and transported to the east during this flow event.

Further information on the easterly directed flow event that occurred in late February 1979 (i.e. 2784 hr $< t <$ 2832) can be gained by comparing progressive vector diagrams for the current meters at all three levels during that period (Fig. 8.6). From 2712 hr $\leqslant t \leqslant$ 2783 hr, the flow is dominated by a rotary circulation with a net westerly drift. For about 11 hours subsequent to $t = 2783$ hr, the water column accelerated rapidly at all levels in response to the passage of a front. Acceleration throughout the water column was rather uniform, as shown by a comparison of the progressive vector plots. The flow was south-

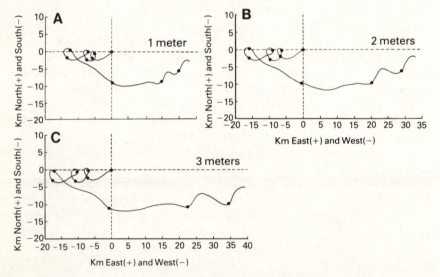

Figure 8.6 Progressive vector diagrams for current meters at heights of 1, 2, and 3 m above the bottom. Note the increased cross-isobath flow with depth during the final 12 hours of the record.

easterly, with a perceptible offshore component during the accelerative event. There was approximately a 5-hour lag between the strong easterly directed wind pulse and the onset of an easterly flowing shelf current. This phase lag was due, in part at least, to the considerable westerly directed momentum in the water column prior to the wind pulse.

During the interval 2794 hr $\leqslant t \leqslant$ 2805 hr, the flow was relatively steady, with a net alongshore motion to the southeast. The mean speed increased from about 50 cm/s at 1 m above the bottom to about 70 cm/s at the upper current meter.

The rotary component again became noticeable for $t >$ 2805 hr, with transshelf motions being superimposed on the drift, which had become essentially easterly. Although not shown on Figure 8.6 the flow again became westerly subsequent to $t =$ 2832 hr approximately.

The progressive vector plots (Fig. 8.6) suggest a turning of the current vector with depth in a counterclockwise sense. Thus the cross-isobath flow component at 1 m is greater than at 3 m. Current veering could be due to the development of a bottom Ekman layer (e.g. Bowden 1978), which would make the sense of the turning correct. The potential effects of local bottom topography on velocity veering have not been examined, however.

Discussion

For more than 4 months of near-bottom current data, it has been possible to characterize the boundary flow and to make inferences concerning sediment transport processes during the winter at a location offshore of the developing Atchafalaya River delta. A distinctive feature of the data record is the presence of eight separate periods during which the bottom stress calculated from the quadratic drag law exceeded the critical stress required to resuspend very fine sand. Seven of those suspension events were characterized by flow of water to the east, and two likely were of sufficient intensity to cause significant bottom scour. The magnitude of the transport and deposition processes induced by one or both of these intense events is suggested by the observations of the depth of burial (0.5 m) of the legs of the instrument platform (see above). The uncertainty of the actual height of the 1-m current meter raises a question as to the accuracy of τ_0 calculated with Equation 8.4. Because C_{100} increases as sediment goes into suspension (Komar 1976b), it is believed that any decrease in U resulting from platform settling should be counterbalanced by this effect, at least during intense resuspension events.

A further indication of sediment resuspension comes from a comparison of velocity profile data from two different flow regimes during the period February 22–26, 1979. The earlier part of this record represents tidally dominated flow that is competent to erode only silts and clays. This earlier period is, to a large extent, representative of the mean flow during the entire $4\frac{1}{2}$-month period. The later regime consists, in part, of the wind-driven easterly flow event (described above), which was of sufficient intensity, as determined by two independent methods, to erode the sand-sized material present. Insofar as the data examined represent mean conditions for the area of interest, there are strong indications that sand-sized material is moved selectively to the east and therefore the bottom sediments west of the source area should be deficient in the coarser components. This is not inconsistent with the few observations of shelf sediment distribution, which generally show a larger sand-silt ratio in winter than in summer.

It is important to note that the analysis does not consider the effects of major storms on the distribution of sediment. It is quite likely that a large storm (a hurricane, for example) would completely obliterate the sediment distribution patterns created by the normal flow patterns. Major storms, however, are sufficiently infrequent that their effects are not likely to be dominant over a time scale of a few tens of years or longer. Such a view is supported by bathymetric data, which show little or no significant change in the position and configuration of major bathymetric features over long time periods.

As a device for examining the relative dispersion of sand-sized sediments in a quantitative manner, current vectors were progressively summed for each period during which $\tau_0 \geq \tau_c = 1.7$ dynes/cm^2. In this case τ_0 was computed from Equation 8.4. Further, it was assumed that, once in suspension, particles would remain in suspension until $\tau_0 < 0.9\,\tau_c$. This progressive transport vector diagram is shown in Figure 8.7. The coordinate system is the same as in Figure 8.6. The net sediment transport, whether calculated with or without wave information, is to the east. This is consistent with the premise that wave motions, although effective in resuspending sediments, do no net transporting. Wind data indicate that movement of the sand-sized component during the period January 23–28, 1979, when current meter data were not available, was nominally eastward. The inclusion of bottom stress generated by wave motion

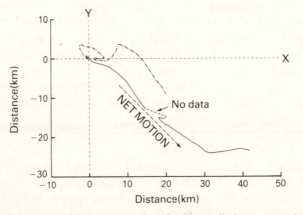

Figure 8.7 Progressive transport vector plot of sediment displacement at the 1-m level for the period November 1, 1978, to March 12, 1979. The dash-dot line is based on combined current and wave stress (from Adams *et al.* 1982. Used by permission of the Port Aransas Marine Lab.).

results in sediment being moved farther to the east, which suggests that wave motion indeed plays a role in sediment transport. In either case, qualitatively the results are the same: an eastward alongshore motion of about 40 km and an offshore displacement of about 20 km. This is in marked contrast to the water displacement (Fig. 8.3) as inferred from the 1-m current meter, which gives a westward and onshore displacement of about 200 km and 150 km, respectively. Thus current meter data, even from very near the bottom, in the absence of information concerning local bottom sediments and resuspension criteria, may be a very poor indicator of sediment motion and dispersion.

The results presented here do not necessarily portray conditions on the Louisiana shelf as a whole. The method of treating the noncohesive sediments (sands) at the study site differs markedly from those that must be used for cohesive sediments, which blanket most of the Louisiana continental shelf. Because the morphology of the delta may be expected to be controlled by the distribution of the coarser sediment components, however, sand transport and distribution should provide a clue to future trends in delta development.

Acknowledgments

We are grateful to H. Frey, K. Hausknecht, and R. Crout for providing the data used in this study. H. Roberts offered valuable suggestions for improving the text. Three anonymous reviewers provided useful suggestions for improving the manuscript. This study was sponsored by the United States Geological Survey under a contract with Louisiana State University.

References

Adams, C. E. and G. L. Weatherly 1981a. Some effects of suspended sediment on the oceanic bottom boundary layer. *J. Geophys. Res.* **86**, 4161–72.

Adams, C. E. and G. L. Weatherly 1981b. Suspended sediment transport and benthic boundary layer dynamics. *Mar. Geol.* **42**, 1–18.

Adams, C. E., J. T. Wells and J. M. Coleman 1982. Sediment transport on the central Louisiana continental shelf: implications for the developing Atchafalaya River delta. *Contr. Mar. Sci.* **25**, 133–48.

Bagnold, R. A. 1966. An approach to the sediment transport problem from general physics. U.S. Geol. Surv. Prof. Pap. 422I.

Brower, W. A., J. M. Meserve and R. G. Quayle 1972. *Environmental guide for the U.S. Gulf Coast*. Prepared for U.S. Army Corps Engrs by National Oceanic and Atmospheric Admin., EOS, NCC, Asheville, N.C.

Businger, J. A. and S. P. S. Arya 1974. Height of the mixed layer in the stably stratified planetary boundary layer. In *Advances in geophysics*, vol. 18A, F. N. Frenkiel and R. E. Munn (eds), 73–92. New York: Academic Press.

Cacchione, D. A. and D. E. Drake 1982. Measurements of storm-generated bottom stresses on the continental shelf. *J. Geophys. Res.* **87**, 1952–60.

Coleman, J. M. 1966. *Recent coastal sedimentation: central Louisiana coast*. Tech. Rpt. 29, Louisiana State University, Baton Rouge: Coastal Studies Institute.

Daddio, E., Wm. J. Wiseman, Jr. and S. P. Murray 1978. Inertial currents over the inner shelf near 30°N. *J. Phys. Oceanogr.* **8**, 728–33.

Fisk, H. N. 1944. *Geological investigation of the alluvial valley of the Lower Mississippi River*. War Department, Corps of Engrs U.S. Army.

Fisk, H. N. 1952. *Geological investigations of the Atchafalaya Basin and the problem of Mississippi River diversion*, Vol. 1. Vicksburg, Ms.: U.S. Army Corps Engrs, Mississippi River Commission.

Frazier, D. E. 1967. Recent deltaic deposits of the Mississippi River: their development and chronology. *Trans. Gulf Coast Assoc. Geol. Soc.* **17**, 287–311.

Frey, H. R. and G. F. Appell (eds.) 1981. *NOS Strategic Petroleum Reserve Support Project: final report, vol. 2—Measurements and data quality assurance*. U.S. Dept. of Commerce, National Oceanic and Atmos. Admin., National Ocean Surv., Rockville Md.

Garrett, B. J., P. Hawxhurst, and J. R. Miller 1969. *Atchafalaya Basin, Louisiana, Lower Atchafalaya River and Wax Lake outlet*. U.S. Army Corps Engrs, New Orleans Dist., 66th Commission on Tidal Hydraulics Conf.

Gust, G. 1976. Observations of turbulent drag reduction in a dilute suspension of clay in sea water. *J. Fl. Mech.* **75**, 29–47.

Hausknecht, K. A. 1980. Biological/chemical survey of Texoma and Capline sector salt dome brine disposal sites off Louisiana. *Work Unit 3.1, Sediments and suspended particulates, final report*. Cambridge, Mass.: Energy Resources Co., Inc.

Huyer, A., R. S. Smith, and E. J. C. Sobey 1978. Seasonal differences in low frequency current fluctuations over the Oregon continental shelf. *J. Geophys. Res.* **83**, 5077–89.

Komar, P. D. 1976a. The transport of cohesionless sediments on continental shelves. In *Marine sediment transport and environmental management*, D. J. Stanley and D. J. P. Swift (eds), 107–25. New York: Wiley Interscience.

Komar, P. D. 1976b. Boundary layer flow under steady, unidirectional currents. In *Marine sediment transport and environmental management*, D. J. Stanley and D. J. P. Swift (eds.), 91–106. New York: Wiley Interscience.

Madsen, O. S. and W. D. Grant 1975. *Sediment transport in the coastal environment*. Ralph M. Parsons Lab., Rept. No.209. Cambridge, Mass.: Massachusetts Inst. Tech.

McCave, I. N. 1971. Sand waves in the North Sea off the coast of Holland. *Mar. Geol.* **10**, 199–225.

McCave, I. N. 1973. Some boundary-layer characteristics of tidal currents bearing sand in suspension. *Mem. Soc. R. Sci. Liege*, 6 ser., **6**, 187–206.

Mellor, G. L. and T. Yamada 1974. A hierarchy of turbulence closure models for planetary boundary layers. *J. Atmos. Sci.* **31**, 1791–1806.

Murray, S. P. 1976. Currents and circulation in the coastal waters of Louisiana. *Sea Grant Publn* 210, Baton Rouge, La.: Center for Wetland Resources, Louisiana State University.

Nihoul, J. C. J. 1977. Turbulent boundary layer bearing silt in suspension. *Phys. Fluids* **10**, 107–203.

Owen, M. W. 1977. Problems in the modeling of transport, erosion, and deposition of cohesive sediments. In *The Sea: marine modeling*. vol. 6, E. D. Goldberg, I. M. McCave, J. J. O'Brien, and J. H. Steele (eds), 515–37. New York: Wiley Interscience.

Roberts, H. H., R. D. Adams and R. H. W. Cunningham 1980. Evaluation of sand-dominant subaerial phase, Atchafalaya Delta, Louisiana. *Bull. Am. Assoc. Petrol. Geol.* **6**, 264–79.

Schlichting, H. 1955. *Boundary layer theory*. New York: McGraw-Hill.

Smith, J. D. 1977. Modeling of sediment on continental shelves. In *The sea: marine modeling*. vol. 6, E. D. Goldberg, I. M. McCave, J. J. O'Brien and J. H. Steele (eds), 539–77. New York: Wiley Interscience.

Smith, J. D. and S. R. McLean 1977. Boundary layer adjustments to bottom topography and suspended sediment. In *Bottom turbulence*, J. Nihoul (ed.), 123–51. New York: Elsevier.

Smith, N. P. 1977. Near-bottom cross-shelf currents in the northwestern Gulf of Mexico; a response to wind forcing. *J. Phys. Oceanogr.* **7**, 615–20.

Sternberg, R. W. 1968. Friction factors in tidal channels with differing bed roughness. *Mar. Geol.* **10**, 113–19.

Turner, J. S. 1973. *Buoyancy effects in fluids*. Cambridge: Cambridge University Press.

Vanoni, V. A. 1953. Some effects of suspended sediment on flow characteristics. *Proc. 5th Hydr. Conf.* Iowa city: State Univ. of Iowa, Studies in Engineering, Bull. 34.

Vanoni, V. A. and G. N. Nomicos 1959. Resistance properties of sediment-laden streams. *J. Hydraul. Div., Proc. Am. Soc. Civ. Engrs.* **85** (HY5), 77–107.

Wells, J. T., S. J. Chinburg, and J. M. Coleman 1982. *Development of the Atchafalaya River delta: generic analysis*. Unpubl. Rept. Baton Rouge, La.: Coastal Studies Inst., Louisiana State Univ.

Winant, C. D. and R. C. Beardsley 1979. A comparison of shallow currents induced by wind stress. *J. Phys. Oceanogr.* **9**, 218–20.

9

Simulation of slope development and the magnitude and frequency of overland flow erosion in an abandoned hydraulic gold mine

L. E. Band

Introduction

A major problem in the analysis of hillslope systems is the pronounced fluctuation in those forces responsible for sediment transport. Gilbert (1877, 1909) first explicitly considered the adjustment of hillslopes to the dominant transport mechanisms and conceptually derived geomorphic laws linking the processes of overland flow, soil creep, and rainsplash to the expected slope form. A persistent question has arisen since that time: How does the geomorphic feature adjust to the great variations of the magnitude of these processes, particularly of hydraulic erosion?

This paper will describe the construction and use of a numerical simulation model of hillslope development to assess the contribution of various runoff magnitudes to soil erosion and the long-term development of hillslopes in an abandoned hydraulic gold mine in northern California. The model is based on detailed field measurements and observations of the geomorphic processes active at the site, and explicitly includes the effects of the fluctuating process rates. This constitutes a unique approach to the magnitude and frequency problem because the geomorphic effects of each runoff magnitude may be directly examined with the use of the simulation model, including the total amount of erosion contributed and the effect on slope shape. Construction of the model is first described along with a discussion of the criteria used for the selection of an appropriate study site and the measurement techniques used in the field analysis. Direct inspection of the erosional contributions of the various runoff magnitudes as simulated within the model is then used to explore the concepts of effective and dominant hydrologic events.

To date, the only direct study of the magnitude and frequency of erosion by overland flow is that of Pearce (1976), who shows that maximum sediment transport is accomplished by intermediate size rainstorms. Unlike river systems, however, on hillslopes there exist no morphologic features such as the height of the channel banks that may be used to suggest the size of an event that dominates the geomorphic shape, or whether, in fact, such an event exists. In addition, the assumption that the bankfull discharge is most responsible for shaping river systems (Wolman & Miller 1960) is based only on consideration of channel cross-sectional shape, and not on longitudinal profile. Therefore the concept of dominance as used in regard to stream systems may not be applicable to hillslopes, and it is necessary to explicitly set up some other criteria to define a dominant event. Such an event should be able to reproduce the geomorphic effects of the full range of erosional events (Carson & Kirkby 1972, p. 104) under the constraint that the total hydrologic input over a period of time (i.e. the annual rainfall) is reproduced at the steady rate. For the case of overland flow on hillslopes, this means that the total annual rainfall (or runoff) occurs at the constant rate.

Obviously it is impossible to test these criteria for any hillslope in the field due to the impracticality of observing the feature change over the character-istically long time scales involved, and the lack of control over the hydrologic inputs. It is for these reasons that combining field work with a simulation model of such a system may provide a profitable approach to this problem by allowing a greater amount of insight than has been gained through field research alone.

Site selection: theoretical and practical considerations

Although a number of simulation models of hillslope development have been presented over the past two decades (Scheidegger 1970, Culling 1963, Ahnert 1973, 1976, Kirkby 1971, Hirano 1975, and others), none has attempted to come to grips with the magnitude and frequency of hydrologic inputs by attempting to reproduce their fluctuations. Rather it is assumed there exist long-term average or representative values that may be used, although no attempt has been made to prove or disprove this assumption. In addition, little attempt has been made to assess the correspondence of these models to real systems by specific application and testing with an actual hillslope.

This may be largely attributed to the complexity of even the simplest hillslope systems. In order to simulate any hillslope it is necessary to estimate (1) sediment transport laws, (2) the magnitude and frequency of the hydrologic events, and (3) initial and boundary conditions. Therefore, any study attempt-ing to construct a simulation model of an actual hillslope requires very careful selection of a site for which these geomorphic factors and processes may be at least roughly estimated. The site chosen for this study fulfills these criteria.

The slopes in the study area have developed over the last century in a large, abandoned hydraulic gold mine in northern California. This area is now

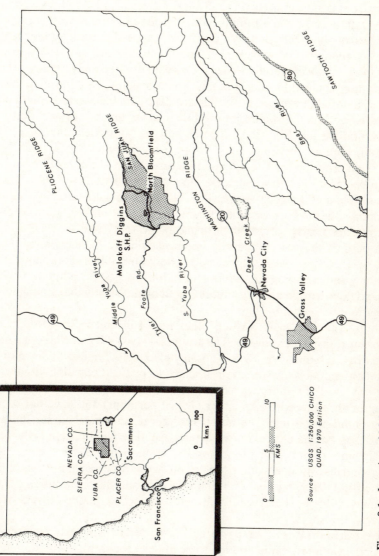

Figure 9.1 Location of Malakoff Diggings State Historic Park.

Malakoff Diggings State Historic Park, located approximately 25 km northeast of Nevada City, California (Fig. 9.1). The site may be accurately located by reference to its position 350 m S 35° W from the intersection of boundaries between T 18 N and T 17 N and R 9 E and R 10 E, shown on the North Bloomfield 7.5 minute topographic quadrangle.

The mine pit is at an elevation of roughly 975 m, with a mean annual precipitation of 1330 mm which is largely received as a series of long, closely spaced frontal storms occurring between October and April. A very small portion falls as snow and is rapidly melted. The slopes are developing in a very weathered phyllite (bulk density 1.6 gm/cm^3) exposed by the hydraulic removal of the overlying fluvial gravels over one hundred years ago. The upper half meter to meter is weathered to a dense clay with isolated pods of more coherent phyllite and some thin quartz seams which leave a lag deposit of residual coarser material developed at the surface. The drainage system is still poorly integrated and the hillslopes are eroding into an 'upland surface' that is still largely undissected (Fig. 9.2.) The slopes are short (less than 6 m), have negligible contour curvature and do not exceed a declivity of 30°–35°.

The first criterion for site selection requires that natural erosion processes should be rapid enough to be adequately measured while still allowing relatively simple mathematical expression. The clay-rich, slow-draining substrate combined with the very wet winter conditions result in a situation where the hillslopes are nearly saturated throughout the rainy season, so that during storm periods they are practically impermeable. This greatly simplifies the hydrologic response of the slope surface, eliminating much of the complexity of describing and modeling the runoff-generation process. In addition, runoff and overland flow erosion are much more frequent and steady than on badlands in more arid environments (Yair *et al.* 1980).

Shifting rills and mass wasting other than soil creep do not seem to be significant transport processes. The winters of 1981–82 and 1982–83 were the wettest two seasons since 1890. The only evidence of minor rill development or shallow mudflows was seen on surface declivities exceeding 35–40°; these were not used for the purpose of simulation. I have been on the slopes during rainfall exceeding 2.5 cm/hr (1 inch per hour) and have not seen any rilling or mass wasting. The absence of rill development under these conditions may be due to the coarser lag deposit. The main benefit is the luxury, for this time, of not having to assess and develop criteria for the growth and decay of rills, a significant problem that was approached by Smith and Bretherton (1972), but one that is still lacking specific solutions as described by Dunne (1980). Sheetwash is, therefore, the only major process of overland flow transport.

Mass transport by soil creep is largely restricted to near surface material and takes place at a rate rapid enough to be at least roughly measured.

Rainsplash may constitute an important portion of the total sediment transport although the lag deposit may reduce some of its effectiveness. Rainsplash erosion, however, was not measured in this study and this lack

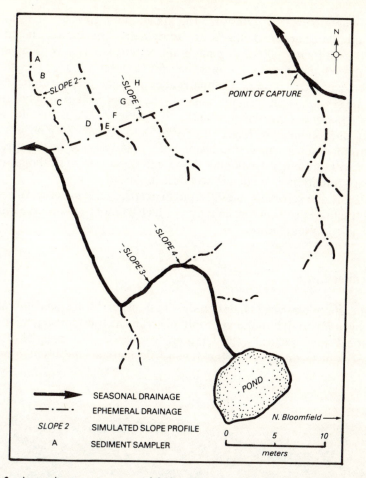

Figure 9.2 Approximate arrangement of drainage lines, with locations of simulated slopes and sediment samplers.

may represent the most significant limitation to a full and accurate description of the total sediment transport on these hillslopes.

The second criterion for a site requires that detailed precipitation records for the area should be available so that the nature of hydrologic events (type, magnitude and frequency) can be assessed. The Sierra Nevada foothill region has available one of the oldest and most detailed precipitation records in the western United States. Daily precipitation has been recorded at the field site and the surrounding region for the full development of the studied slopes, and continuous recording rain gauges in the region provide an adequate database for the assessment of the distribution of short duration intensities.

As a final criterion, initial and boundary conditions should be reasonably simple and be approximated by available evidence. Initial and boundary conditions, which are generally unknown for even simple badland systems, may be reasonably estimated at this site. The Malakoff Diggings are an historically important mining area and numerous photographs of the pit were taken in the late 1800s and early 1900s. A photograph taken by C. E. Watkins in 1871 during the Whitney Survey of California (Fig. 9.3) shows that by then the area had been exposed to erosion. Comparison of the photograph with the present landscape (Fig. 9.4) suggests the slopes in the foreground of Watkins' photograph are the slopes instrumented in this study (Fig. 9.2). Consideration of the photograph, combined with tree-ring dating of the Ponderosa pine growing in the current drainage lines, the history of stream capture in the area, and analogy with presently developing slopes allows estimation of the initial and boundary conditions of some of the slopes in the site.

While the site selection rules are rather restrictive, their observation provides a tractable problem for simulation, which has not been the case for many other badland sites studied.

Data collection

Field measurements were made of sediment transport by overland flow and soil creep in order to determine appropriate sediment transport equations. The theoretical and practical considerations involved in designing the specific measurement techniques and the actual field methods are briefly described

Figure 9.3 Instrumented hillslopes (slopes 1 and 2) in foreground, 1871. View towards prominent north wall of the Malakoff pit. Courtesy of the Stanford University Historical Archives.

Figure 9.4 Instrumented hillslopes (slopes 1 and 2) in foreground, 1980.

here. A more detailed account, however, is beyond the scope of this paper and may be found elsewhere (Band 1983, Weirich, in preparation).

Overland flow generation and sediment transport
A number of workers (Zingg 1940, Musgrave 1947, Kirkby 1969, Kilinc & Richardson 1973) have determined that soil transport by overland flow may be adequately described as a power function of form:

$$q_h = K_h q_w^m S^n \qquad (9.1)$$

where q_h is the soil transport rate, k_h is a rate constant, q_w is the water discharge, S is the local declivity and m and n are empirically determined exponents. The strong non-linearity of hydraulic sediment transport dictates that an adequate sampling scheme must operate with near instantaneous discharge measurements in order to capture the major fluctuations of sediment transport corresponding to short-term variations of the surface water discharge.

For this study, specially designed automated samplers were constructed that could simultaneously collect surface water and sediment for short time periods (five minutes or less) at various points on a slope, a number of times through a storm. More detailed descriptions of the design, operation, and difficulties encountered with these samplers are given elsewhere (Band 1983, Weirich, in preparation). During a storm event several of these samplers were operated along with a simple rain gauge, providing measurements of the total rainfall, runoff, and sediment transport for very short duration events. This allowed collection of a large data set describing these processes over a relatively short period of time spent in the field.

The sediment transport equation derived from this data set by regression analysis is:

$$q_h = 0.000025 \, q_w^{2.07} \, S^{0.84} \qquad \qquad (9.1a)$$

where q_h is given as gm/min/cm, q_w as cm³/min/cm and $S = \tan \theta$ where θ is the local declivity in degrees. A high level of prediction is given by Equation 9.1a with these parameters, $r = 0.90$, $r^2 = 0.82$. A plot of observed and predicted values of sediment discharge (Fig. 9.5) shows this to be an acceptable form to describe surface wash transport on these hillslopes.

As discussed above, the slopes are quite short, have negligible contour curvature, and are practically impermeable during storm events. Under these special conditions, a very simple rainfall-runoff relation may describe the runoff generation:

$$q_w = rx \qquad \qquad (9.2)$$

where r is the rainfall rate and x is the horizontal distance to the divide.

Soil creep
Culling (1963, 1965), Kirkby (1967), and others provide theoretical and empirical evidence suggesting that soil creep may be described as:

$$q_c = K_c S \qquad \qquad (9.3)$$

where q_c is mass transport, S is the declivity and K_c the site and soil specific creep coefficient.

Field measurement of soil creep in this study was approached with two

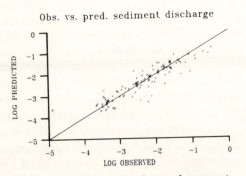

Figure 9.5 Observed and predicted rates of surface wash transport.

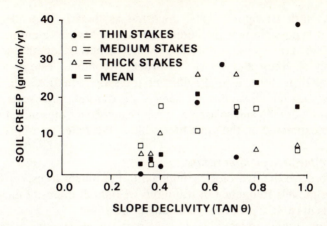

Figure 9.6 Soil creep rates measured by downslope displacement and rotation of stakes of varying diameters.

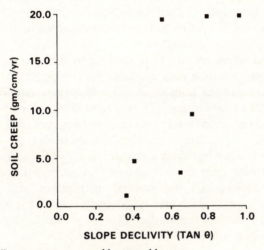

Figure 9.7 Soil creep rates measured by traceable cores.

methods. The first method involved the use of wooden dowels of various widths inserted vertically into the soil to different depths in order to record downslope rotation and displacement through the soil column, and the second method involved the filling of augered cores with a traceable material. Unfortunately, reliable readings could be taken only during the first year due to the site's adjacency to a sight-seeing trail that is heavily used during the summer months.

Despite this drawback, creep rates in the clay-rich substrate were sufficiently rapid to allow a rough estimate of the annual creep rate. Plots of soil creep rates recorded by the two techniques (Figs 9.6 and 9.7) show considerable scatter but suggest an annual creep coefficient between 15–20 gm/cm/yr. A value of K_c is set at 16 gm/cm/yr as a first approximation, although for reasons discussed elsewhere (Band 1983) this may represent an underestimate.

As rainsplash transport is also reported as linearly dependent on the declivity, it is possible that the addition of a quantity representing the time-averaged rate of rainsplash to the creep coefficient may adequately describe the joint effect of creep and splash. In order to combine Equation 9.3 with Equation 9.1a to form a total sediment transport equation, the joint rate coefficient must be adjusted to the same short-duration time scale. As a first approximation, creep and splash are spread evenly over the entire time rain is falling and expressed on the same time scale as the water transport.

Analysis of the precipitation regime

In order to drive the hydrologic component of sediment transport within the simulation model, frequency distributions of rainfall intensity must be provided. This may be approached in a number of ways.

The simplest method would be to construct a single frequency distribution for the entire length of simulation (the past century) and run the model with successive precipitation intensity increments for the amount of time each is estimated to have occurred. Unfortunately this precludes an analysis of the year to year variations in erosion induced by natural fluctuations of annual precipitation amounts and distributions.

The method used allows an analysis of the year to year variations and takes advantage of the precipitation data available for the region. These are: (1) annual precipitation and the number of rainy days (precipitation exceeding 0.01 inch) from 1871 to the present; (2) daily rainfall rates for a portion of the last century; and (3) a limited number of continuously recorded rain charts.

From these data, annual frequency distributions of discrete precipitation intensities are constructed based on a mixture density (Feller 1965, p. 51) of distributions describing daily rainfall amounts for each year, modeled as an exponential distribution, and the internal distribution of 5-min rainfall amounts within a rain day which may be adequately modeled by a truncated Poisson distribution with the randomized Poisson parameter statistically estimated from the daily rainfall.

Boundary and initial conditions

Although the use of hillslopes for which historic photographs exist allows a better estimate of initial conditions, the relatively short amount of time over which the erosional processes have been active leaves the final (present) slope shape not significantly different from the initial state. The total amount of ground lowering at a point on the slope is often not significantly out of the range of the probable error of estimate of the initial conditions. A trade-off is therefore gained in opting for recently developed slopes for although initial conditions may be more easily determined, the precision with which they need to be determined so that meaningful comparisons of computed and observed profiles may be made is dramatically increased. Due to this, the initial conditions set up below, although based on as detailed evidence as is available, must be considered only as approximations to the true initial conditions.

Figure 9.8   Slope profile 2.

By combining and comparing the information drawn from the historic photographs, cores taken from trees growing in and adjacent to drainage lines, the apparent beheading of the major drainage line by a steeper gully (Fig. 9.2), and analogy with presently developing slopes the initial and boundary conditions for slope profiles 2 and 4 (Figs 9.2, 9.8, & 9.9) are constructed. Boundary conditions for slope 2 are set as a slope toe fixed in vertical and horizontal position, and a symmetrical slope divide, fixed in horizontal position and across which there is no flux of sediment or water. Boundary conditions for slope 4 are set as two points fixed in vertical and horizontal position at the bases of two slopes with a common divide. While a no flux condition is also imposed at the divide, it is free to migrate horizontally in accordance with differing rates of erosion on either side of the crest. The divide is interpolated after each time step with the use of a second order polynomial. Initial conditions for both slopes are shown in Figures 9.10 and 9.11.

Development of the numerical model

A numerical simulation model is constructed incorporating the above described hydrologic inputs, sediment transport processes, and slope boundary conditions. We begin by setting up a slope diagram that portrays the major characteristics of a slope profile of interest to the modeling effort (Fig. 9.12).

At the slope toe the model slope is bounded by a stream that removes all

Figure 9.9   Slope profile 4.

sediment delivered from the hillslope and the slope divide. The horizontal dimension, x, is given as the distance from the slope crest, which may be horizontally mobile, while the vertical dimension, h, is given as the height above the stable slope toe. The declivity on the main slope may then be expressed $S = -\partial h / \partial x$.

The net change of elevation at any point on the hillslope is a function of the mass balance as determined by the sediment transport process. A basic continuity equation expressing this relationship may be written:

$$\rho \, \partial h / \partial t = -\partial q_s / \partial x \qquad (9.4)$$

where ρ is the bulk density of the substrate, h is the height above the slope toe, and x is the horizontal distance from the divide. Substituting Equations 9.1 and 9.3 into Equation 9.4, and substituting $S = - h / x$ forms:

$$\rho \; \partial h / \partial t = - \partial / \partial x [K_h q_w^m (- \partial h / \partial x)^n - K_c \partial h / \partial x] \qquad (9.5)$$

Prior to solution, Equation 9.5 is transformed into dimensionless form by introducing the following variables:

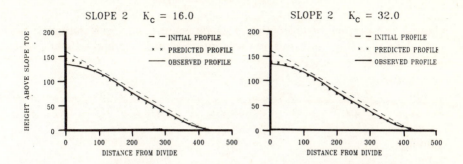

Figure 9.10 Initial, observed and predicted profiles for slope 2. Annual creep coefficients (K_c) are 16 and 32 gm/cm/yr. The fit of observed to predicted slopes improves with the larger K_c which may model the effect of rainsplash.

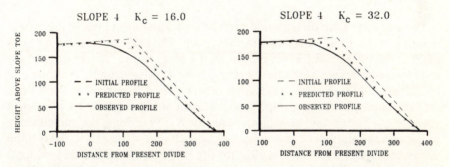

Figure 9.11 Initial, observed and predicted profiles for slope 4. K_c values are 16 and 32 gm/cm/yr. Again the fit is better for the larger value.

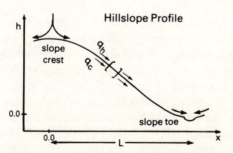

Figure 9.12 Conceptual diagram of hillslope profile system.

$$\zeta = x/L, \ \eta = h/L, \ \tau = t/t_c$$

where L is a characteristic length (i.e. slope length) and t_c is a characteristic time, defined below. Substituting these into Equation 9.5 and recalling that $q_w = rx$ forms:

$$\frac{\partial \eta}{\partial \tau} = \frac{-t_c K_h r^m L^m}{\rho L^2} \ \frac{\partial}{\partial \zeta} \ (\zeta^m(-\frac{\partial \eta}{\partial \zeta})^n - \frac{K_c}{k_h r^m L^m} \frac{\partial \eta}{\partial \zeta}) \qquad (9.6)$$

Defining

$$t_c = \frac{\rho L^2}{k_h r^m L^m} \ \text{ and } \ \beta = \frac{K_c}{k_h r^m L^m}$$

Equation 9.6 is simplified as:

$$\frac{\partial \eta}{\partial \tau} = -\frac{\partial}{\partial \zeta} \ (\zeta^m(-\frac{\partial \eta}{\partial \zeta})^n - \beta\frac{\partial \eta}{\partial \zeta}) \qquad (9.6a)$$

The purpose of this dimensional exercise is to allow an analysis of the system in terms of this simplified, dimensionless form. The major use of the development of the parameter β is the information it gives on the relative rates of the soil creep and surface wash terms by comparing its value to unity. As described by Gilbert (1909), Kirkby (1971), and others, the relative sizes of these rates are crucial to the resulting slope shape. For the simulated hillslopes, taking a value of $L = 500$ cm and an average value for K_c (by dividing the annual value by the average amount of time rain is falling each year, as discussed above), the values of β range between 2×10^{-1} for the least intense (and most frequent) rainfall, and 2×10^{-3} for the most intense (and rarest) rainfall. This indicates that, although the balance between hydraulic erosion and soil creep will vary up and down the slope with fluctuations of rainfall intensity, erosion by overland flow is always significantly larger than soil creep, and will dominate the slope's development. By this reasoning, over a significant period of time the slopes should become increasingly concave, reflecting the dominant transport process.

The definition of the characteristic time, t_c, gives a process time scale based on the dominant transport process, hydraulic erosion. For the studied hillslopes, again taking $L = 500$ cm, t_c calculated for that runoff transporting the most sediment over time (discussed below) is about one thousand years. This represents the time scale needed for the hillslopes to be significantly changed (or shaped) by the dominant transport process (R.L. Shreve, pers. comm. 1980). This bears out the previous observation that the last century has not seen a significant change in shape of the hillslopes when compared to their initial conditions and exemplifies the very long time scales over which even these rapidly eroding geomorphic forms develop and change.

In terms of the magnitude and frequency of overland flow erosion, we investigate the role of r, the variable rainfall rate. An important observation is that the particular value of r is only involved in determining the relative magnitude of overland flow erosion and soil creep, β, and in the overall rate of erosion, t_c, but is not involved in the hydraulic erosion term. This results directly from the simple rainfall-runoff relation described for the hillslopes and does not necessarily occur on hydrologically more complex slopes without uniform, consistent increases of runoff with distance from the divide.

For these and other similar hillslopes, however, this indicates that the pattern of hydraulic erosion does not change with r; each rainfall–runoff has a similar geomorphic effect, varying only in magnitude. Therefore the concept of a dominant runoff, one that dominates the development of hillslope shape, is not meaningful on these slopes as the same pattern of hydraulic erosion is set up by all precipitation–runoff events. This is discussed further with regard to the simulation results.

Equation 9.6a may now be solved, given initial and boundary conditions, and values of the parameters comprising t_c and β, to yield time-dependent hillslope profiles. β may be set as a constant, representing a steady, or dominant runoff event, or as a variable, modeling fluctuations of precipitation and runoff conditions. This latter scheme is chosen so that an analysis of the contributions to the total sediment transport and hillslope development by various magnitudes of precipitation and runoff intensity is possible.

Equation 9.6a is solved by converting to a finite difference form using a Crank-Nicolson central differencing scheme and writing the finite difference equation for a number of fixed points along the slope. This nonlinear set of equations is then iterated to a solution using a Newton–Raphson method (see Gerald 1977, or other texts on numerical computing). The fluctuations of rainfall intensity are incorporated by running the model each year with the different, discrete rainfall rates for the amount of time that rate is calculated to occur. The numerical techniques are written as a FORTRAN program that includes algorithms for the calculation of statistics describing the amount and distribution of ground lowering along the slopes for each precipitation intensity and for each year, and the relative frequency distribution of precipitation intensities over the full length of simulation to provide the data necessary for the magnitude and frequency analysis.

Interpretation of simulation results

Figures 9.10a and 9.11a show observed, predicted, and estimated initial profiles of the two simulated slopes with the annual $K_c = 16$ gm/cm/yr. As previously stated, an increase in the value of K_c may model the effect of rainsplash. Although any such increase in the absence of empirical evidence indicating an appropriate value must be considered speculative, we double the value of K_c used and solve again for the slope profiles (Figs 9.10b and 9.11b).

As expected there is a greater net erosion in the vicinity of the divide convexities and a slight decrease of net erosion in the concavities near the slope toes, leading to an overall improved correspondence of predicted and observed forms.

An evaluation of the model's validity in terms of the correspondence between observed and predicted profiles is difficult to make because the initial conditions are not known with enough precision, considering the relatively small amount of net erosion (less than 20 cm) that has occurred over the past century. As has been noted, despite the dominance of overland flow erosion, pronounced concavities have not yet developed due to the short amount of time over which these slopes have developed relative to the process time scale. The overall patterns of erosion predicted, however, seem consistent with the inferred geomorphic development of the hillslopes, and the model does provide the opportunity to investigate the magnitude and frequency of erosion by overland flow.

In order to assess the individual contributions of each precipitation intensity and creep to net erosion and slope development, slopes 2 and 4 are modeled by running each precipitation intensity and soil creep/rainsplash separately each year. The amount and distribution of net erosion for each is recorded and summed for the full length of simulation. Profiles produced in this manner are virtually identical to those produced with simultaneous creep and wash, showing that the feedback between the two processes is not sensitive to changes in their timing at this scale.

Analysis of the magnitude and frequency over the past century of surficial wash erosion may be accomplished by examining the distribution of the total erosion carried out among the different precipitation intensities. Figure 9.13 shows the relative frequency distribution of precipitation intensities and the proportion of the total wash erosion accomplished by each of the discrete intensity events for slope 2. The distributions for slope 4 are identical because both slopes share common precipitation frequency distributions and sediment transport laws. The mode for the erosion distribution is in the 0.010 cm/min class and the shape of the distribution suggests the actual value corresponding to the maximum ground loss lies between an intensity of 0.010 and 0.015 cm/min. This value may be approximated by polynomial interpolation at 0.011 cm/min. This represents the peak of the magnitude and frequency curve: the event that transports and erodes the most sediment and by Wolman and Miller's definition does the most geomorphic work. This value is in accord with observations in river systems that the most effective (maximum work) event is a fairly common event (Wolman & Miller 1960, Andrews 1980).

We now attempt to locate an event that alone can reproduce the geomorphic effects of the full spectrum of events when the entire runoff takes place at that intensity. This may be termed a dominant event and, as previously discussed, we expect that such a runoff can be found because the pattern of hydraulic erosion apparently does not vary with different runoff intensities. In this case all that is necessary is to find the particular runoff that will reproduce the

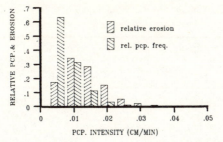

Figure 9.13 Percentage of the total net soil erosion and frequency distribution of precipitation over discrete intensity increments.

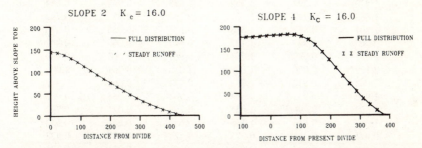

Figure 9.14 Slope profiles 2 and 4 produced by steady runoff and by the full spectrum of runoff magnitudes.

magnitude of erosion generated by the full distribution when the entire runoff takes place at that rate.

This runoff is also found close to 0.011 cm/min, very close to the magnitude of the maximum work event, defined above. This value is now used as the sole precipitation-runoff intensity for the simulation of slopes 2 and 4. An excellent match is gained with the profiles produced with the full range of precipitation intensities for both cases (Figs 9.14a & b) as predicted by the dimensional analysis. These results indicate two important properties regarding the shape and erosional development of these particular slopes:

(1) A steady precipitation intensity that can reproduce the effects of the full range of hydrologic events when run for the total annual precipitation does exist, and

(2) this value is very close to, and may be indistinguishable from, the maximum work event.

The invariance of the relative geomorphic effect through the full spectrum of runoff events is now demonstrated by examining the distribution of net erosion along the hillslope contributed by each discrete intensity. As the total magnitude of erosion varies as discussed above, the distributions are standardized for comparison by dividing the net erosion at each point on the slope by the

Table 9.1 Distribution of hydraulic erosion along slope 2 at different precipitation (and runoff) intensities.

Distance from the divide x/L	Erosion for precip. int. of 0.005 cm/min.		Erosion for precip. int. of 0.020 cm/min.		Erosion for precip. int. of 0.035 cm/min.	
	Net (cm)	Net/max.*	Net (cm)	Net/max.*	Net (cm)	Net/max.*
0.00	0.00	0.00	0.00	0.00	0.00	0.00
0.05	0.141	0.066	0.121	0.067	0.004	0.067
0.10	0.380	0.179	0.324	0.180	0.011	0.179
0.15	0.613	0.289	0.523	0.289	0.019	0.289
0.20	0.830	0.391	0.709	0.392	0.025	0.391
0.25	1.049	0.494	0.895	0.495	0.032	0.494
0.30	1.263	0.594	1.079	0.596	0.038	0.595
0.35	1.473	0.693	1.258	0.695	0.045	0.694
0.40	1.683	0.792	1.437	0.794	0.051	0.793
0.45	1.861	0.877	1.588	0.877	0.056	0.876
0.50	2.000	0.943	1.707	0.943	0.060	0.942
0.55	2.087	0.989	1.781	0.989	0.063	0.989
0.60	2.122*	1.000	1.809*	1.000	0.064*	1.000
0.65	2.063	0.972	1.759	0.972	0.062	0.972
0.70	1.916	0.903	1.632	0.902	0.058	0.902
0.75	1.692	0.798	1.435	0.793	0.051	0.792
0.80	1.355	0.633	1.145	0.639	0.041	0.632
0.85	0.921	0.433	0.773	0.428	0.027	0.428
0.90	0.419	0.197	0.321	0.177	0.011	0.174
0.95	−0.311	−0.147	−0.153	−0.085	−0.006	−0.092
1.00	0.000	0.000	0.000	0.000	0.000	0.000

maximum net erosion (the net erosion at that point experiencing the greatest ground removal).

Comparison of these data for different intensities on slope 2 (Table 9.1) shows that the relative distribution of surface wash erosion is quite similar, as expected. The constraint that gives the identified dominant intensity its significance is that it will reproduce the overall magnitude when run for the full annual precipitation, thus providing the mean magnitude of hydraulic erosion per increment of rainfall. However, unless this constraint can be justified in geomorphic terms, the significance of this particular value may be more statistical than physical. The proximity of this mean magnitude to the erosion mode is dependent on the shape of the magnitude-frequency curve and would not hold as well for more asymmetrical erosion distributions.

It is interesting to speculate whether a similar relation holds for river channels. Andrews (1980) computed sediment transport for alluvial streams for different discrete discharge intervals, along with their frequency of occurrence. Using his data, the mean sediment transport per increment of the annual discharge is computed. The discharge that produces this transport rate is very close to the bank full discharge. Although this analysis should be repeated for other data sets before a general relationship can be described, these results suggest this statistical property may apply to certain river systems as well as to certain hillslopes.

Discussion and summary of results

The numerical simulation model described here represents the first attempt to explicitly construct and apply a simulation model of naturally eroding hillslopes by first determining and representing the significant forms of sediment transport, the boundary and initial conditions, and the nature, magnitude, and frequency of hydrologic events. Application of a theoretical model to a specific field situation forces consideration of a number of factors and details whose significance are often not obvious when sitting at a desk or computer terminal. Without careful determination of these geomorphic factors before the simulation, these quantities are often chosen with no other criteria than that of mathematical convenience, or the generation of output, or results that are attractive for either theoretical or empirical considerations.

At this point the correspondence of the model to the natural slopes is difficult to assess because the total amount of erosion that has occurred over the past century is not significantly larger than the potential errors in the estimation of the initial conditions. It is possible that an accurate test may only be possible in a laboratory model in which initial conditions could be accurately surveyed and boundary conditions carefully controlled or monitored.

The model does allow an analysis of the individual contributions of the discrete rainfall intensities to the total soil loss and the development of hillslope forms by overland flow erosion. This constitutes a unique approach to the magnitude and frequency problem. Use of the model output and consideration of the wash transport equation and frequency distribution of runoff events allows the determination of the 'maximum work' event, and also shows that the pattern of erosion by surface wash differs only in magnitude between the different precipitation intensities.

The significance of the event, identified above, that will reproduce both the magnitude and pattern of the full spectrum of runoff events when run for the full annual runoff, is based on the statistical property of providing the mean magnitude of erosion per increment of rainfall, rather than controlling the pattern of erosion. If the pattern of wash erosion is not changed by variations in runoff intensity, then slope evolution must be determined wholly by the variations in the relative magnitudes of hydraulic, creep, and rainsplash sediment transport.

On actual hillslopes the annual and instantaneous rates of these processes are more interdependent than represented in this model. The most significant of these interactions is probably that between overland flow and rainsplash. Explicit inclusion of this element into the model requires a much more thorough assessment of the problem than the simple declivity-dependent relation used, including the effects of the rainfall intensity and overland flow depth. An accurate incorporation of these interactions would allow a more detailed investigation of the temporal and spatial variability of the relative magnitudes of the slope-forming processes along the hillslope. Dunne (1980) has pointed out that this interaction is crucial to the growth and decay of rills along a hillslope which will bear heavily on both the rate and pattern of hydraulic erosion, as well as on the development of new drainage lines as approached by Smith and Bretherton (1972).

In conclusion, it has been demonstrated that simulation modeling of specific field systems can provide a flexible and powerful tool for discrimination of the relative geomorphic contributions of the hydrologic inputs. Construction of a theoretical framework in which to work also serves to illuminate the significant aspects of a field problem. Model building is a sequential operation, with each stage demonstrating remaining limitations and suggesting improvements to be incorporated. In this case a more detailed representation of the interactions and effects involving rainsplash appears to be the next significant step.

Acknowledgments

The paper represents a portion of a doctoral thesis prepared under the direction of A. R. Orme of the Department of Geography UCLA. Many of the original concepts and techniques used were developed in a seminar on theoretical geomorphology offered by R. L. Shreve, Department of Earth and Space Science, UCLA and in conversation with Professor Shreve. The author assumes full responsibility for the text.

References

Ahnert, F. 1973. COSLOPE2—A comprehensive model program for simulating slope profile development. *Geocom. Programs* **8**, 99–119.

Ahnert, F. 1976. Brief description of a comprehensive three-dimensional process–response model of landform development. *Zeit. Geomorph*. Suppl. **25**, 29–49.

Andrews, E. D. 1980. Effective and bankfull discharges of streams in the Yampa River basin, Colorado and Wyoming. *J. Hydrol.* **46**, 301–10.

Band, L. E. 1983. Measurement and simulation of hillslope development. Ph.D. thesis, Dept. of Geography, University of California, Los Angeles.

Carson, M. A. and M. J. Kirkby 1972. *Hillslope form and process*. Cambridge: Cambridge University Press.

Culling, W. E. H. 1963. Soil creep and the development of hillside slopes. *J. Geol.* **71**, 127–61.

Culling, W. E. H. 1965. Theory of erosion on soil covered slopes, *J. Geol.* **73**, 230–54.

Dunne, T. 1980. Formation and controls of channel networks. *Prog. Physical Geog.* **4**, 211–40.

Feller, W. 1965. *An introduction to probability theory and its applications* vol. 2. New York: John Wiley.

Gerald, C. F. 1977. *Applied numerical analysis* 2nd edn. Reading, Mass.: Addison Wesley.

Gilbert, G. K. 1877. *Report on the geology of the Henry mountains*. Washington: U.S. Geol. Survey.

Gilbert, G. K. 1909. The convexity of hilltops. *J. Geol.* **17**, 344–51.

Hirano, M. 1975. Simulation of developmental process of interfluvial slopes with reference to graded form. *J. Geol.* **83**, 113–23.

Kilinc, M. and E. V. Richardson 1973. *Mechanics of soil erosion from overland flow generated by simulated rainfall.* Colorado St. Univ. Hydrol. Pap. 63.

Kirkby, M. J. 1967. Measurement and theory of soil creep. *J. Geol.* **75**, 359–78.

Kirkby, M. J. 1969. Erosion by water on hillslopes. In *Water, earth and man.* R. J. Chorley (ed.), 229–38. London: Methuen.

Kirkby, M. J. 1971. Hillslope process-response models based on the continuity equation. *Inst. Brit. Geogr. Spec. Pub.* **3**, 15–30.

Musgrave, G. W. 1947. Quantitative evaluation of factors on water erosion—a first approximation. *J. Soil and Water Conserv.* **2**, 133–38.

Pearce, A. J. 1976. Magnitude and frequency of Hortonian overland flow erosion. *J. Geol.* **84**, 65–80.

Scheidegger, A. 1970. *Theoretical geomorphology*, 2nd edn Englewood, NJ: Prentice-Hall.

Schumm, S. A. 1956. The role of creep and rainwash on the retreat of badland slopes. *Am. J. Sci.* **254**, 693–706.

Smith T. R. and F. P. Bretherton 1972. Stability and the conservation of mass in drainage basin evolution. *Water Resources Res.* **8**(6), 1506–29.

Wolman, M.G. and J. P. Miller 1960. Magnitude and frequency of forces in geomorphic processes. *J. Geol.* **68**, 54–74.

Weirich, F. in preparation. *Automated system for rapid sampling of hillslope runoff.*

Yair, A., H. Lavee, R. B. Bryan and E. Adar 1980. Runoff and erosion processes and rates in the Zin Valley badlands, northern Negev, Israel. *Earth Surf. Proc.* **5**, 205–25.

Zingg, A. W. 1940. Degree and length of land slope as it affects soil loss in runoff. *Agric. Engng* **21**, 59–64.

10

A model for the evolution of regolith-mantled slopes

M. J. Kirkby

Introduction

The intention of this paper is to explore the relationships between rock type and slope form, mediated by the soil. It puts forward as simple an explanatory theory as is considered reasonable, and examines its assumptions and geomorphological implications through a series of related models. Knowledge of these relationships is a major gap in our understanding of how landforms evolve, and has received less attention than its importance warrants. Perhaps the most serious attempts have been those by Ahnert (1964) and Huggett (1975) but the former, although an important step, has insufficient basis, while the latter does not extend far enough toward landform development. The present model builds most directly on Kirkby (1976b), but with a more explicitly formulated soil model and with lithological differences downslope. The slope hydrology here is somewhat simplified, and procedures are also incorporated for slope evolution by landslides (Kirkby, in press), with rates also responding to lithology.

The assumptions underlying the present approach are that rock type has two, or perhaps three, major influences that are able to act through its regolith or soil cover. Rock type is thought to influence (a) rates of solutional denudation, (b) geotechnical properties of the soil and (c) rates of percolation to groundwater through the rock mass and its voids network. Climate is seen as a second, independent, control on slope hydrology, which in turn determines the partition between overland and subsurface flow on the hillside. The current slope morphology, acting largely through slope gradients, influences rates of landslide, creep/solifluction and wash transport. With soil depth it also has an influence on slope hydrology, by setting an upper limit on subsurface discharge. The combination of mechanical and solutional denudation controls the evolution of both the slope profile and the soil on it. The interactions considered are summarized in Figure 10.1, and the selection of these links as predominant is a first crucial step in explanation. The second level is to examine some of the interactions shown in greater detail, and examine suitable

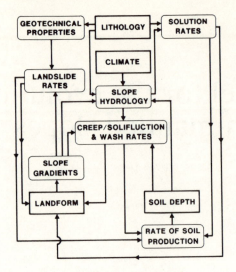

Figure 10.1 Linkages considered in the models discussed.

functional forms for each within an overall model, as well as to explain why some possibly relevant linkages have been omitted. Before doing so, however, the governing mass balance equations should be briefly stated in a consistent notation.

Mass balance for a section of a slope profile may be expressed in the form:

$$-\partial z/\partial t = \partial(S+V)/\partial x \qquad (10.1)$$

where z = elevation above an arbitrary datum measured in rock equivalent
         depths (*i.e.* elevation of a rock column of equal mass),
    t = time elapsed,
    x = horizontal distance from the divide,
    S = total mechanical sediment transport downslope,
and V = total solute transport downslope.

Expressions of this general form have been widely used since at least the 1960s (e.g. Ahnert 1967, Culling 1963). In this formulation it is assumed that the slope profile is of uniform width which is not changing over time, and that there is negligible direct aerial input or output of sediment.

In a similar way a mass balance may be defined for soil, using the concept of 'soil deficit' (Kirkby 1976b). The degree of weathering at depth, y, within the soil may be described by the proportional substance remaining, p. Thus $p = 1$ for bedrock and decreases with weathering. The soil deficit is then defined as

$$w = \int_{y=0}^{\infty}(1-p)\,dy \qquad (10.2)$$

This quantity has the dimension of depth and normally increases monotonically with it, but is more precisely defined where the weathering front is diffuse. The balance for soil deficit is then:

$$\partial w/\partial t = \partial(V - S\pi_s)/\partial x \tag{10.3}$$

where $\pi_s = (1 - p_s)/p_s$ and p_s is the proportional substance near the surface (beneath the A horizon). This equation expresses the balance between addition to the soil by weathering and loss by mechanical denudation, which is assumed to take place entirely at the surface.

Lithology and solute concentration

The first linkage to be considered in detail is that between lithology and solution rate, assuming a known subsurface flow. The rate of solute pick up (i.e. volumetric concentration) should tend asymptotically to a concentration characteristic of rock type downslope from a lithological boundary. In addition the total solute load should be zero at the divide and should not change discontinuously at a lithological contact. It is argued that equilibrium thermodynamic models for solution are more relevant than kinetic models because residence times in the soil matrix are long enough (>100 hours) to approach equilibrium with the current soil solids. It is also argued, partly from exploration of soil profiles, that most solution occurs from the base of the soil within a zone where soil and bedrock compositions differ little. Three alternative formulations were considered for the solute pick up with a zone of constant lithology. They are:

$$dV/dq = c \tag{10.4a}$$

$$V = cq \tag{10.4b}$$

$$dV/dq = (cq - V)/q_* \tag{10.4c}$$

where q is the subsurface flow discharge which is assumed to come into chemical equilibrium with the rock/soil; c is a characteristic concentration for the rock; and q_* is a characteristic discharge.

Equation 10.4a assumes that local rates of pick up depend on local lithology and that the overlying soil can maintain solute load from upslope. Equation 10.4b assumes that the total load reaches instantaneous equilibrium with local lithology, thus giving discontinuities in load at a lithological boundary. Equation 10.4c assumes that some load carrying capacity is lost as the soil from upslope is lost by mechanical erosion, and that a compensating load is picked up from bedrock to approach equilibrium: the constant q_* here is related to the distance downslope for which soil survives before it is lost by erosion. Although in most ways the most conceptually attractive model, it gives for transport near the origin:

$$V = Ux - Ux_*[1 - \exp(-x/x_*)]$$

where $U = c[dq/dx]_{x=0}$ and x_* is the characteristic distance cq_{x*}/U.
  This expression behaves like

$$V \simeq Ux^2/(2x_*)$$

which would give no weathering at divides, in contradiction to observation. Equation 10.4c is therefore rejected and equation 10.4a, which seems to have the least disadvantages, is adopted here.

Soil deficits and soil weathering

In order to solve the mass-balance equation for soil deficit, an additional relationship is required. Even when the geomorphic environment, as defined by the transport rates S and V is known, π_s and w still represent independent soil variables. The simplest approach is to note that the course of soil weathering should produce consistent increases in both soil deficit and the ratio π_s. Explorations of a soil profile model (Kirkby 1976a, 1985) suggest that this relationship is substantially unique for a given parent material and climate, and is substantially unaffected by changes in the rate of mechanical denudation. Thus π_s may be expressed as a function of soil deficit, w. It appears further that for small deficits the function is essentially linear, corresponding to an exponential decay curve of the form:

$$p = 1 - Ae^{-by}$$

for constants A, b. This form is related to the dominance of ionic diffusion for transferring solutes upward from bedrock into the soil. It has also been shown that for large deficits, π_s tends to a constant upper value, π_*. This corresponds to the observed rather constant composition of lateritic and other deeply weathered profiles at $\pi_* = 1$ to 1.5. A simple and suitable functional form for the relationship is:

$$\pi_s = \pi_* w/(w + w_1) \tag{10.5}$$

for constants π_* and w_1 which describe respectively the limiting value and the slope of the linear relationship for small deficits. Appropriate values for w_1 are thought to be about 1–2 m. The form of Equation 10.5 and the set of soil profiles associated with it is shown in Figure 10.2.
  The general course of soil development at a site may readily be seen from Equations 10.3 and 10.5, if it is assumed that denudation rates are constant over time, and that the variation in soil deficit downslope may be neglected. In that case Equation 10.3 reduces to:

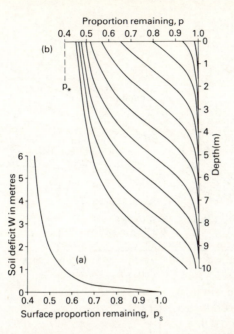

Figure 10.2 (a) The relationship between the degree of surface weathering, p_s and total soil deficit, w given by Equation 10.5. (b) The implicit sequence of soil profile evolution (at approximately equal time intervals). Curves drawn for $p_* = 0.4$; $\pi_* = 1.5$; $w_1 = 1\,\text{m}$.

$$\frac{dw}{dt} = U - T\frac{\pi_* w}{w + w_1} \tag{10.6}$$

where $U = dV/dx$ and $T = dS/dx$.

The relevant solution is then:

$$t = \frac{w}{u - T\pi_*} - \frac{T\pi_* w_1}{(U - T\pi_*)^2}\ln\left[\frac{w(U - T\pi_*)}{w_1 U} + 1\right] \tag{10.7}$$

The soil tends towards an equilibrium if $T\pi_* > U$. At the equilibrium, soil deficit

$$w = \frac{Uw_1}{T\pi_* - U}$$

and the proportional weathering at the surface

$$\pi_s = U/T.$$

If $T\pi_* \leq U$ then equilibrium is never attained, and the soil gets deeper for an indefinite period. This expression is illustrated in dimensionless form in Figure

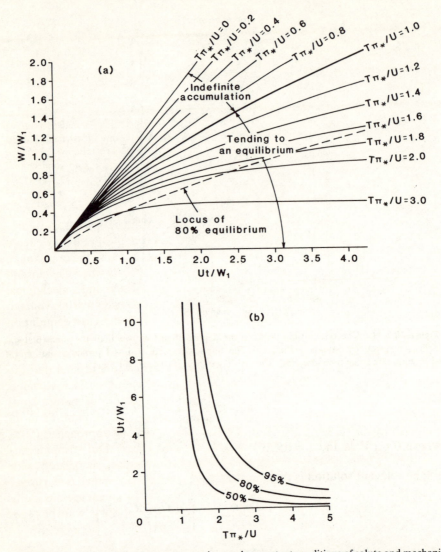

Figure 10.3 Model soil development over time under constant conditions of solute and mechanical denudation. For values of constants assumed see Table 10.1. (a) Soil deficit as a function of time elapsed. Broken line shows locus of points for which soil deficit has reached 80% of its equilibrium value. (b) Times taken to reach different percentages of equilibrium deficit, in terms of relative rates of mechanical and solutional denudation.

10.3. Reasonable values for chemical denudation range from $0.1 - 10\,\mu m/yr$ for igneous rocks and from $0.1 - 100\mu m/year$ for limestone, in both cases increasing with rainfall. The constants in Figure 10.3 are given in Table 10.1.

Figure 10.3a shows the general course of soil deficit evolution. Taking a central value of $U = 5\,\mu m/yr$, it may be seen that negligible divergence occurs until $Ut/w_1 = 0.25$; that is for the first 50 000 years, except under very rapid

Table 10.1 Values of the constants used in Figure 10.3. These values assume that the soil weathering constants are $\pi_* = 1$ and $w_1 = 1$ m.

Igneous rocks		Limestones		
Arid	Humid	Arid	Humid	
0.1	10	0.1	100	U = rate of solutional denudation (μm/yr)
0.1	10	0.1	100	T^1 = minimum rate of mechanical denudation for equilibrium (μm/yr)
10^7	10^5	10^7	10^4	w_1/U = scale multiplier to connect time axis to years

mechanical stripping. Postglacial soil evolution is therefore not expected to show much sensitivity to site factors that control mechanical stripping except perhaps on the steepest slopes ($T>50$ μm/yr). Figure 3b shows the times taken to approach equilibrium, which are considerable, especially when T is close to its critical value. Thus, for the example value of $U = 5$ μm/yr and a mechanical denudation of twice the critical value (10 μm/yr), the soil takes 490 000 years to attain 80% of its equilibrium deficit. In this time the total landscape denudation is $490\,000 \times [(5+10)\times10^{-6}] \approx 7.5$ m. Although appreciable, this erosion loss normally represents a small fraction of the available relief (150 m on average for a denudation of 15 μm/yr (Ahnert 1970)). It is therefore argued that the assumption of constant denudation during soil development is sufficiently met, and that equilibrium soils should occur as a normal feature of mature landscapes. Indeed on landscapes eroding at very rapid rates of 1 mm or more annually, equilibrium soils (admittedly rather thin) will be approached (50% level) in a period of only 1000 years, when only 1 m of total stripping has occurred. This model forecasts a downslope soil catena associated with weathering. For example, if constant solutional loss, U is assumed downslope, then for a normal mature slope down which mechanical loss decreases, a catena of soils deepening downslope will develop. Equilibrium will be first approached near the divide, where denudation is most rapid, and progressively later, if at all, at successive sites downslope.

Soil thickness and mechanical transport rates

Soil thickness is thought to have some influence on mechanical transport rates (Ahnert 1964), and a simple model for the relationship might assume truncation of the shear rate profile at the 'base' of the soil. Thus for an exponential shear profile truncated at depth y_1:

$$dv/dy = -v_0 \exp(-y/y_0)/y_0$$

$$v(y) \quad = v_0[\exp(-y/y_0) - \exp(-y_1/y_0)]$$

$$S = \int_0^{y_1} v(y)dy = v_0 y_0[1-(1+y_1/y_0)\exp(-y_1/y_0)]$$

where v is the velocity of soil movement, y_1 is the soil depth, and y_0 is a depth characteristic of the alteration of velocity with depth for the slope process. This tends to an upper limit, $v_0 y_0$ at large $y_1 y_0$, and behaves like $(v_0 y_0) \times (y_1/y_0)^2/2$ for small y_1/y_0. Other plausible velocity profiles give comparable analyses (Figure 10.4), suggesting that the ratio of actual to maximum transport rate should be treated as an increasing function of the ratio y_1/y_0.

In the present context it seems appropriate to use soil deficit in place of soil depth even though the notion of truncation is less well-defined. Thus the potential transport rate has been reduced by a factor, for which one simple for is:

$$\phi(w) = \frac{w^2}{w^2 + 2w_0^2} \tag{10.8}$$

where w_0 is a soil deficit given by the depth of influence of each slope process. It is expected to take values of 100–500 mm for soil creep or solifluction. For surface wash, soil thickness influences transport rates via grain size. Although the process operates within the surface 10–20 mm, thin soils tend to be coarse grained, and transport rates are therefore considerably reduced below

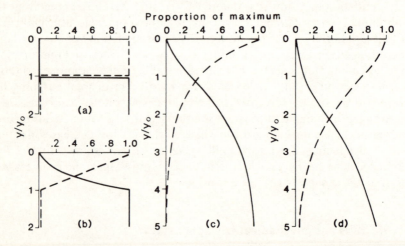

Figure 10.4 The influence of soil truncation on sediment transport rates for four assumed velocity distributions with depth.
(a)   Constant: $v = v_0$ for $y \leqslant y_0$
          $v = 0$ for $y > y_0$
(b)   Constant shear: $v = v_0 (1-y/y_0)$ for $y \leqslant y_0$
          $v = 0$ for $y \geqslant y_0$
(c)   Exponential: $v = v_0 \exp(-y/y_0)$
(d)   Modified exponential: $v = v_0 (1+y/y_0) \exp(-y/y_0)$
Broken curves show velocity distributions. Solid curves show sediment transport for truncation at depth y.

maximum values. The influence of soil thickness will be seen to be strongest when soils are very thin, but in many cases it has little impact.

Hillslope hydrology

For long-term soil and slope evolution, a highly simplified hydrological basis is needed. In this case a constant vegetation cover and climate are assumed. The vegetation cover and the organic soil derived from it are considered to establish a critical storage capacity, h, above which daily rainfall amounts produce Horton overland flow. Annual overland flow produced in this way, summed over the daily rainfall distribution, totals:

$$H_0 = P \exp(-h/r_0) \tag{10.9}$$

where P = annual precipitation and r_0 = mean rain per rain-day.

The remaining rainfall is considered to percolate into the soil and be available for evapotranspiration, up to a climatically determined maximum, E_P. An estimate of actual evapotranspiration is then made as:

$$E_A/E_P = (x^n - x)/(x^n - 1) \tag{10.10}$$

where $x = (P - H_0)/E_P$ and n is a constant $\simeq 4$.

This expression (see Fig. 10.5) follows Langbein's (1949) generalized rainfall-runoff curves for USA.

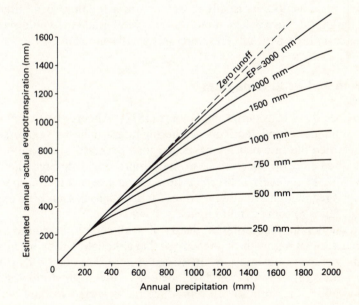

Figure 10.5 Generalized estimates for actual evapotranspiration derived using Equation 10.10. Overland flow has not been subtracted from precipitation to obtain these curves.

The remaining rainfall is next considered to be available for percolation to groundwater, at a rate, Q, which is characteristic of the bedrock. Any surplus remaining is then available for subsurface flow:

$$J_0 = P - H_0 - E_A - Q \qquad (10.11)$$

Subsurface flow is limited by the capacity of the weathered soil to carry it, which may be considered to depend on the soil deficit. For simplicity the total subsurface flow capacity has been taken in the form:

$$q_{max} = Kwi \qquad (10.12)$$

where K is an appropriate hydraulic conductivity and i is the local surface gradient. Where

$$\frac{dq_{max}}{dx} < J_0$$

then not all the available flow can be carried within the soil, and the saturation excess is added to the accumulated overland flow. The procedure described above establishes the partition between overland flow, which is considered capable of producing wash erosion but too rapid to pick up solutes; and subsurface flow and percolation to groundwater, both of which are considered to pick up solutes at concentrations related to lithology. The impact of these hydrological influences is usually most important for thin soils; although a second important effect is seen for areas of very low gradient, where even deep soils can carry little flow and therefore only small solute loads.

Soil depth and weathering rates

It has been widely considered, by Gilbert (1877), Ahnert (1964) and others, that the influence of soil depth on weathering rates can best be expressed by a curve in which the rate first increases and then decreases with depth. In the present discussion the increase is strongly represented through the role of shallow soils in reducing subsurface flow and hence solute transport. The subsequent reduction in weathering rates for deep soils has not been explicitly modeled, but is to some extent implicit.

The reduction in weathering for deep soils is thought to be due to the very slow circulation of waters through them; so that although solute concentrations are high, removal rates are low. It is argued here that deep soils generally require gentle gradients for their development, so that the hydrological effects described above are sufficient to reproduce the expected gradual decline in weathering rates where soils are deep, so that an explicit formulation is not appropriate. At a single site, the proposed hydrological model provides for a linear increase of solutional loss with soil deficit until the soil is thick enough to carry the available component of rainfall, J_0. For thicker soils, solution then remains constant at $U = cJ_0$.

Mechanical transport processes

Following most previous work, soil creep and solifluction are considered here to transport material at a rate proportional to tangent slope gradient. Appropriate constants are thought to be $0.001 \, m^2/yr$ for soil creep and about $0.01 \, m^2/yr$ for solifluction (Carson & Kirkby 1973, Finlayson 1976). The rates obtained are then corrected for soil thickness as described above.

Wash processes have been modeled in a wider variety of ways, although transport rates (S) are most commonly forecast by an expression of the form:

$$S \propto q^m i^n$$

where q = overland flow discharge and m, n are constant exponents, usually in the range 1–2.5.

In many cases distance from the divide is used in place of discharge on the assumption of uniform overland flow production. Here the form adopted is a convenient one within the range indicated, namely:

$$S = 0.02q^2 i \tag{10.13}$$

The constant has been obtained by calibration from erosion data for substantially unvegetated surfaces (Carson & Kirkby, 1973, p. 216) and assumes that q and S are measured in cubic meters per unit width per year. A lower value for the constant is appropriate for vegetated surfaces due to the substantial flow resistance of plant stems. In the absence of a satisfactory model for variation with soil depth, the correction used has provisionally been taken to be the same as for soil creep and solifluction.

Landsliding has been modeled in a rather different way, which is discussed at greater length elsewhere (Kirkby, 1984), and is only briefly summarized here. As a largely 'weathering-limited' process, it seems best to specify denudation rates initially, in terms of lateral slope retreat at gradients above an ultimate threshold. Thus:

$$D = 0 \qquad \text{for} \quad i \leqslant i_*$$

$$D = R(i - i_*) \qquad \text{for} \quad i \geqslant i_* \tag{10.14}$$

where D is the rate of lateral retreat, i_* is the ultimate threshold gradient and R is a rate constant.

In modeling the evolution of coastal cliffs, a single threshold appeared fully

Table 10.2 Empirically derived constants which could be used to predict the lateral retreat of cliffs for the Old Red Sandstone and the London Clay. See Equation 10.14.

	R(m/yr)	i_*
ORS sandstone/shale sequence	0.001	0.40
London Clay	2.5	0.1

adequate and therefore preferable to the use of a series of thresholds. Values obtained from this study on Old Red Sandstone sequences in South Wales, and by reworking Hutchinson's (1967) historical data on London Clay Cliffs gives some appropriate values for the constants. These are shown in Table 10.2. Denudation rates measured vertically may be obtained from D by multiplying by slope gradient.

Because deposition may occur at gradients above the threshold, i_*; and because landslides have appreciable travel distances relative to total slope length, a simple weathering-limited model is not sufficient. Instead an erosion-limited model is proposed (Kirkby 1971, Bennett 1974, Foster & Meyer 1975). Denudation is considered to take place at a rate proportional to the difference between transporting capacity and actual transport rate. The link from the free denudation rate, D, to transport capacity is via the travel distance, h; capacity being equal to their product.

The local vertical denudation rate is then:

$$-dz/dt = Di - S/h \tag{10.15}$$

In this expression travel distance is also thought to increase with gradient, becoming infinite for talus slopes. The form proposed is:

$$h = h_0/(i_0 - i) \quad \text{for} \quad i < i_0$$

$$1/h = 0 \quad \text{for} \quad i \geqslant i_0 \tag{10.16}$$

for constants h_0, i_0 with i_0 usually greater than i_*.

For 35° talus slopes, $i_0 \simeq 0.7$. The use of a constant, h_0, is equivalent to an assumption that material is dropped freely from a height, h_0, that its total impact momentum is transferred downslope without loss, and that it then comes to rest by sliding on a surface whose angle of friction is $\tan^{-1}(i_0)$. Travel distances as modeled then reflect the difference between maximum and residual angles of internal friction and the effect of cohesion in producing macroscopic slides. The derivation above does not represent the processes acting, but nevertheless indicates the magnitude of the potential energy available in a landslide. Realistic values for h_0 are thought to lie in the 5–30 m. range.

Between the two critical gradients, i_* and i_0, the slide process, as modeled,

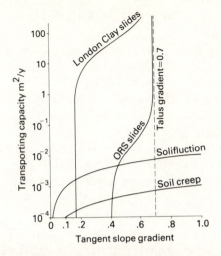

Figure 10.6 Transporting capacity for creep, solifluction and, where defined, for landsliding for London Clay and Old Red Sandstone parent materials. Values used as in text.

has a meaningful transporting capacity, equal to Dh, which has been compared with rates for creep/solifluction in Figure 10.6. It may be seen that there is only a narrow range of gradients within which the two sets of processes are comparable in magnitude. (A wider range may be applicable where wash is significant). Outside this range, one or other process is overwhelmingly dominant. The landslide process as modeled here applies best to shallow slides. For deep-seated slides, debris avalanches and flow-slides its relevance declines sharply.

Slope profiles and soils under landsliding

The influence of Equations 10.14 to 10.16 may be seen most simply by assuming a constant rate of lateral slope retreat, G (which may be $+$ or $-$). Then from Equation 10.15:

$$S = hi(D-G) = G(z_0-z)$$

for some constant z_0.

Substitution in Equations 10.14 and 10.16 then gives:

(a)

$$0<i<i_*<i_0 : \quad i = i_0 \frac{(z_0-z)}{(z_0-z) -h_0} \qquad (10.17)$$

This gives meaningful values for

$$0 > (z_0 - z) > -i_* h_0/(i_0 - i_*)$$

which corresponds to the concave toe of a talus slope, with z_0 at the slope base.

(b) $i_* < i < i_0$: $Rh_0 i^2 - i[Rh_0 i_* + Gh_0 - G(z_0 - z)] - G(z_0 - z)i_0 = 0$ (10.18)

This gives two solutions, only one of them valid.

For slope retreat ($G > 0$) a convex slope is formed, which steepens towards a uniform talus angle downwards. For accumulation slope ($G < 0$), a concave slope develops, straightening upward toward a talus angle.

(c) $i_0 < i$: $i = i_* + G/R$ (10.19)

which is a straight slope at an angle that steepens with the rate of retreat. This solution is illustrated in Figure 10.7.

Figure 10.8 shows a slightly different solution in which the rate of retreat is constant over time, but decreases from a positive maximum at the crest to an equal but negative minumum at the base. The slopes derived thus show a simplified version of slopes undergoing free degradation by landslides. For rapid rates of retreat a cliff is formed with accumulation on a slope close to the talus gradient. At low rates of retreat, profiles are smoothly concave throughout. In all cases, a run-out concavity forms at the slope base at gradients below the threshold angle. The scale of the entire slope is largely determined by choice of the travel distance parameter, h_0. The long-term evolution of an initial cliff may be approximated by arranging these profiles in sequence. Even though the assumptions of constant retreat over time and a change from erosion to deposition at mid height are thereby violated, the more exact numerical solution closely parallels this sequence (with profiles crossing at mid height).

It may be seen, from the formulation of Equations 10.14 to 10.16 above and from Figure 10.1, that soil thickness and/or weathering is not treated as an influence on the rate of sliding, but only as a result of sliding. The argument for making this assumption is that soils are thin, and not necessarily derived from the bedrock on which they are lying (increasingly downslope), so that the crucial geotechnical variables relate to the weathering front zone and so primarily to the bedrock rather than to the soil properties. Nevertheless if the *vertical* denudation rate is assumed to be constant over time, and the solutional denudation loss, U, is taken as fixed, then there is a clear relationship between equilibrium soil thickness, indicated by:

$$\pi_s = (i - p_s)/p_s$$

and the slope gradient. This relationship is shown in Figure 10.9, and is given

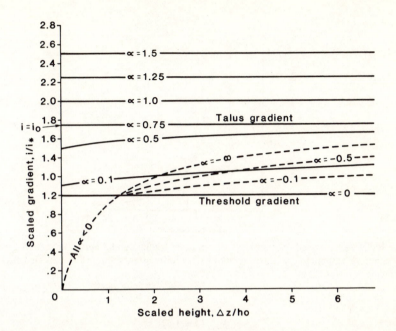

Figure 10.7 Equilibrium gradients for constant rates of retreat (α>0:solid curves) or accumulation (α<0:broken curves). Height difference Δz measured below crest of retreating slopes and above base of accumulation slopes. α = G/(R_* where G is rate of lateral retreat. Other notation as in Equations 10.14 and 10.16.

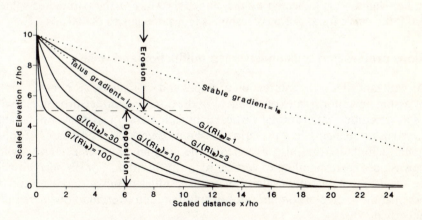

Figure 10.8 Model slope profiles in equilibrium with landsliding, where retreat is at rate G at top of slope, and decreases linearly to $-G$ at base. Curves correspond to retreat for slopes of various gradients. Dotted lines show assumed talus (i_0) and stable (i_*) gradients.

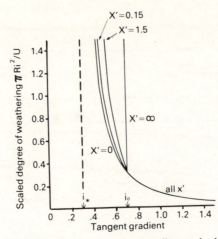

Figure 10.9 The modeled relationship between slope gradient under landsliding and degree of soil weathering, assuming constant downcutting rates and constant solution loss U. Scaled distance from divide, $x' = i_*x/h_0$.

by:

$$\pi_s Ri_*^2/U = 1/B$$

$$i_0 \leqslant i{:}i{-}i_* = [1+\sqrt{(1+4B)}]/2 \qquad (10.20)$$

$$i_* < i \leqslant i_0{:}i/i_* = \alpha + \sqrt{[\alpha^2+(i_*xh_0+1)B]}$$

where $\alpha = (1-i_*xB/h_0)/2$.

For thin soils and gradients at least as steep as talus, the relationship is single valued, but shows a dependence on position, x for lower gradients, due to the influence of material deposited from upslope. In most cases this influence is rather slight, as is indicated by the curves shown. For the parameter values used, the curve for $x' = 1.5$ corresponds to a slope length of 100 m.

Slope profiles and soils under creep solifluction and wash

As with landslides, the interaction of slopes and soils can be seen most simply by assuming uniform rates of change. For the transport limited processes it will be assumed that rates of total vertical denudation are constant over time; and constant or changing in a prescribed fashion downslope. Local rates of solutional denudation are also assumed to be constant over time, and varying downslope, corresponding to vertical rock stratification. For simplicity it will be assumed that the soil can everywhere carry sufficient sub-surface flow and/or percolation to give maximum rates of solution, and overland flow increasing linearly downslope.

At a distance x from the divide:

$$S = (\bar{Y}-\bar{U})x = \beta s[\phi+(x/x_1)^2] \qquad (10.21)$$

where $\bar{Y}$ is the average rate of total denudation from the divide,
$\bar{U}$ is the average rate of solutional denudation from the divide,
β is the rate constant for soil creep/solifluction,
s is slope gradient at x,
ϕ is the correction factor for creep/solifluction in thin soils (cf. Equation 10.8), and
x_1 is the distance from the divide at which wash rates become equal to those for creep/solifluction (cf. Equation 10.13).

Slope gradient is then given by:

$$\beta s = (\bar{Y} - \bar{U})x/[\phi + (x/x_1)^2] \tag{10.22}$$

Differentiating with respect to distance:

$$\frac{x}{s}\frac{ds}{dx} = \frac{x_1^2 - x^2}{x_1^2 + x^2} - \frac{\bar{Y} - Y}{\bar{Y} - \bar{U}} - \frac{x^2 d\phi/dx}{x_1^2 + x^2} + \frac{\bar{U} - U}{\bar{Y} - \bar{U}} \tag{10.23}$$

where Y and U are the local rates of total and solutional denudation respectively [i.e. $Y = d/dx(\bar{Y}x)$].

The four terms on the right-hand side of Equation 10.23 may be described as follows:

(a) The first term shows the normal change in response to the downslope action of slope processes. For creep/solifluction alone, $x_1 \to \infty$ and this term takes the value 1.0, representing a constant convexity. Otherwise, gradient increases for $x < x_1$, and decreases for $x > x_1$, giving a convexo-concave profile. The magnitude of this first term expresses the 'standard' response of the slope for a uniform lithology and constant downcutting, and subsequent terms are superimposed upon it.

(b) The second term (mainly) shows the influence of variations in total denudation rate. For example if, as is normal for mature slopes, the local rate and average decline downslope, then this term is negative, indicating less convexity or greater concavity than under conditions of constant downcutting. The magnitude of this term commonly increases downslope, corresponding to a monotonic reduction in local downcutting rates.

(c) The third term shows that wherever soils thin ($d\phi/dx < 0$) downslope, slopes will steepen relatively. This effect is greatest over the range of soil deficits where $\phi(w)$ responds most strongly: that is when $w < 5w_0$ and especially near $w = w_0$.

(d) The final term shows the effect of lithological differences acting via solution rate. Where the local parent material is less soluble than the average from the divide (i.e. $\bar{U} > U$), slopes will steepen relatively; and where the local rock is more soluble than average, slopes will become relatively gentler. Thus if the standard response to process were a straight slope, hard rock bands would produce convexities and soft rock would produce concavities in the profile.

In summary, therefore, the analysis of Equations 22 and 23 shows that the effect of lithology is to produce relative steepening of the profile where the underlying rock is less soluble than average and/or where soils thin downslope. It will be seen that these effects commonly reinforce one another. However because differences in soil thickness have most influence at sites where the soil is generally thin, such sites show the greatest response of slope gradient to lithology.

The response of the soil to changes in lithology can be indicated roughly by assuming that soil has reached equilibrium. Even though time spans are commonly insufficient to justify this assumption, an equilibrium analysis is able to show the direction of expected differences, while overestimating the magnitudes likely to be observed in practice. Figure 10.10 illustrates the kind of forms to be expected, assuming that soils have reached equilibrium with the constant downcutting rate, and that the gradients obtained are less than landslide thresholds for the materials in question. In the example solifluction/creep is considered as the only mechanical process, so that where the resistant bands are the same as the remainder of the rock ($u'/u = 1.0$), the soil is of constant

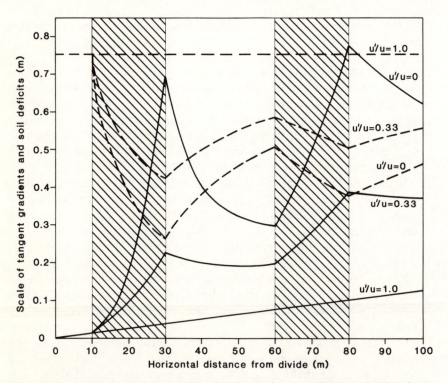

Figure 10.10 Modeled response of gradient (solid curves) and equilibrium soil deficit (broken curves) on a hillside undergoing constant total denudation at 20 μm/year. Rock solution in unshaded zones at $U = 15$ μm/year, and in shaded zones at $U^1 = 0$–15 μm/year. Note the convexity of the less soluble bands, and the weakening response downslope.

depth and slope gradient increases linearly downslope. As the resistant bands are considered to be progressively less soluble, each band shows an exaggeration of the convexity with soil thinning downslope over the hard band, followed by a less marked concavity with thickening soils as gradient and soil recover toward their previous values.

Where soils become thin locally, diversion of subsurface flow overland will tend to reduce local solutional denudation, and still further accentuate the response to lithological differences indicated by Equation 10.23 above. However this diversion will also increase overland flow, and the slopes will tend to respond by evolving towards lower gradients. The combined influence of the flow diversion is therefore not immediately apparent. On steep slopes solifluction etc. are still considered to operate, even though the dominance of more rapid processes will prevent the development of any morphological expression.

An integrated slope evolution model

The various processes described above have all been combined in a digital computer program to simulate some examples of slope profile and soil evolution. The model solves the partial differential Equations 10.1 and 10.3 using an explicit method, and selects the iteration time-step so that gradient and soil thickness change by less than a specified proportion at all points.

The boundary conditions at $x = 0$ is specified as a divide so that $S = 0$ and $V = 0$ at this point. At the basal point the boundary condition is specified by expressions for elevation and soil thickness at a fixed value of x. In the examples used here elevation has been held constant at the basal point, an assumption that normally leads to Davisian decline or 'characteristic form' solutions in simple cases. A corresponding condition for soil is less obvious. Fixed basal elevation is usually supposed to correspond to a basal river removing all debris supplied to it at a base level that is fixed in height. Under such circumstances it is artificial to set soil thickness to a fixed value, since weathering may proceed beneath the river bed. Nor are symmetry conditions of zero transport, like those used for the divide, appropriate at a basal river, although they would be appropriate for the base of an infilling, undrained valley. The condition adopted has been for symmetry with non-zero transport, that is for

$$\mathrm{d}w/\mathrm{d}x = 0$$

at the slope base. This condition implies negligible interference by the basal river on the base of the soil profile, and appears to be at least a neutral choice, appropriate in the absence of other constraints.

Lithology has been represented as a sequence of parallel strata at any specified inclination. Each stratum is assigned a thickness, and process parameters for solute concentration (c in Equation 10.4), landslide threshold and rate of sliding above it (i_* and R respectively in Equation 10.14). All other parameters have been given the same values for all strata. Example runs have been carried out for slopes divided into 40 horizontal increments. This resolution limits the number of distinct strata that can usefully be included to three or four.

All of the linkages shown in Figure 10.1, and discussed above, have been incorporated into the model. Climate is explicitly represented by a set of hydrological parameters that define the initial partition of flow, but it is recognised that climate will also have some influence on a number of process parameters. Perhaps the most important of such implicit effects are those on the solute concentration (via vegetation and soil CO_2 levels); on the landslide rate parameter, R (via solute rates); and on the creep/solifluction rate parameter (via freeze-thaw frequency and vegetation cover). Similarly lithology implicitly influences all soil parameters, which may consequently vary downslope, and through them most process parameters.

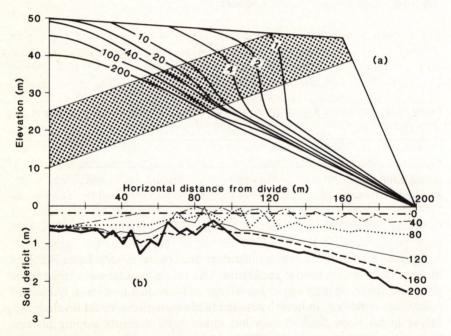

Figure 10.11 Simulation of slope development over 200 years for initial form shown, with a hard band dipping at 10° into slope. Simulation assumes values from Table 10.3. Properties of the strata are:

	C	R	i_*
shaded (hard) band	1×10^{-6}	0.005m/yr	0.25
unshaded (soft) bands	50×10^{-6}	0.5m/yr	0.25

(a) Slope profile evolution. (b) Soil deficit evolution.

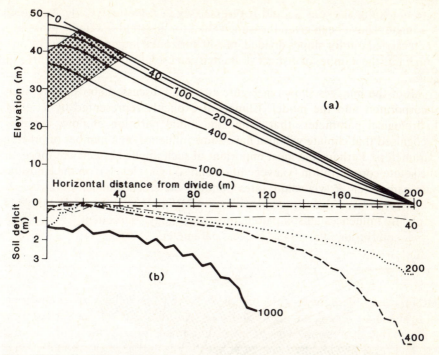

Figure 10.12 Simulation of slope development over 1000 years with hard band dipping at 20°. Simulation assumes values from Table 10.13. Properties of the strata are:

	C	R	i_*
shaded (hard) band	1×10^{-6}	0.01m/yr	0.25
unshaded (soft) bands	50×10^{-6}	0.01m/yr	0.25

(a) Slope profile evolution. (b) Soil deficit evolution.

Figures 10.11 and 10.12 show illustrative runs of the model. Table 10.3 lists the parameter values used in these runs. The values used lie within the ranges proposed above on the basis of knowledge of individual processes, but would need some optimization before detailed field comparisons could be made.

The initial slope form chosen has either been a gently sloping plateau terminating in a steep bluff (Figure 10.11) or a uniform gradient at the landslide-stable gradient i_* (Figure 10.12). In all cases an initial uniform soil deficit of $0.1-0.5$ m has been assumed. A single more resistant layer has been included in the slope strata, dipping at $10°-20°$ into the slope. Each figure shows the evolution of the overall slope topography (a) and of soil deficits (b).

Figure 10.11 shows a hard band which is less soluble and has a lower rate of landslide retreat, R, than the softer strata on either side, but the same ultimate threshold gradient for landsliding ($i_* = 0.25$). Slope development (Fig. 10.11a) shows rapid development of gradients close to the threshold for the softer

Table 10.3 Parameter values used in illustrative numerical simulations of Figures 10.11 and 10.12.

Parameter	Value and units	Equation
creep solifluction rate	$0.01 \, m^2/yr$	–
wash rate	$0.02 \, m^2/yr$	13
landslide travel distance	$h_0 = 20 \, m$	16
talus gradient	$i_0 = 0.7$	16
groundwater percolation	$Q = 150 \, mm/yr$	11
soil permeability parameter	$K = 5000 \, m^2/yr$	12
limiting degree of weathering	$\pi_* = 1.5(@p_5 = 0.4)$	5
rate of change of soil deficit		
with weathering	$w_1 = 2 \, m$	5
scale depth to allow full transport	$w_0 = 0.5 \, m$	8
annual rainfall	$P = 1000 \, mm/yr$	9, 11
number of rain-days	$P/r_0 = 175/yr$	9
potential evapo-transpiration	$E_P = 500 \, mm/yr$	10
soil 'A' horizon storage for		
daily rainfall	$h = 80 \, mm/day$	9

strata, with undercutting of the harder rock, which initially steepens considerably and then very gradually declines, so that after 200 000 years it has little topographic expression. The steep section comes to coincide with the hard strata after a few hundred years (within the resolution of the model). After approximately 20 000 years, a summit convexity begins to replace the threshold slope above the hard band, and this process is substantially complete after 200 000 years. At the end of this period, however, the lower slope, although no longer subject to sliding, still stands at an almost constant gradient (0.22) just below the landslide threshold. Under the assumed climate little wash is evident. Soils initially thicken due to weathering on the plateau, and are stripped by landslides below. Because slides occur locally, the soil distribution is very uneven. The size of slides shown is an artefact of the computational time-step rather than a true representation of individual events, but nevertheless appears to mimic reality. Once slopes begin to stabilize, soil is again able to thicken everywhere. It shows the expected increase downslope and over time, but with a decrease within the less soluble hard band, which is in the direction indicated in Figure 10.10, and becomes more marked over time.

Figure 10.12 shows the effect of solubility differences alone, with an initial form at the limit of landslide stability. The overall slope evolution consists of the development of a convexo-concave slope on which wash eventually predominates over the lower half of the slope, and soils thicken progressively downslope. Within the insoluble band, however, there is local steepening so that landsliding occurs there to a small extent. Beneath the insoluble band a compensating concavity develops before the slope reverts to its 'normal' trend of convexity in the upper half of the profile. Soils thin downslope over the insoluble band, and thicken elsewhere, with the trend to thickening developing from the divide downward. The simulation thus shows the effect of a less soluble band which is illustrated in Figure 10.10. It also shows the tendency to approach soil equilibrium first where downcutting is most rapid (near the divide in this case).

Conclusions

A number of provisional deductions may be made from the model simulations, and from the analyses of individual processes in sections that discussed slope profiles and soils under landsliding and under creep/solifluction and wash. These deductions appear to correspond to features of real landscapes, and so encourage detailed comparisons between model and field data, after due allowance has been made for evidence of slope base and climatic history.

Geotechnical properties, especially the rate of decline towards the threshold R, influence landforms much more strongly than do solutional properties. The latter only begin to play a dominant role on slopes of low gradient and after long times. In common stratigraphic associations, like limestone/shale or sandstone/shale sequences, the rock that is considered to be more resistant appears to be geotechnically stronger but more soluble than the less resistant shales. Gradients close to landslide thresholds commonly outlive landslide activity for many thousands of years, and so play perhaps the dominant role in determining regional relief in a tectonically stable area.

Soils are generally thin under active landsliding and wash. Thick soils thus tend to indicate the predominance of solution and creep/solifluction in landscape denudation. In a humid climate, and where downcutting is least near the slope base and greatest near divides, it is to be expected that soils will eventually thicken downslope. They commonly approach a quasi-equilibrium with current rates of denudation, and approach it sooner at upslope than downslope sites. In this way pronounced soil catenas are developed due to the balance between mechanical processes and solution. Over time, as elevations decline, mechanical denudation falls steadily while solution remains relatively constant, so that the quasi-equilibrium is steadily shifting toward deeper soils over time. Under semi-arid conditions the greater importance of wash produces thin soils except on very low gradients. The catenas developed are likely to be reversed, with deeper soils upslope and shallower soils downslope.

There is no evidence for any significant hangover from previous positions, as erosion changes outcrop location on the slope (except for vertical strata). Within the resolution of the model, gradient steepening and soil thinning downslope over resistant strata is thus strictly associated with the current outcrop location. Small proportions of resistant strata appear to have a disproportionate effect in steepening the slope profile overall. The resistant bands, by maintaining locally steep gradients, tend to hold the less resistant strata close to their threshold gradients, and so increase gradients everywhere.

In conditions where wash is important simulated slope retreat commonly consists of a steep slope at a gradient determined by sliding. Basal replacement (dominated by wash) from an early stage produces a pediment-like feature which is progressively regraded, and has little soil developed on it as long as its evolution remains active (i.e., for example until incision).

Under generally thin soils associated with rapid sliding or wash, the model shows periods during which soil thickness fluctuates appreciably downslope, and travels as a wave over time. Although it is recognized that the scale of these effects in the model results from the dynamic choice of the simulation time step, the sediment waves are thought to provide a helpful analog of episodic landsliding and discontinuous soil erosion. In the case of landslides, soil accumulation builds up local gradients to trigger slides downslope in a cyclic fashion. For wash, thin soil areas produce larger volumes of overland flow, and so localize erosion and produce local deposition downslope. Denudation reduces the local gradient until it is too low to produce erosion and its locus then tends to shift upslope.

The model is thought to reflect reality in that slope and soil evolution is highly sensitive to the exact criteria for overland flow and wash transport. Particular care would be needed in optimizing relevant parameters before making particular field comparisons. Such comparisons have not been attempted here, but clearly represent the next stage in validating this model. The conclusions are not thought to be radical, but to confirm existing views of hillslope development. Such qualitative comparisons provide the initial steps in model validation, and lead towards the more quantitative comparisons.

References

Ahnert, F. 1964. Quantitative models of slope development as a function of waste cover thickness. *Abstr. of Pap. 20th I.G.U. Congress, London*, 188.

Ahnert, F. 1967. The role of the equilibrium concept in the interpretation of landforms of fluvial erosion and deposition. *I.G.U. Slopes Commission Report* 6, 71–84.

Ahnert, F. 1970. Functional relationships between denudation, relief and uplift in large, mid-latitude drainage basins. *Am J. Sci.* **268**, 132–63.

Bennett, J. P. 1974. Concepts of mathematical modelling of sediment yield. *Water Resources Res.* **10**, 485–92.

Carson, M. A. and M. J. Kirkby 1973. *Hillslope form and process*, Cambridge: Cambridge University Press.

Culling, W. E. H. 1963. Soil creep and the development of hillside slopes. *J. Geol.* **71**, 127–62.

Finlayson, B. L. 1976. Measurements of geomorphic processes in a small drainage basin. Ph.D. Thesis, University of Bristol.

Foster, G. R. and L. D. Meyer 1975. Mathematical simulation of upland erosion by fundamental erosion mechanics, In *Present and prospective technology for predicting sediment yields and sources*. U.S. Dept. Ag. Sedimentation Lab. Report ARS-S-40, 190–207. Oxford, Miss.

Gilbert, G. K. 1877. *Report on the geology of the Henry Mountains*. Washington: U.S. Geol. Survey.

Huggett, R. J. 1975. Soil landscape systems: a model of soil genesis. *Geoderma* **13**, 1–22.

Hutchinson, J. N. 1967. The free degradation of London Clay cliffs. *Proc. Geotech. Conf. Oslo* **1**, 113–18.

Kirkby, M. J. 1971. Hillslope process response models based on the continuity equation. *Inst. Brit. Geogr. Spec. Pub.* **3**, 15–30.

Kirkby, M. J. 1976a. Soil development models as a component of slope models. *Earth Surf. Proc.* **2**, 203–30.

Kirkby, M. J. 1976b. Hydrological slope models—the influence of climate. In *Geomorphology and Climate*, E. Derbyshire (ed.), 247–67. Chichester: John Wiley.

Kirkby, M. J. 1984. Landslide models in the context of cliff profiles in S. Wales. *Zeits. für Geomorph.* 28(4), 405.

Kirkby, M. J. 1985. The basis for soil profile modelling in a geomorphic context. J. Soils Sci. 36(1), 97–122.

Langbein, W. B. 1949. Annual runoff in the United States. *U.S. Geol. Surv. Circular* 52.

11

Topologic properties of delta distributary networks

Marie Morisawa

Introduction

Many studies, both qualitative and quantitative, have been made of drainage basin network topology (e.g., Shreve 1966, 1978, Ranalli & Scheidegger 1968, Smart 1969, Woldenberg 1969, and Abrahams 1975). Such networks are catchment patterns where runoff is collected in many channels. These feed into ever fewer and larger main channels and, finally, into one master outlet (Fig. 11.1). Almost no studies have been done on the topology of networks of discharge-disposal systems with multiple bifurcations and rejoining of waterways, such as occur in braided rivers and on distributary patterns of deltas and alluvial fans. The topology of braided channels was studied by Howard *et al.* (1970); followed by Krumbein and Orme (1972). Although many researchers studying the sedimentology of deltaic systems have briefly noted and discussed distributary patterns (Russell 1967, Coleman & Wright 1975, Coleman 1976) the first topologic study of such networks was that of Smart and Moruzzi (1972). Their work was followed much later by Morisawa and Montgomery (1983) who used their techniques of analysis and elaborated upon them.

This paper presents an analysis of the topologic configuration of the distributary patterns of twenty deltas using the approach of Smart and Moruzzi (1972). In addition, a similar analysis is performed on 13 randomly generated distributary nets. A comparison is made between the topology of the natural networks and those of the models.

General characteristics

Most deltas are fan shaped or can be broken up into subareas that are fan shaped (Russell 1967). This inverted 'V shape' has its apex at or near the first distributary fork and the sides are delineated by the landward margins. The angle formed by the binding landward margin will, in this study, be called the exterior angle of the delta (Fig. 11.2). The exterior angle is a subjective

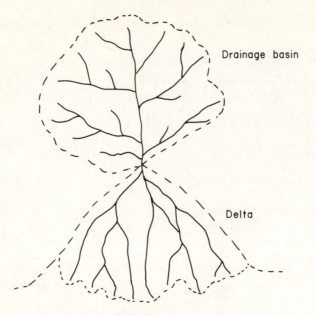

Figure 11.1   Drainage basin and delta networks.

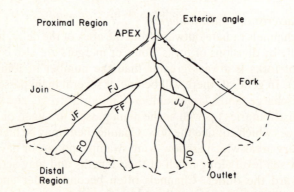

Figure 11.2   Nomenclature used in this paper. Channels bifurcate at forks (*F*), merge at joins (*J*) and enter the sea at outlets (*O*). Links are defined by the vertices which bound them.

measure of the average trend of the irregular delta sides. It may be difficult to determine the exact position of the margin where detailed topographic maps are not available. In this case, the edge was traced in this study by using the outermost distributary channels as markers.

Usually, discharge enters the delta via a single channel at the apex. The water is then carried outward through numerous bifurcations and merging channelways, criss-crossing the delta surface and entering the receiving basin through a number of outlets (Fig. 11.2, Smart and Moruzzi 1972). Coleman

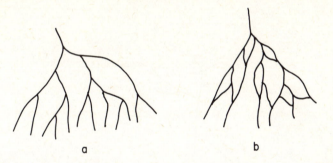

a   b

Figure 11.3   Coleman's classification of distributary nets: (a) bifurcating; (b) rejoining.

(1976) classified distributary nets as bifurcating, rejoining, or single (Fig. 11.3). According to his definition, a bifurcating network is one where channels continuously fork and the number of channels increases exponentially, resulting in a large number of outlets. A rejoining network is one where channels merge after bifurcating. In this case, the number of outlets would not be so numerous as in a bifurcating net. A single channel pattern occurs when discharge from a river reaches the receiving basin primarily through one channelway, and there are none or very few distributaries or outlets. Single channel deltas are not considered in this paper.

Physical characteristics of distributaries
Distributary channels radiate outward from the apex of the delta, increasing in number distally by divisions or forks. Joins or mergers are also common, but the number of forks is always greater than the number of joins (Smart & Moruzzi 1972). In general, there will be one or several main passes that carry most of the discharge, and a large number of minor distributaries with smaller discharges. Distributaries (as well as the main passes) are lined on both banks by natural levees which confine the flow. Channel beds may be higher than the interdistributary areas which are low-lying swamps, ponds, or bays. The beds may lie below sea level along much of their lengths and may actually rise in elevation toward the outlet. Channels often become wider and shallower toward the sea. One or more bars (distributary mouth bars) commonly occur at the outlets.

The growth of the delta is accompanied by formation of new distributaries and abandonment of old ones. Thus the network is dynamic and highly unstable in its configuration. The temporary nature of deltaic networks has been illustrated by many studies. Kolb and Van Lopik (1966) noted hundreds of channel-fill sediments that marked vanished stream courses on the Mississippi delta. Van Andel (1967) found many oxbow lakes and point bar deposits on the Orinoco delta, indicating recent channel changes. Using air photographs, Kanes (1970) documented differences over time in the distributary network on the Colorado River delta, Texas.

Bifurcation may result from a number of causes. Some branching is braiding,

Figure 11.4   A braided river reach.

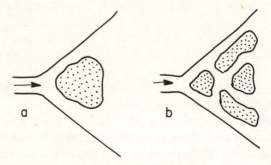

Figure 11.5   Distributary mouth bars. (a) simple; (b) complex.

that is, the division of a single channel into two or more stems by deposition of midchannel bars (Russell 1967) because of a large bedload (Fig. 11.4). It is thought that multiple channels provide more competence to carry the bedload than does a single channel (Church 1972). Branching may also occur during progradation or elongation of a distributary as mouth bars form (Fig. 11.5). Deposition of a bar causes separation of flow around the bar, and each part of the flow may be split again by further deposition. Such bifurcations during growth would result in a geometric increase of bifurcation seaward according to Russell (1967).

Multiple distributaries can also be caused by crevassing. Because the height and width of natural levees decrease distally, the banks can be fairly easily breached seaward by high discharge. Such crevassing is accompanied by creation of many new channels as the water breaks through the levee and splays over the interdistributary area (Fig. 11.6). In some cases the old channel is abandoned but not necessarily so. If the crevassing creates a shorter reach to the sea with a steeper gradient it may become the main pass. However, such an instantaneous increase in branching may be only temporary, as demonstrated by crevassing of West Bay on the Mississippi delta (Coleman 1976). And not only is the topologic structure of the network variable over longer periods of time but also from season to season as river discharge changes. Flooding during the rainy season may cause formation of multiple channels that disappear during lower stages.

Topological characteristics
Smart and Moruzzi (1972) defined a number of parameters characterizing distributary networks. There are three types of vertices: forks, *F*, where a

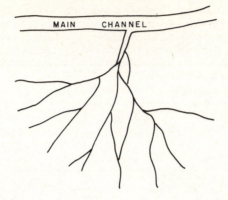

Figure 11.6   A crevasse splay.

channel bifurcates, joins, J, where channels merge, and outlets, O, where distributaries enter the receiving basin. A segment of channel between two vertices is called a link, L. Links are classified on the basis of the vertices at their two ends (Fig 11.2). There are six types of links—FF, FJ, JF, JJ, JO and FO. The ratio of the number of joins to the number of forks was defined as the recombination factor, α, which varies from one to zero in value.

$$\alpha = J/F \tag{11.1}$$

Smart and Moruzzi assumed that all vertices were connected at random and developed a random connection model for link distributions. In this model the fraction of FJ links, for example, is the product of the respective probabilities of finding a fork and a junction.

$$f_{FJ} = \frac{FJ}{L} = \frac{(2F)(2J)}{L^2} = \frac{(2F)(2J)}{(2F+J)^2} = \frac{4\alpha}{(2+\alpha)^2} \tag{11.2}$$

where f_{FJ} is the fraction of FJ links; FJ is the number of FJ links; $L = 2F+J$ is the number of all links; $2F$ and $2J$ are the number of fork links downstream and join links upstream for each fork and join. Using analogous relations, the probabilities of each type of link were determined as functions of α (Table 11.1).

Thus, in the random connection model, link distribution depends on α. If α is zero, there are no joins and the net represents Coleman's true bifurcating pattern (Fig. 11.3). If α is one, the pattern is that of a braided stream where every fork rejoins, i.e., the number of forks and joins are equal and there will be only one outlet. When α equals zero, fractions of FF and FO links are at a maximum (each is 50% of the links) and there are no other kinds of links. As α increases in value, the proportion of joins to forks increases and the fractional

Table 11.1 Probability of links in a randomly connected delta (Smart & Moruzzi 1972, p. 275). The marginal sums are included here. If the marginals are known, knowledge of any two link probabilities allows one to calculate the others by subtraction. To compute the predicted values for an actual network, multiply the probabilities by the number of links.

Upstream vertex	Downstream vertex F	J	O	Totals
F	$2/(2+\alpha)^2$	$4\alpha/(2+\alpha)^2$	$2(1-\alpha)/(2-\alpha)^2$	$2/(2+\alpha)$
J	$\alpha/(2+\alpha)^2$	$2\alpha^2/(2+\alpha)^2$	$\alpha(1-\alpha)/(2+\alpha)^2$	$\alpha/(2+\alpha)$
Totals	$1/(2+\alpha)$	$2\alpha/(2+\alpha)$	$(1-\alpha)/(2+\alpha)$	1

values of FF and FO links decrease while those of the other links (*FJ, JJ, JF, JO*) increase. The fraction of *JO* links rises and then falls as α varies from zero to one simply because for small values of α there are not many junctions, while for large values of α there are not many outlets.

Smart and Moruzzi (1972) found that for the deltas they studied the random connection model seemed to describe the link distribution fairly well, although both the Colville and Niger deltas showed discrepancies in the probability of *FO* and *JO* links between the natural networks and those of the random model. Morisawa and Montgomery (1983) showed that, whereas the distribution of most link types generally conforms to the random connection model, a higher than expected proportion of joins occurred just before the distal edge of the deltas. This means that there are fewer *FO* links and more *JO*s than predicted. They attributed this to energy and space constraints.

The natural deltas

Distributary patterns of twenty deltas were studied (Table 11.2). These include deltas in a wide range of sizes and from a variety of settings representing different climates and different energy regimes, and having watersheds and receiving basins with differing physiographic and geologic characteristics. Networks and some data for five of the deltas were taken from the paper by Smart and Moruzzi (1972) and data for two deltas was obtained from Morisawa and Montgomery (1983). Other networks were traced from topographic and political maps of differing scales (Fig. 11.7). Thus, some networks are probably more complete than others. However, for any one delta and scale it is assumed bifurcations and joins would be omitted in the same ratio as for the network as a whole so that errors in these counts would balance out. Moreover, since recent maps were used, it is possible that configuration of the channel network of some deltas may have been altered by draining, diversion, or dredging. To eliminate these man-made effects, channels cutting straight across at right angles between two other channels were omitted from the net. The high

Table 11.2 Deltas studied.

Delta	Location	Area (km^2)
Colville*	Alaska	1687
Irrawaddy*	Burma	20 571
Yukon*	Alaska	3613
Niger*	West Africa	19 135
Paraña*	South America	5440
Sungai Kajang	Borneo	474
Batang Kajang	Borneo	2256
Song Hong Ha	Vietnam	3486
Hsi Chiang	China	1610
Colorado River, Texas	USA	9.1
Cedar Lake	Manitoba, Canada	24.8
St. Clair	Ontario, Canada	290
Goose Bay	Labrador, Canada	1.55
Grand Pass	Mississippi River, USA	165
Raphael Pass	"	222
Southeast Pass	"	213
Cubits Gap†	"	167
Orinoco†	South America	20 642
Volga	USSR	27 224
Godavari	India	6322

* Networks taken from Smart & Moruzzi (1972).
† Networks from Morisawa & Montgomery (1983).

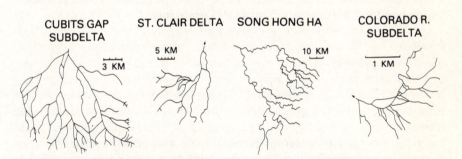

Figure 11.7 Representative deltaic networks.

possibility of human effects on the nets was one reason why such deltas as the Nile, Danube and Rhone were not used in this study.

Recombination factors were calculated. In addition, the ratios of links to outlets, and outlets to vertices were determined, because in analyzing the data it was noticed that the number of outlets tended to remain fairly conservative in range with changing delta size. Data obtained and calculations from these data are given in Table 11.3.

Table 11.3 Topologic properties of natural deltas.

Delta	N_V	N_F	N_J	N_O	N_L
Sungai Kajang	41	20	14	7	54
Batang Kajang	47	23	16	8	62
Song Hong Ha	29	14	5	10	33
Hsi Chiang	35	17	9	9	43
Godavari	59	29	14	16	72
Colorado R., Texas	45	22	11	12	55
Cedar Lake	41	20	7	14	46
St. Clair	31	15	2	14	32
Goose Bay	69	34	13	22	81
Grand Pass	21	10	4	7	24
Raphael Pass	51	25	11	15	61
Southeast Pass	57	28	7	22	63
Cubits Gap	109	54	14	41	122
Orinoco	131	65	43	23	173
Volga	89	44	22	23	110
Colville	107	53	34	20	140
Irrawaddy	71	35	30	6	100
Yukon	135	67	44	24	178
Niger	131	65	51	15	181
Paraña	71	35	18	18	88

Frequency of links

The predominant types of links are those joining two forks, FF, and those with a fork upstream and a join down, *FJ*. The lowest frequency occurs with a link from a join to an outlet, *JO*, (Table 11.4). These result because forks are more numerous in all the study networks than are joins. As the common deltaic form is that of a fan, there is a tendency for more bifurcations to take place distally as the 'V' widens and gives more space. Bifurcations are more common just above outlets also, because of the deposition of distributary mouth bars (Fig. 11.5) which often results in numerous *FO* links. An inordinate number of joins occurred before outlets in four deltas: the Niger, Colville, Grand Pass, and Godavari.

Of the several parameters suggested by Smart and Moruzzi (1972), the recombination factor, α, was chosen for study here since it was thought to be more likely to express differences in the pattern of the distributaries. Thus the behavior of each type of link frequency was examined as α increased (Fig. 11.8). As expected, the number of *FF* and *FO* links decrease, and the number of *JJ, JF,* and *FJ* links increase as α increases. As the relative number of joins increase, one would naturally expect more *J* involved links. However, there seems to be no relation of the frequency of *JO* links to α. It is interesting to note that the proportion of *FF* links decreases quite rapidly as α increases. The number of *JF* links rises more quickly than does the number of *FJ* links with increasing α. This indicates a tendency for channels to divide again after merging.

Table 11.4 Link frequencies.

Delta	FF		FJ		JF		JJ		JO		FO	
	Obs.	Calc.	Obs.	Calc.	Obs.	Calc.	Obs.	Calc.	Obs.	Calc.	Obs.	Calc.
Sungai Kajang	0.28	0.27	0.37	0.38	0.07	0.10	0.15	0.13	0.04	0.03	0.09	0.08
Batang Kajang	0.24	0.28	0.42	0.38	0.11	0.09	0.10	0.13	0.05	0.03	0.08	0.08
Song Hong Ha	0.30	0.36	0.30	0.26	0.09	0.06	0	0.04	0.06	0.04	0.24	0.23
Hsi Chiang	0.27	0.31	0.34	0.33	0.09	0.08	0.07	0.09	0.05	0.04	0.16	0.15
Godavari	0.33	0.32	0.35	0.31	0.06	0.08	0.04	0.08	0.10	0.04	0.12	0.17
Colo. R., Texas	0.34	0.32	0.31	0.32	0.06	0.08	0.09	0.08	0.07	0.04	0.12	0.16
Cedar Lake	0.35	0.36	0.28	0.25	0.04	0.06	0.02	0.04	0.06	0.06	0.24	0.23
St. Clair	0.41	0.44	0.12	0.12	0.03	0.03	0	0.08	0.03	0.02	0.41	0.38
Goose Bay	0.33	0.35	0.28	0.27	0.07	0.07	0.04	0.05	0.03	0.04	0.22	0.22
Grand Pass	0.29	0.35	0.29	0.28	0.08	0.07	0	0.06	0.08	0.04	0.21	0.22
Raphael Pass	0.33	0.34	0.29	0.30	0.06	0.07	0.06	0.06	0.08	0.04	0.20	0.19
SE Pass	0.37	0.39	0.22	0.21	0.05	0.05	0.01	0.03	0.05	0.04	0.28	0.28
Cubits Gap	0.38	0.39	0.18	0.19	0.04	0.05	0.06	0.03	0.02	0.04	0.32	0.30
Orinoco	0.24	0.28	0.42	0.37	0.11	0.10	0.10	0.12	0.05	0.04	0.17	0.16
Volga	0.29	0.32	0.34	0.32	0.10	0.08	0.06	0.08	0.04	0.04	0.08	0.10
Colville	0.26	0.29	0.43	0.37	0.11	0.10	0.05	0.12	0.04	0.03	0.06	0.10
Irrawaddy	0.25	0.25	0.43	0.42	0.09	0.09	0.19	0.18	0.04	0.02	0.02	0.03
Yukon	0.26	0.28	0.39	0.37	0.11	0.09	0.11	0.12	0.03	0.03	0.02	0.10
Niger	0.27	0.26	0.43	0.40	0.08	0.10	0.13	0.16	0.07	0.02	0.02	0.06
Paraña	0.31	0.32	0.32	0.33	0.08	0.08	0.09	0.08	0.03	0.04	0.17	0.15

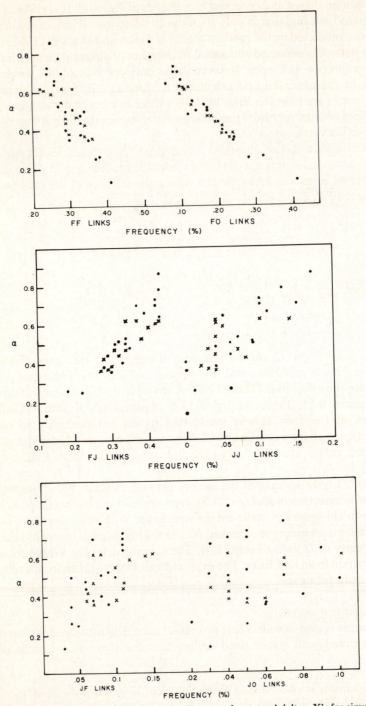

Figure 11.8 Plots of link frequencies and α. Dots are for natural deltas, X's for simulations.

When predicted and observed frequencies of various links for each delta are compared, we find that in only 10 of the 20 deltas are all link frequencies close to those predicted by the random model of Smart and Moruzzi (Table 11.4). In three networks expected and actual frequencies of different links differ by 0.04 or more in two link types. In three deltas only one type of link varies 0.04 or more in the observed and calculated frequencies. In four deltas predicted frequencies are found to differ by 0.04 or more in more than two types of links. All types of links vary from predictions, but those for the JF link vary by 0.04 or more in only one case.

A chi square test[1] of observed versus expected frequencies for the six link types can reveal significant deviations from the random model. If we count the number of joins and forks we can derive the number of outlets (Smart and Moruzzi 1972, 271–2).

$$O = F - J + 1 \qquad\qquad (11.3a)$$

because

$$V = F + J + O \qquad\qquad (11.3b)$$

and

$$V = 2F + 1 \qquad\qquad (11.3c)$$

where V, F, J and O are the numbers of vertices, forks, joins and outlets. Knowing these quantities and the numbers of only two link types, one can calculate the other four (Table 11.5). A chi-square table for each delta can be constructed from Tables 11.1 and 11.5. Again, such a table has only two degrees of freedom. It was found that it was not necessary to combine categories as all the networks which have a significant chi-square statistic have expected values above 1.0 for each link type (Snedecor & Cochran 1967, p. 235).

Table 11.6 shows that of the twenty networks tested, five differ from the random connection model ($p < 0.05$). In every case the most important contribution to chi square is made by the join links. With the exception of Cubits Gap, there is an excess of observed JO links. This excess is compensated for by a deficiency of JJ links (Table 11.4). The Colville delta has a large chi-square contribution from fork links. The expected values were taken from a table with four significant figures.

Variations with delta size
Each factor varied so widely that no regular change in measured characteristics can be noted with larger sized deltas. Because there was a great deal of

---

[1] In his capacity as editor, Michael Woldenberg added the section on chi-square. He modified Table 11.1 and added Tables 11.5 and 11.6. He is responsible for any errors relating to the presentation and discussion of the chi-square test.

Table 11.5 Constraints on the number of observed links (after Smart & Moruzzi 1972, p. 274). The marginal sums are expressed in terms of vertices. Note that the number of links beginning with a fork is given by 2F; the number beginning with a join is given by J. The number of links ending with a fork is given by F-1, the number ending with a join is given by $2J$ and the number ending with an outlet is given by O. Since $O = F-J+1$, the grand sum of the columns equals that of the rows, i.e., $2F+J$. Assuming the marginal sums are known, knowing the number of any two link types makes it possible to calculate the other four. Thus the matrix has two degrees of freedom.

Upstream vertex	Downstream vertex F	J	O	Totals
F	FF	FJ	FO	2F
J	JF	JJ	JO	J
Totals	F−1	2J	O	2F+J

Table 11.6 Contributions to χ^2 summed for JJ, JF and JO links and for FF, FJ and FO links. There are only 2 degrees of freedom (Tables 11.1 and 11.5; Smart and Moruzzi 1972, p. 271–4). * indicates significance at the 0.05 level, ** at the 0.025 level and for *** $p<0.005$.

Delta	$\Sigma\chi^2$ of join links	$\Sigma\chi^2$ of fork links	Total χ^2
Godavari	7.29	1.07	8.36**
Cubits Gap	7.58	.75	8.33**
Orinoco	3.68	2.43	6.11*
Colville	11.82	10.67	22.49***
Niger	17.83	5.51	23.34***
all others (15)			not sign.

variation of each parameter with area, trends were determined by separating the deltas into size groups: those with an area less than 100 km^2; $100–1000 \text{ km}^2$; $1000–10\,000 \text{ km}^2$; and larger than $10\,000 \text{ km}^2$. Figure 11.9 shows that despite the great range of any one parameter within a size group, there is a general tendency for the average number of vertices, forks, joins, and links to increase with larger sized deltas. However, the average number of outlets tends to remain constant even though the delta area is greater. Hence, the L/O ratio increases and the O/V ratio decreases with deltas of increasing size. The number of links per unit area decreases with larger deltas.

Although there is no regular change with size, the overall trend and the average α of each areal group increases with the larger area (Fig. 11.10). If α increases with increased delta area, the ratio of joins to forks increases. Thus, the number of joins increases faster than the number of forks with larger sized deltas. This is not what one would expect. One would expect that lifting of space constraints would result in comparatively fewer merges as surface area increases and distributaries are allowed to spread out.

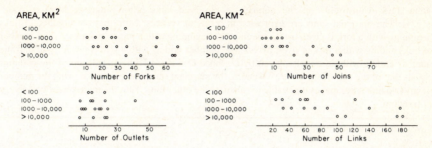

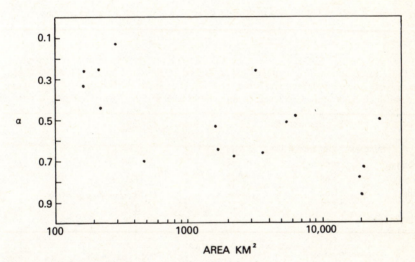

Figure 11.9 Plots of topologic parameters against area, showing overlap of points from group to group.

Figure 11.10 Graph showing increase of α with delta size.

Variations within a delta

In their study Smart and Moruzzi (1972) proposed that the α value became quickly established in the distributary network and remained constant thereafter. This suggested that the ratio of joins to forks tends to be constant over the delta surface. Morisawa and Montgomery (1983) tested this hypothesis on the deltas they studied. Concentric arcs were drawn at 10% intervals of the distance from the apex of the delta to the farthest distributary mouth (Fig. 11.11). α was determined for successive arcs by accumulating the number of joins and forks moving outward from the apex. A graph was then drawn showing the α value with cumulative distance (in percent) from the apex (Fig. 11.12).

The present data, combined with those reported by Morisawa and Montgomery, indicate that α is variable from apex to distal edge of the delta, but not regularly so. The α value does rise distally and is greatest at the outlets, but not

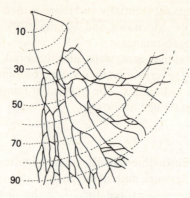

Figure 11.11   Illustration of method of drawing equal concentric arcs from the delta apex.

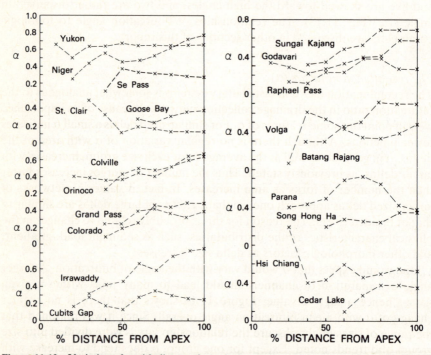

Figure 11.12   Variation of α with distance from the apex on natural deltas.

in all cases. Only in the Yukon delta does α 'quickly' assume a constant value. And only in nine cases does it become fairly constant at 80% of the distance from apex to mouths, i.e. far down-delta. Usually α does not even become a real value until about 20–30% of the distance toward the outlets is traversed, because a number of forks must form before joins can occur. Moreover, quite often there is a decrease in α after an increase in its value. This occurs when the number of joins jumps substantially and there is then a reaction to produce a large number of forks. Morisawa and Montgomery (1983) called this alternation in the α value, resonance.

Links within successive arcs were also counted. The number of links increases outward for several arcs, then varies, becoming lower then higher as forks and joins vary in number. The maximum number of links is generally found toward the distal edge of the delta.

Angle of bifurcation

Because it seemed reasonable that the angle of bifurcation should affect the configuration of the network and influence the value of topologic parameters, bifurcation angles were measured for 14 of the distributary patterns. Frequency distributions of the angles are normal in only five of the 14 natural deltas analyzed. Two distributions are strongly skewed toward the low angles and five are skewed toward the high angles, and two are rather dispersed in angle size (Fig. 11.13). The relationships of bifurcation angle to topologic parameters are discussed in other sections of this report.

The recombination factor (α)

The recombination factor in distributary systems is somewhat analogous to the bifurcation ratio in the drainage collection network. Because of its importance we will point out some features of α. For similar sized deltas a small α indicates fewer junctions. Although there is no regular variation of α with area of the deltas, α in general, as well as the average α for each size group, increases with larger deltas, as previously stated. Thus the number of merges increases faster than the number of forks as area increases. In fact, α values of networks of larger sized deltas approach one, so that patterns of large deltas are similar to that of braided rivers. The reason for this is probably related to changes in the physical characteristics of the distributaries, such as sediment load, gradient, and other morphologic factors as delta size increases.

One would expect that α would vary with the angle of bifurcation. A larger angle of separation of channels should lead to many joins within a given space, hence a high α value. Figure 11.14 shows that there is no regular change of α with mean bifurcation angle overall. Since it was reasoned that the range of sizes might obscure the relationship, areas were divided into size groups and trends noted. Except for one group, α does tend to increase with bifurcation angle within size groups, but at differing rates. Although this observation agrees with a priori reasoning, its validity needs to be verified by further research. The results are also obscured by the range of values of bifurcation angles in each net and the validity of a mean under such conditions.

The exterior angle, since it determines the width of the delta, might affect the value of α. A wide exterior angle should lead to more forks and fewer merges since distributaries could spread out over a wide surface. Again, the wide range

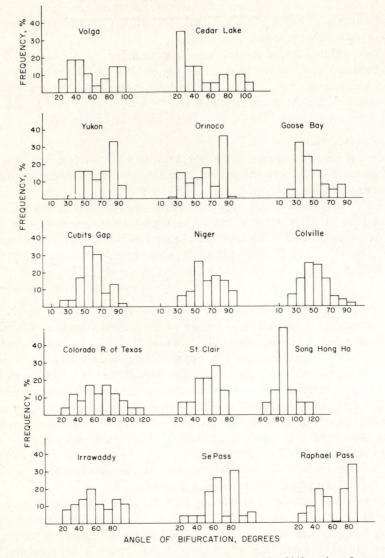

Figure 11.13 Histograms of frequency distributions of the angle of bifurcation of some deltaic networks.

of delta sizes may obscure relationships, but division into size groups does not aid in pointing out trends (Fig. 11.15). Although α does increase in value with increasing external angle in two size groups, the trend is indeterminate in the other size groups. One concludes that neither the external angle nor bifurcation angle are very influential in determining the value of α and that there are other factors that have more of an effect on α than either of these. For example, discharge and sediment load which are not considered here must be as

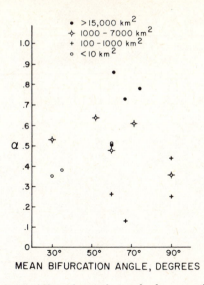

Figure 11.14   Relation of mean bifurcation angle to α for four area classes.

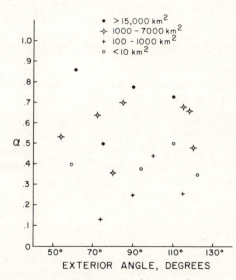

Figure 11.15   Relation of external angle to α for four area classes.

important in distributaries as they are in determining river channel character-istics.

Finally we would point out that digitate deltas have a low α value, i.e. few joins and many forks. This seems to be related to the observation that digitate delta networks are often formed by crevassing, e.g. subdeltas of the Mississippi River. In general, crevassing produces networks with many forks and few joins (see Fig. 11.6).

Table 11.7 Exterior and bifurcating angles of model nets.

Model	Exterior angle, degrees	Bifurcation angle, degrees	Real delta analog
I	74	67	St Clair
II	54	30	Hsi Chiang
III	80	90	Song Hong Ha
IV	75	60	Volga
V	110	60	Colorado
VI	100	90	Raphael Pass
VII	90	90	SE Pass
VIII	90	30	Goose Bay
IX	120	60	Godavari
X	60	60	Irrawaddy
XI	115	60	Cubits Gap
XII	118	70	Yukon
XIII	110	67	Orinoco

Summary of topology of natural deltas

To summarize the topology of natural distributary networks in the twenty deltas studied, forks are more numerous than joins and the *FF* link is therefore predominant; *FO* links are much more common than *JO* links, although in many nets there are more *JO* links than might be expected; and the random network model for predicting frequency of link types does not closely fit the actual frequencies. The random model can be rejected at the .05 level in five of twenty cases.

There seems to be no regular change in topologic parameters with size of delta, although average trends can be noted. The recombination factor, α, varies irregularly (resonates) with distance from the apex and generally reaches its highest value distally. The angle of bifurcation may have an effect on α, but the sample is too small and scatter too large to be certain.

The models

To permit statistical analysis of random distributary networks and to enable comparison with natural deltaic distributary patterns, 13 random walk models were constructed. The rules laid down for creating the models are as follows:

(1) An exterior angle and a bifurcation angle equal to that of one of the natural deltas studied are selected. The exterior angle represents the edge of the shoreline bounding the model delta. The bifurcation angle chosen was that which represents the mean angle of bifurcation of the natural delta with the chosen exterior angle (Table 11.7).

(2) At the apex of the exterior angle it is decided whether the feeder stream continues straight or bifurcates by using a random number table; the

probability of a continuation or a fork is decided by specifying the numbers between 0 and 9 that indicate a continuation and those that indicate a bifurcation.

(3)   If the decision is to extend the channel without a fork, the stream continues for a fixed distance which is arbitrarily set and used through the modeling. If the decision is to bifurcate, both branches are extended this unit distance with the chosen bifurcation angle between the branches.

(4)   After the first fork, the network is continued in the same manner with links either being extended straight or bifurcating with each decision. The distributaries are played in order from left to right each time. All models are extended the same distance, i.e. with the same number of plays.

(5)   Any distributary that intersects another joins it. If one distributary intersects another upstream from the second's end point, the first joins the second which is unaffected in its path. If two distributaries meet at their end points, they join and form a single channel which is on the next play either extended one unit distance along a line bisecting the angle between the junction or which bifurcates into two channels each a unit distance at the given bifurcation angle.

(6)   No distributary can move toward the apex or enter another against the direction of flow. Any link that does so is erased.

(7)   If a distributary encounters the extension of the bounding exterior angle, it turns to flow along it. When it bifurcates, one branch is lost and the other leaves the edge moving inward at half the bifurcation angle.

Figure 11.6 illustrates several of the model deltas. All thirteen models were analyzed in the same way as were the actual delta networks. Results are given in Tables 11.8 and 11.9

Frequency of links
As with the natural deltas, the highest frequency of links occurs with the *FF* and *FJ* types. In contrast to natural deltas, in the models the lowest fraction of links appears to be the *JJ* type rather than the *JO* outlets which have the second

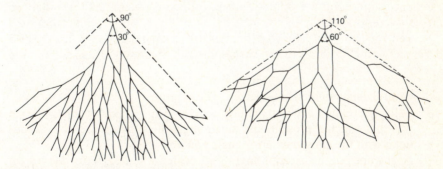

Figure 11.16   Some illustrative model delta networks.

Table 11.8 Topologic properties of model deltaic networks.

Model	N_V	N_F	N_J	N_O	N_L
I	57	28	17	12	73
II	65	32	21	12	85
III	65	32	19	14	83
IV	67	33	14	20	80
V	95	47	25	23	123
VI	85	42	26	17	110
VII	43	21	8	14	50
VIII	117	58	27	32	143
IX	69	34	13	22	81
X	91	45	29	17	119
XI	77	38	18	21	94
XII	81	40	25	16	105
XIII	55	27	12	16	66

lowest frequency. The *FF* and *FO* links decrease in number as α increases (Fig. 11.8). The relationship between the frequency of *FO* links and α is close, with the scatter being much less than it is for natural deltaic networks. The frequencies of *FJ*, *JJ*, and *JF* types increase with increasing α values. Unlike the natural networks, model nets tend to increase the frequency of *JO* links with α. Finally, both natural and model networks produce more *JO* links than expected.

A comparison between the observed and predicted numbers of links of various types may be made by examining Table 11.10. A chi-square test,[1] using a frequency table with four significant figures, indicates that four of the thirteen simulated networks have significant deviations from random expectation (Table 11.10). Most of the contribution to chi square again comes from the join links, and in particular from *JJ* and *JO* links. The proportion of the networks whose six link types do not correspond to the random model appears to be slightly, but not significantly greater for model networks than for the real delta networks—four of thirteen versus five of twenty.

The pattern of correspondence (and deviance) of actual and predicted frequencies for real and simulated delta networks indicates that the simulations are fairly representative of the real deltas in terms of agreement between observed and expected values for link types for each delta as a whole. They do differ, however, in terms of the variations with increasing distance from the apex.

Variations in alpha with increasing distance from the apex
The model networks were divided into arcs at equal intervals of 10% of the distance from apex to outer edge of the net and forks, joins, and links were

[1] Michael Woldenberg added this section on the chi-square test and Table 11.10. He is responsible for any errors.

Table 11.9 Link frequencies.

Models	FF		FJ		JF		JJ		JO		FO	
	Obs.	Calc.	Obs.	Calc.	Obs.	Calc.	Obs.	Calc.	Obs.	Calc.	Obs.	Calc.
I	0.23	0.29	0.42	0.36	0.14	0.09	0.04	0.11	0.06	0.04	0.11	0.12
II	0.34	0.35	0.29	0.26	0.06	0.07	0.03	0.05	0.06	0.04	0.21	0.22
III	0.25	0.30	0.40	0.35	0.12	0.09	0.05	0.04	0.07	0.04	0.10	0.12
IV	0.34	0.34	0.28	0.29	0.06	0.07	0.08	0.06	0.04	0.04	0.21	0.19
V	0.28	0.31	0.38	0.34	0.10	0.08	0.04	0.09	0.07	0.04	0.12	0.14
VI	0.22	0.29	0.43	0.36	0.15	0.09	0.04	0.11	0.04	0.03	0.11	0.11
VII	0.30	0.35	0.30	0.27	0.10	0.07	0.02	0.05	0.04	0.04	0.24	0.22
VIII	0.29	0.33	0.34	0.31	0.11	0.08	0.03	0.05	0.05	0.04	0.18	0.18
IX	0.36	0.35	0.34	0.31	0.07	0.07	0.04	0.07	0.04	0.04	0.23	0.23
X	0.31	0.29	0.27	0.26	0.07	0.08	0.14	0.05	0.04	0.04	0.18	0.18
XI	0.32	0.33	0.31	0.35	0.07	0.08	0.07	0.10	0.04	0.04	0.11	0.11
XII	0.28	0.29	0.37	0.36	0.08	0.09	0.10	0.11	0.04	0.03	0.10	0.17
XIII	0.29	0.33	0.32	0.29	0.11	0.07	0.04	0.07	0.03	0.04	0.21	0.19

Table 11.10 Contributions to χ^2 summed for *JJ, JF* and *JO* links and for *FF, FJ* and *FO* links. The observed data are from simulation models. There are 6 categories but only 2 degrees of freedom (Tables 11.1 and 11.5. Smart & Moruzzi 1972, p. 271–4). * indicates significance at the 0.05 level, ** at the 0.025 level and for *** p.<0.005.

Model	Analog	$\Sigma\chi^2$ of join links	$\Sigma\chi^2$ of fork links	Total χ^2
I	St. Clair	5.72	1.90	7.62**
V	Colorado	6.19	2.43	8.62**
VI	Raphael Pass	8.05	3.42	11.47***
VIII	Goose Bay	6.01	1.19	7.20*
	all others (9)			not sign.

counted within each arc. In the models α generally increases with distance from the apex, reaching a maximum distally and often remaining constant after reaching the maximum value (Fig. 11.17). Thus, the models support the statement of Smart and Moruzzi (1972) that α reaches a constant value on the delta surface even though it does so part of the way down the delta. That the maximum constant value of α is not reached until toward the distal area results from the fact that joins do not occur until midway down the delta surface. In other words, a critical space filling has to occur before distributaries merge. Resonance is not observable in most of the model networks. This is a crucial difference between the models and the real deltas. In general, the number of links in each successive arc increases toward the distal edge and then decreases at the final outlet arc.

Variations with angles
Contrary to what might be expected since the only constraints are spatial, the models show no relation of either bifurcation angle or exterior angle to any of the parameters measured or counted. However, if one averages counts for each bifurcation angle, then the number of vertices, forks, joins, and outlets all decrease with increasing bifurcation angle. A possible explanation for this is that as space is filled with forks of smaller bifurcation angle, there is a higher chance of one link encountering another.

Again, although individual delta networks are quite variable, if one averages ratios in networks of the same bifurcation angle, the number of outlets compared to number of vertices is approximately 25%; the number of outlets compared to forks is approximately 50%; and the number of outlets in relation to number of links is about 20%.

Increase of vertices and links with alpha
Although the scatter is great (Fig. 11.18) there is a slight tendency for the number of vertices, forks, joins and links to increase with increasing values of α. There is little variation in number of outlets whatever the value of α. The relationship of the ratios *O/F, O/V*, and *O/L* with α is very good (Fig. 11.19), all

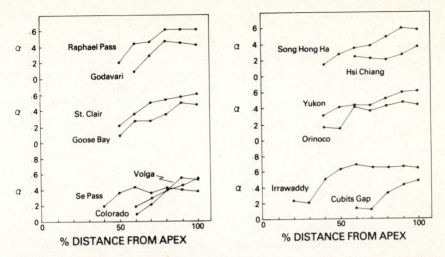

Figure 11.17 Variation of α with distance from the apex on model networks. Names refer to rivers whose deltas are being modeled. See Table 11.6.

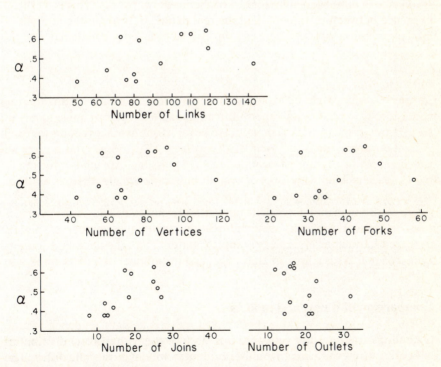

Figure 11.18 Relation of topologic factors to α on model networks.

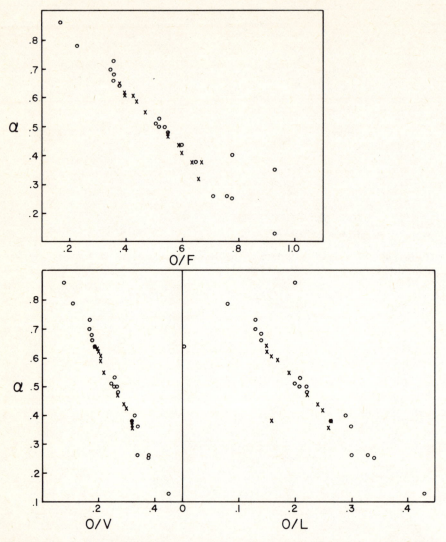

Figure 11.19   Variation of topologic ratios to α for model networks (X) and natural deltas (open circle).

of these factors decreasing as α decreases. This is expected, of course, because of the conservative nature of outlet frequency.

Comparison of deltas and models

Overall, *FJ* links are most numerous in 55% of the natural delta distributary networks, whereas *FJ* links are most frequent in 46% of those delta distributary patterns for which analogous models were made. The models clearly show that

FJ links are the most frequent type of link; *FJ* links are most numerous in 62% of the models. In all the nets analyzed *FF* links are the next most frequent. *JO* links tend to occur least often in all natural deltas. However, in the model networks *JJ* links tend to be lowest in frequency. In both natural and model patterns frequencies of all types of links behave in a similar manner as α increases or decreases (Fig. 11.8). Predicted frequencies of link types by the random connection model seem to be closer to the actual observed frequencies in the model networks than in patterns of the natural deltas, although both actual and simulated networks tend to have too many *JO* links and too few *JJ* links, and about the same proportion of networks are significantly different from random (Tables 11.6 and 11.10).

There is no correspondence in α values between each model and its analogous natural delta, either overall or as α changes with distance from the apex (Figs. 11.12 & 11.17). Variability in all parameters is greater in the delta patterns than in the corresponding model. For example, the range of number of vertices in the models is 24, that of the natural deltas is 106; the range of the number of outlets in the models is 20, that of the natural networks is 35. This could be a result of the size limit imposed on generation of the models, but in any case does illustrate the space filling constraints in random models. Trends in changes in parameters with α are more defined and regular in the random models than in the natural networks. Most imporantly, in real networks α varies with distance from the apex; in model networks it does not.

Although the topology of the random models generated by the rules stipulated is similar to that of the natural networks, including the tendency to produce too many *JO* links and too few *JJ* links, there are enough differences between the real and simulated random patterns to conclude that natural deltaic networks are not random.

Discussion

A fluvial system is composed of two parts. One is the drainage basin, a collection agency for water and sediment, and the other is the delta or fan, a dispersal agency for water and sediment (Fig. 11.1). The network configuration of the drainage basin reflects the interaction of the geologic and topographic character of the watershed. The network pattern of the deltaic distributaries also depends, in part, on the character of the upper drainage basin and in part on coastal processes.

As we consider the distributary channels as components of this total fluvial system, the branches are subject to the same principles as those that govern the dynamics and morphology of rivers. Studies have shown that a river adjusts its form to attain an equilibrium between its energy and the work it has to do. Hence, one would expect distributaries to act in the same way—that is, to establish the most efficient conditions for transporting its discharge and sediment load.

Multiple channels are narrower and shallower, with a smaller channel cross-section area and have steeper gradients and higher velocities than a single channel (Church, 1972). Bifurcation (braiding) is thus an adjustment made by a river in order to increase its competence. So when a distributary channel lacks the power required to transport its load, it will divide. A merge to a single channel indicates the distributary has enough power to carry the load. I suggest that the resonance observed in the α values is a result of a phenomenon similar to that of the pool/riffle sequence in channels of the collecting watershed. That is, resonance is caused by the efforts of the distributaries to attain an equilibrium of power expenditure along the network by dividing and then merging several times along the length of the delta.

Braided rivers also are characterized by a highly variable discharge and a heavy caliber bedload. Coleman (1976) noted that distributary nets classified as rejoining (those similar to braided rivers which have an α value of 1) have erratic discharges and coarse sediment loads. Those deltas he typed as bifurcating have less variable discharge and carry a fine sediment load. Thus we conclude that branching in deltaic networks is an expression of the adjustment of the channel to its discharge and load.

These conclusions are supported by the work of Chang (1982). He found that in experimental, flume-built deltas, distributaries increased in number as aggradation (braiding) took place because of decreased discharge or increased sediment load. During degradation brought about by increased discharge and/or decreased sediment load, distributaries merged into a single channel. He attributed each of these pattern changes to an attempt by the distributary to achieve equal power expenditure along its length.

At the mouth subaqueous processes such as deposition of distributary mouth bars are important during delta progradation in the absence of strong wave, current, or tidal action. For example, formation of a single bar (Fig. 11.20a)

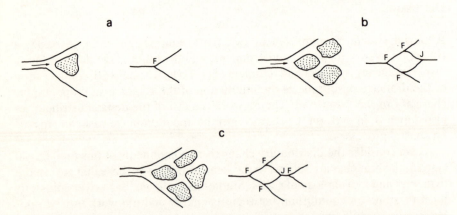

Figure 11.20 Distributary mouth bars and corresponding topologic pattern of links and vertices. (a) Single bar deposited at the distributary mouth resulting in two *FO* links. (b) Multiple bars at the mouth giving two *FO* links and one *JO* link. (c) Multiple bars at the mouth forming four *FO* links.

will lead to a fork. This can be seen at the mouths of many distributaries, and gives rise to many *FO* links at the distal edge of the delta. Multiple bar deposition will create a more complex situation (Fig. 11.20b & c) where a number of forks and joins will form as channels are extended around bars. Note the formation of *JO* links in this manner. This is probably the reason for the inordinate number of joins at the distal edge of the Niger, Colville, Godavari and Grand Pass networks.

Of the parameters suggested by Smart and Moruzzi (1972), it is considered that the recombination factor, α, may be most helpful in topologic studies since it describes the distributary pattern in terms of joins and forks. The data here indicate that of the geometric factors that might influence the value of α, the angle of bifurcation may be important whereas the exterior angle of the shoreline enclosing the delta appears to be of little importance. Other factors, not considered in this study, that may significantly affect α values are wave dynamics, current velocities, tidal range, distributary discharge, frequency and magnitude of floods, amount and caliber of load, bank vegetation, and the morphology of the channels. They represent a fertile field of investigation for further work on delta-distributary topology.

What do the α variations shown in Figure 11.12 and previously discussed suggest about the relation of forks to merges on the delta surface? First, the initial height or rise of the α curve near the apex is, in many cases, followed by a decline in the value of α and then a rise again (resonance). This results partly because of geometry; for joins to occur there must be a number of channels filling the space. Near the apex the single channel entering the delta divides. Consequently there are one or more forks near it and a low α value, or in actuality a zero α. As joins occur downstream from these forks and farther from the apex, α increases in value. Hence, the value of α tends to rise toward the outlets rather than stabilizing near the apex. The number of merges increases faster than bifurcations occur so that especially on large deltas the network approaches the braided configuration with an equal number of forks and joins. On natural deltas, factors such as energy constraints may be the cause, as has been previously discussed. As gradients decrease toward the distal edge of channels, less energy is available. Also as forks multiply the power (i.e. discharge) available to each branch lessens. To conserve the power the branches merge. The fact that α values almost always rise toward the outer delta edge in the models suggest that there are also spatial constraints on the number of forks.

Finally, we can relate the topology of the networks to Coleman's (1976) classification of distributary networks using data from those studied here. Table 11.11 indicates that those nets he classed as bifurcating have lower values of α than those he regarded as rejoining. The Volga, which he placed in both categories, has an α value of 0.50. Because α is the ratio of number of joins to number of forks, it seems valid to use 0.50 as the threshold between the two classes, i.e. when $\alpha > 0.50$ there is a rejoining pattern. Hence on an objective basis, the Paraña and the Volga would be on the borderline, the Godavari

Table 11.11 Classification of distributary networks by Coleman.

Delta	α	No. of outlets	Classification (Coleman's)
Godavari	0.483	16	bifurcating
Paraña	0.514	18	bifurcating
Colville	0.642	20	bifurcating
Volga	0.500	23	bifurc./rejoining
Orinoco	0.728	23	rejoining
Niger	0.785	15	rejoining
Irrawaddy	0.857	6	rejoining

would be bifurcating and the other networks in the table would be rejoining. Of the 20 networks studied here, nine are bifurcating: Song Hong Ha, Godavari, Cedar Lake, St. Clair, Goose Bay, Grand Pass, Cubits Gap, Raphael Pass, and Southeast Pass. Three are on or close to the borderline: the Volga, Paraña, and the Colorado River of Texas. The remaining eight are rejoining networks (Table 11.3).

Table 11.11 also demonstrates that number of outlets is not dependent upon type of network; bifurcating patterns may not necessarily have a larger number of outlets than do rejoining networks.

Summary

We have extended the topologic analysis of Smart and Moruzzi to 20 natural delta distributary networks and to 13 random models. Observations indicate that:

(1) Deltaic networks are considered to be non random, being controlled by space and energy constraints. In addition, there are probably other factors affecting distributary patterns, factors not studied here.

(2) The most frequent type of links in the distributary system are those joining two forks, *FF*, and those with a fork at the upstream end and a join at the downstream vertex, *FJ*. The least frequent link type is that with a join before an outlet, *JO*.

(3) The recombination factor, α, is variable over the delta surface. Its value generally rises distally and often displays resonance. Its values are thought to be governed by the fluvial principles of equal power expenditure.

(4) There is an indication that α values are affected by bifurcation angle such that α increases with increasing angle of bifurcation. Other physical factors of fluvial and coastal morphology that might influence α were not considered here.

(5) The number of *FO* and *JO* links at the outlets are primarily determined by deposition of simple or complex distributary mouth bars.

(6) There is a general tendency for the average number of vertices, forks, joins and links to increase with larger sized deltas. The average number of outlets remains fairly constant. The α value of large deltas tends toward 1.

(7) Classification of distributary networks as bifurcating or rejoining can be made on the basis of α values.

In this study we have been primarily concerned with the description of the topologic properties of delta distributary systems. We have tried to explain constraints on some of these parameters but much more study needs to be done on the physical characteristics of distributary channels (such as their hydraulic geometry) before an adequate explanation for topologic properties can be obtained. Moreover, although we have accented the fluvial aspects of delta distributaries, it will be necessary in future research to measure the effects of coastal processes on network configuration. This paper represents only a small beginning to a tremendously large subject.

References

Abrahams, A. D. 1975. Topologically random channel networks in the presence of environmental controls. *Geol. Soc. Am. Bul.* **86**, 1459–62.

Chang, H. H. 1982. Fluvial hydraulics of deltas and alluvial fans. *Am Soc. Civil Eng.,* **HY11**, 1282–95.

Church, M. 1972. Baffin Island sandurs; a study of arctic fluvial processes. *Can. Geol. Surv. Bull.* **216**.

Coleman, J. M. 1976. *Deltas: processes of deposition and models for exploration.* Champaign, Ill.: Cont. Educ. Pub. Co.

Coleman, J. M. and L.D. Wright 1975. Modern river deltas: variability of processes and sandbodies. In *Deltas, models for exploration*, M. L. Broussard (ed.), 99–150. Houston: Houston Geol. Soc.

Howard, A. D., M. E. Keetch and C. L. Vincent 1970. Topological and geometrical properties of braided streams. *Water Res. Res.* **6**, 1674–88.

Kanes, W. H. 1970. Facies and development of the Colorado River delta in Texas. In *Deltaic sedimentation, modern and ancient*, J. P. Morgan (ed.), 78–106. Soc. Eco. Paleo. Min. Spec. Pub. No. 15.

Kolb, C. R. and J. R. Van Lopik 1966. Depositional environments of the Mississippi River deltaic plain—southeast Louisiana. In *Deltas in their geologic framework*, Shirley, M. L. (ed.) 17–61. Houston: Houston Geol. Soc.

Krumbein, W. C. and A. R. Orme 1972. Field mapping and computer simulation of braided-stream networks. *Geol. Soc. Am. Bull.* **83**, 3369–79.

Morisawa, M. and W. Montgomery 1983. Delta distributary networks: some quantitative aspects. In *Facets of Geomorphology*, Delhi, India.

Ranalli, G. and A. E. Scheidegger. A test of the topological structure of river nets. *Int. Assoc. Sci. Hydrol. Bull.* **13**, 142–53.

Russell, R. J. 1967. *River plains and sea coasts.* Los Angeles: Univ. Calif. Press.

Shreve, R. L. 1966. Statistical law of stream numbers. *J. Geol.* **74**, 17–37.

Shreve, R. L. 1978. Infinite topologically random channel networks. *J. Geol.* **75**, 178–186.

Smart, J. S. 1969. Topological properties of channel networks. *Geol. Soc. Am. Bull.* **80**, 1757–73.

Smart, J. S. and V. L. Moruzzi 1972. Quantitative properties of delta channel networks. *Zeit. Geomorph.* **16**(3), 268–82.

Snedecor, G. W. and W. G. Cochran 1967. *Statistical methods*, 6th edn Ames: Iowa State Univ. Press.

Van Andel, T. H. 1967. The Orinoco delta. *J. Sed. Petrol.* **37**, 297–310.

Woldenberg, M. J. 1969. Spatial order in fluvial systems. Horton's laws derived from mixed hexagonal hierarchies of drainage basin areas. *Geol. Soc. Am. Bull.* **80**, 97–112.

12

Optimal models of river branching angles

André G. Roy

Introduction

It has long been recognized that branching angles are a fundamental component of river network morphometry. At a junction, branching angles are defined as the angles formed by the tributaries and the axial prolongation of the main (receiving) stream (Fig. 12.1a). These angles (θ_1 and θ_2) are often called the angles of entry and their sum is simply the junction angle (ψ).

Assuming that stream channels are relatively straight in the vicinity of a junction and that the locations of the end-points of the channels are known, branching angles will be determined by locating the junction point. Displacements of that point will obviously change the angles of branching (Fig. 1b). Thus, one way of modeling the geometry of a junction is to see the problem as an exercise in the locational analysis of a junction point.

In order to find the location of the junction point, one has to put forward locational criteria. In that respect, optimality can be viewed as a governing principle in the fluvial system. The fundamental postulate of the optimality principle is that a river will perform its function with maximum efficiency. The net result of this efficiency is that the costs involved in the operation and/or maintenance of the river system will be minimized. Thus, the geometry of a junction will be optimally adjusted and will help the river to bring its costs to a minimum. In this paper, I will first derive an optimality model for the geometry of stream junctions. I will then discuss the effects on angles of the two parameters that compose the general optimal angular geometry models. Finally, I will present empirical data to show that the model is successful in predicting average angles for a whole network but not so successful in predicting angles for individual junctions.

Derivation of the optimal model

The application of the optimality principle to the geometry of branching systems was first introduced in biology (Hess 1903, Murray 1926b, 1927). In

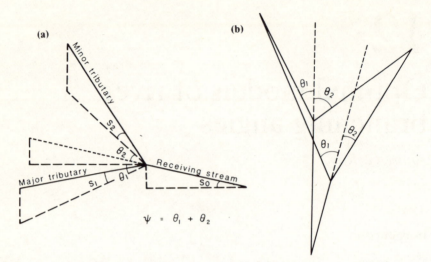

Figure 12.1   (a) Definition of branching angles and channel slopes at a junction. (From Roy 1983; used by permission of Ohio State Univ. Press). (b) The effect of displacing the junction point on the angular geometry of a junction.

this field, natural selection and adaption provide a strong rationale for the optimality point of view (Rosen 1967) and as pointed out by Murray: 'The concept of adaptation has been treated in a quantitative manner by experimental morphologists ... who have shown again and again the tendency toward perfect fitness between structure and function in all sorts of plants and animals' (Murray 1926a, p. 207). Developments in the morphological study of organisms have been heavily dependent upon the optimality principle (see for example, Rashevsky 1960, Cohn 1954, 1955, Zamir 1976). In fluvial geomorphology, research on optimal angular geometry has followed the same path developed in biology and for this reason I will first derive the general model and then specify it for the fluvial system.

The general optimal angular geometry model (i.e. a model that applies equally well to all types of bifurcating trees) assumes that the geometry will minimize total operation and/or maintenance costs. In order to derive the model it must be possible to evaluate what those costs are. If one assumes that costs increase linearly with channel length, then total costs (T) are given by

$$T = \sum_{i=0}^{2} w_i l_i \qquad (12.1)$$

where w_i is the cost per unit length of operating and/or maintaining channel i and l_i is the length of channel i. The subscript i denotes the receiving stream (0) and major (1) and minor (2) tributaries. Given that channels are straight, equation (1) becomes

$$T = \sum_{i=0}^{2} w_i \left((X_i - X)^2 + (Y_i - Y)^2 \right)^{\frac{1}{2}} \qquad (12.2)$$

where X_i, Y_i are the coordinates in Euclidean space of the end-points of channel i and X, Y the coordinates of the junction point. To find an optimum, one has to minimize Equation 12.2 and therefore find a solution for the unknown location of the junction point by differentiating T with respect to X and Y. This problem is the famous three-point Weber problem in location theory (Weber 1909, Miehle 1956, Wesolowsky 1973). The geometrical counterpart of this locational problem is obtained by substituting the appropriate trigonometric expressions into the solution for X and Y (Zamir 1976, Roy 1982). The resulting angular geometry model is

$$\cos \theta_1 = (w_0^2 + w_1^2 - w_2^2)/2w_0 w_1 \qquad (12.3a)$$

$$\cos \theta_2 = (w_0^2 - w_1^2 + w_2^2)/2w_0 w_2 \qquad (12.3b)$$

$$\cos \psi = (w_0^2 - w_1^2 - w_2^2)/2w_1 w_2 \qquad (12.3c)$$

Although they represent a more general form of the model, these equations were first derived by Murray (1926b, 1927) for a minimum power losses model.

The general model is thus derived from a length minimization procedure where length of each channel is weighted by a cost per unit length. One interesting feature of this branching angle model is its independence from the original set of coordinates and its strict reliance on the weights (w_i). Therefore, the resulting angular geometry will not be affected by displacements in the end-points and, as long as the weights remain constant, the junction point will always adjust to maintain the branching angles.

Given the general model of optimal branching angles, it is readily apparent that the problem is to specify the weights. How do we evaluate the costs of maintaining and/or operating a stream channel? One has to determine an optimality criterion which will govern the angular geometry of stream networks. Howard (1971) first proposed a minimum power losses criterion. In this case the weight is

$$w = \rho g Q S \qquad (12.4a)$$

where g is the gravitational constant, ρ is fluid density, Q is water discharge and S is slope. I have argued that several other optimality criteria could be useful in evaluating the costs involved in the operation of a channel reach (Roy 1982, 1983). I proposed three other ways of specifying the weights and derived a minimum total flow resistance model where

$$w = \rho g W D S / v^2, \qquad (12.4b)$$

a minimum resisting force model where

$$w = \rho g W D S \qquad (12.4c)$$

and a minimum volume model where

$$w = W D. \qquad (12.4d)$$

In these equations W, D and v are the width, average depth and velocity of the water.

Pursuing an approach used by Zamir (1978) and Roy and Woldenberg (1982) for arterial branching, I generalized the model by substituting the appropriate hydraulic geometry relationships into the weight equations (Roy 1982, 1983).

Assuming that W, D, v and S vary downstream with discharge (Q) as power functions with exponents b, f, m and z (Leopold and Maddock 1957), then the weights become

$$w = j Q^{1+z} \qquad (12.5a)$$

for minimum power losses,

$$w = j Q^{1+z-3m} \qquad (12.5b)$$

for minimum total flow resistance,

$$w = j Q^{1+z-m} \qquad (12.5c)$$

for minimum resisting force, and

$$w = j Q^{1-m} \qquad (12.5d)$$

for minumum volume. Thus, all the weights have the same general form

$$w = j Q^{1+k} \qquad (12.6)$$

where j is a constant if fluid density is a constant and k is a function of the downstream hydraulic geometry relationships involved in the specification of the optimality criterion. Since z is always negative and m is positive, k is always negative.

The model can be simplified even further by using the continuity equation of flows

$$Q_0 = Q_1 + Q_2 \qquad (12.7)$$

and defining a symmetry ratio

$$\alpha = \frac{Q_2}{Q_1} \quad \text{(where } Q_2 \leqslant Q_1\text{)} \tag{12.8}$$

that measures the relative sizes of the tributaries. By combining α and the continuity equation, it is possible to express all the weights in terms of Q_1, the discharge of the major tributary, and to derive the general optimal angular geometry model for rivers

$$\cos \theta_1 = \frac{(1+\alpha)^{2x}+1-\alpha^{2x}}{2(1+\alpha)^x} \tag{12.9a}$$

$$\cos \theta_2 = \frac{(1+\alpha)^{2x}-1+\alpha^{2x}}{2\alpha^x(1+\alpha)^x} \tag{12.9b}$$

$$\cos \psi = \frac{(1+\alpha)^{2x}-1-\alpha^{2x}}{2\alpha^x} \tag{12.9c}$$

where $x = 1+k$ and $k < 0$. (For a complete derivation, see Roy 1982 or 1983).

The effect of α and k on optimal angles
According to the model given by Equations 12.9, optimal branching angles are a function of the symmetry ratio (α) and of the negative exponent k. The effect of these parameters on branching angles must first be assessed by looking at them one at a time. On the one hand, for any given value of k, the symmetry ratio controls the relative size of θ_1 with respect to θ_2. When two streams of similar sizes ($\alpha = 1$) merge the angles of entry are equal. As branching becomes asymmetrical in size (i.e. low α) it also becomes asymmetrical in angles and θ_1 diminishes while θ_2 increases. The relationships between α and the angles of entry are shown in Figure 12.2 for $k = -0.4$. As much as the values of θ_1 and θ_2 are affected by the symmetry ratio, the junction angle (ψ) is nearly constant over the whole range of α. Only when α is very small (0.0001) do we see an effect on ψ. Calculations show that for k larger than -1.0, the limiting junction angle as α tends to 0 is 90°. Thus, a very small stream enters a much larger one at a right angle. If k is smaller than -1.0, the limiting value of ψ is 180°. Thus we see that the role of the symmetry ratio is chiefly to determine how the junction angle is partitioned into θ_1 and θ_2.

The junction angle, on the other hand, varies with k. As k declines, the junction angle widens. Since the value of k is determined by the rates of change in slope and velocity with discharge, it implies that ψ becomes wider when slope and/or velocity changes very rapidly downstream. The effect of k on ψ is displayed in Figure 12.3 for symmetrical and asymmetrical branching. Note how the two curves are nearly superimposed, illustrating again the lack of effect of the symmetry ratio for $k > -1.0$. Junction angles in this range of k are therefore mainly controlled by the exponent k.

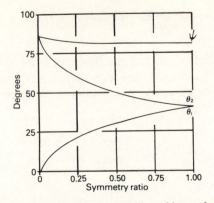

Figure 12.2 The relationship between α and optimal branching angles for $k = -0.4$. (From Roy 1983; used by permission of Ohio State Univ. Press).

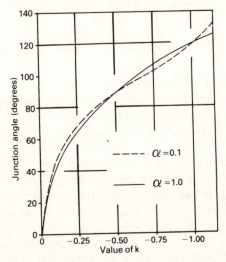

Figure 12.3 The relationship between k and the optimal junction angles for symmetrical and asymmetrical branching. (From Roy 1983; used by permission of Ohio State Univ. Press).

The fact that α does not play an important role on junction angles when $0.1 < α < 1$ is very puzzling because it has been noticed that junction angles increase as branching becomes asymmetrical (Lubowe 1964, Pieri 1979, Pieri in press). Does this failure of the model to predict the widening of junction angles as α declines imply that the model is not readily applicable to river branching? To answer this question I will rework Lubowe's results for the Bell Watershed in the San Dimas Forest in California. I will show that α and k are interrelated and that their combined effect explains the widening of the junction angle.

Lubowe's data (1964, p. 331) along with the average α are reported in Table 12.1. It is clear that as the order-difference between tributaries increases

Table 12.1 Analysis of Lubowe's (1964) stream junction data for Bell Watershed.

Tributary orders	Average junction angle (degrees)	Average symmetry (ratio of tributary drainage basin areas)	Average slope change
11	44.3	0.65	1.43
12	55.7	0.23	1.70
13	70.0	0.07	2.01
14	74.1	0.02	2.74

average junction angles increase and that the order difference is a surrogate measure of the average symmetry ratio. Of course, it is tempting to conclude from this table that there is a direct relationship between α and ψ.

Let us assume, for the sake of argument, that a minimum power losses model is adequate and thus k is equal to z. At a junction, if the continuity equation of flows holds and if the relationship between slope and discharge is given by

$$S = tQ^z \tag{12.10}$$

then, slope change will be expressed by

$$S_0^a = S_1^a + S_2^a \tag{12.11}$$

where

$$a = 1/z \tag{12.12}$$

(Roy and Woldenberg, in press). Thus, a is a negative exponent equal to the reciprocal of the hydraulic geometry exponent. A value of a close to 0 means a high rate of change in slope at a junction. The relationship between α and $k = z$ is shown in Figure 12.4. For any given slope ratio S_0/S_1, z is always smaller when α is small thus indicating that asymmetrical branching is associated with higher rates of change in slope at a junction. This result is intuitively correct because when a very small channel joins a large mainstream, changes in slope are generally drastic. Thus, a low α theoretically implies a low z.

One problem in using Equation 12.11 is that given S_0, S_1 and S_2 at a junction it is not possible to solve for a when S_0 is equal to S_1 or S_2 or lies between them (Roy & Woldenberg, in press). As a consequence, the relationship between α and z for several junctions is difficult to substantiate empirically. To overcome this problem, I have chosen to illustrate the relationship between α and slope change at a junction by using an average slope change measure (ΔS).

$$\Delta S = \frac{S_1 + S_2}{2S_0} \tag{12.13}$$

As ΔS increases, the rate of change in slope (z) becomes more negative. I have

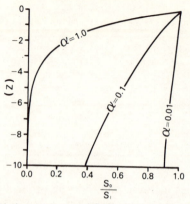

Figure 12.4 Interrelationship between α z and S_0/S_1.

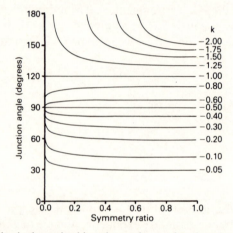

Figure 12.5 Variation in the optimal junction angles as a function of α and k (after Woldenberg and Horsfield, in preparation).

already suggested that ΔS increases with branching asymmetry (low α). The results in Table 12.1 demonstrate this point very clearly.

The apparent relationship between α and ψ has to be understood through the effect of α on z and therefore on k. The combined effect of α and k is displayed in Figure 12.5. Among other things, Figure 12.5 suggests the following. For any α, ψ increases as k becomes more negative. ψ is affected by α only for small α, except when k is more negative than, say, -1.25; here the impact of declining α is pronounced, and causes ψ to increase to a limit of 180°. In our case, -1.25 is an unrealistic value for $z = k$, since z is usually greater than -1 (Leopold *et al.* 1964, p. 244). Thus in the domain $k = z > -1$, for $α > 0.1$, any tendency for ψ to increase can only be explained by a decrease of k. In the interval $-.5 > k=z > -1$, when α approaches zero ψ decreases to 90°. When $k=z > -.5$, ψ increases to 90° as α approaches zero. Lubowe's mean ψ values (Table 12.1) are less than 90° and increase with decreasing α suggesting $z = k > -0.5$. Since these mean ψ values increase at a much greater rate than is predicted by α alone, I argue that the main reason for the increase in ψ is the decrease in $z = k$, as reflected in the increasing average slope change.

Comparison between optimal and observed river branching angles

Before testing the optimal angular geometry models an investigator has to address several questions. First, he has to determine which networks will be used in the comparison of observed and optimal angles. Network selection must be guided by several criteria which greatly reduce the possible number of regions (Roy 1982). For instance, the networks must represent fluvially-eroded landscapes that have evolved for a long period of time and that are devoid of structural controls. Large-scale and accurate topographic maps must also be available for the area. Following these criteria, I have selected four drainage networks from the south-west of England. They were delineated from the blue lines of the 1:25000 Second Series Ordnance Survey topographic sheets. Their drainage areas range from 20 to 26 km² and their magnitudes from 49 to 79. Two networks (Dart and Little Dart) are underlaid by massive sandstones while the others (Great Torrington and Duntz) are developed on evenly bedded sandstones, slates and shales.

Secondly, how do we measure branching angles? Several definitions of angles may be put forward. For instance, we could define angles using the end points of the channel links forming the junction. The main disadvantage of this method is that branching angles become dependent upon the accuracy of the channel network delineation. To avoid this problem, angles should be measured around the junction point. Using a circle centred on the junction point, the angles are given by the lines joining the center to the intersections of the circle and the stream channels (Fig. 12.6). This method yields highly reproducible results and is independent of network delineation techniques. Angles measured in the vicinity of the junction point, however, are generally wider than those obtained from the end points of channel links (Roy 1982). For

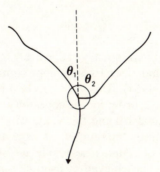

Figure 12.6   Operational definition of branching angles.

Table 12.2 Means (and standard deviations) of the branching angles observed in four river networks from Devon.

		Angles (degrees)		
	n	θ_1	θ_2	ψ
Dart	78	24.1 (20.6)	55.0 (21.7)	79.1 (17.7)
Little Dart	48	21.8 (18.6)	62.2 (25.3)	84.0 (16.9)
Great Torrington	53	23.2 (22.4)	53.9 (18.6)	77.2 (18.7)
Duntz	70	23.2 (19.9)	58.3 (24.8)	81.5 (22.2)

this analysis, I set the radius of the circle equal to 4 mm which represents 100 meters in the field. The means and standard deviations of the observed branching angles are presented in Table 12.2. The maximum difference in the means of a given angle is 9 degrees. The variability among the means is thus very small and there is no effect of lithology on angle means. The standard deviations, on the other hand, are very large thus indicating that within a drainage network, angles are highly variable.

In rivers, size differences between merging tributaries are usually large and branching tends to be asymmetrical. As pointed out by Woldenberg and Horsfield (1983) in the case of blood vessels, if an optimality principle is involved in determining branching angles, then one should expect θ_1 to be smaller than θ_2. This is confirmed by the large difference in the means of θ_1 and θ_2 (Table 12.2). Also, θ_1 is smaller than θ_2 for 81% of the junctions; the probability that this could be due to chance is less than .0001.

A third problem arises from the fact that discharge measurements have to be available for each junction. This, of course, is unrealistic and one has to rely on surrogate variables in order to specify the parameters of the model. Drainage network magnitude and drainage area can both be used to estimate discharge. Given that magnitude is related to the accuracy of network delineation and that it does not allow discrimination between sources of different sizes, drainage areas should be preferred. Within an homogeneous region, drainage basin area (A) and water discharge are related by a power function:

$$Q = cA^p \tag{12.14}$$

The value of p varies with the frequency of the discharge considered and ranges from 1.0 for mean annual discharge to approximately 0.75 for bankfull discharge (Leopold et al. 1964). This variation in p is important since it will affect the symmetry ratio. In order for the continuity equation of flows to hold at a junction, p has to equal 1.0 and this value will be used here. Thus, the symmetry ratio (α) is given by the ratio of drainage areas.

Finally, the model may be tested at several levels. An overall assessment of the adequacy of the model is achieved by using the model backwards. Given the symmetry ratios and the observed angles at each junction, it is possible to determine the value of k which yields the best prediction. Then, one has to determine whether these 'optimal' k values are meaningful and interpretable with respect to different optimality criteria. A second approach deals with the prediction of the average observed angles for a drainage network. The predictions are obtained by substituting average values of α and k into the model. At the opposite end of this approach one could look at the prediction of the angles for individual junctions based upon junction-specific values of α and k. Thus, the parameters are determined from the morphometry of the three branches merging at the junction. I will present results for each type of analysis and show that the model appears to be more adequate for the macro-scale than for the micro-scale predictions.

Assessment of the adequacy of the model: is the best k meaningful?
The first test of the optimal angular geometry model is to assess whether it has any capacity for predicting the actual angles. The procedure used here is to find the value of k that will yield the best prediction. Thus the problem is to minimize

$$\sum_{i=1}^{n} |O_i - E_i| \tag{12.15}$$

where n is the number of junctions in the network, O_i and E_i are the observed and expected angles for junction i. The expected angles are a function of α, a known value, and k, an unknown. By allowing k to vary and recomputing the sum in Equation 12.15 as a function of k, one is able to determine the best values of k. These differ for each network and angle (Table 12.3). Except for θ_1, the values of k appear to be quite consistent from river to river. The k values associated with the junction angles always fall between the k values for θ_1 and θ_2.

Using the symmetry ratios and the best value of k, optimal angles may be computed for each junction. For each network, the best k was set equal to the one computed from the junction angles. The comparison of optimal and observed angles shows that the model accounts for a substantial amount of the variance in the observed angles (Table 12.4). This is particularly true for θ_2 where the proportion of variance explained is always higher than 24%. Predictions of the junction angle, however, do not seem adequate, especially

Table 12.3 Best values of the exponent k in four stream networks from Devon.

	θ_1	θ_2	ψ
Dart	−0.36	−0.30	−0.33
Little Dart	−0.55	−0.37	−0.42
Great Torrington	−0.33	−0.31	−0.32
Duntz	−0.47	−0.30	−0.37

Table 12.4 Proportions (%) of the variance in the observed angles explained by the optimal model in four stream networks from Devon.

	θ_1	θ_2	ψ
Dart	4.0	24.0	14.4
Little Dart	31.4	32.5	4.4
Great Torrington	19.4	28.1	0.0
Duntz	13.7	38.4	7.8
all	13.8	31.7	5.3

Table 12.5 Average absolute value of the residuals (in degrees) associated with the best value of k for four stream networks from Devon.

	θ_1	θ_2	ψ
Dart	17.9	15.1	13.3
Little Dart	12.9	15.0	13.8
Great Torrington	15.3	13.1	14.2
Duntz	15.6	15.5	16.6

for the Great Torrington and Little Dart rivers. This result is partly explained by the fact that a single value of k is used for all α. (Figure 12.4 shows that $z = k$ becomes more negative as α decreases). Hence, the optimal junction angles are constrained to vary only within a small range. Nonetheless, one should expect the proportion of variance explained to be larger than 0. Despite this variability in the proportion of explained variance, the best predictions yield a very consistent mean absolute value of the residuals (Table 12.5). Generally, the average misprediction is on the order of 15 degrees. Because I used the same value of k to compute the optimal values of θ_1, θ_2 and ψ for a network, the predictions tend to overestimate slightly θ_1 and underestimate θ_2. This bias is on the order of 2 to 3 degrees and it disappears if the best k value used to compute the expected angles is derived from the observed θ_1 and θ_2. Predictions of the junction angles, on the other hand, are not biased since the best k was found from the observed ψ. Considering the great variability in angles within a drainage network, optimal predictions are good.

At best, the model is capable of predicting a great deal of variability in θ_1 and θ_2 but not in ψ. But is the best k a meaningful value? To address this question, I will use only the best k derived from the junction angles. Those k values are quite low and this would readily eliminate the minimum volume model because velocity does not increase downstream at a 0.3 and higher power. Thus, a low k implies that slope is a dominant factor. In Table 12.6 I compare the best k value and the exponent of the channel slope-drainage area relationship. Except for the Dart river, the two sets of exponents agree very well, suggesting that k is equal to z. Thus, the minimum power losses model appears to be highly plausible. Of course, one cannot dismiss the two remaining models especially if velocity increases only very slightly downstream. This interpretation of the exponent k clearly suggests that, in this case, the model is adequate.

Predictions of average observed angles

In this application of the model, the parameters α and k are set equal to average values describing the whole drainage network. The purpose of this application is to derive a single set of optimal θ_1, θ_2 and ψ for a given network and to compare these optimal values with the means of the observed angles. Optimal angles found by substituting the average symmetry ratios and the exponent of the channel slope-drainage area relationship into Equations 12.9 are in very close agreement with the observed mean angles (Table 12.7). The most important discrepancy is on the order of 7 degrees (Dart River) but generally the predicted values are within 2 degrees of the observed average angles. Thus, the model, calibrated as a minimum power losses model, is very successful in predicting average angles.

Predictions using junction-specific parameters

Optimal angles can also be derived for individual junctions by using a value of k that describes the junction-specific hydraulic geometry relationships. An example of such relationships is given by Equation 12.11:

$$S_0^a = S_1^a + S_2^a \tag{12.11}$$

If slopes are known at a junction, then a can be found and the junction-specific hydraulic geometry exponent z is the reciprocal of a. Applying this procedure

Table 12.6 Comparison between the best k value and the exponent z from the slope-drainage area relationship in four stream networks from Devon.

	k	z
Dart	−0.33	−0.45
Little Dart	−0.42	−0.43
Great Torrington	−0.32	−0.36
Duntz	−0.37	−0.39

Table 12.7 Comparison between average observed (O) angles and predicted (P) angles using the average symmetry ratio and the exponent of the slope-drainage area relationship.

	θ_1		θ_2		ψ	
	O	P	O	P	O	P
Dart	24.1	28.0	55.0	58.1	79.1	86.1
Little Dart	21.8	23.4	62.2	61.3	84.0	84.7
Great Torrington	23.2	24.4	53.9	53.9	77.2	78.3
Duntz	23.2	24.6	58.3	56.5	81.5	81.1

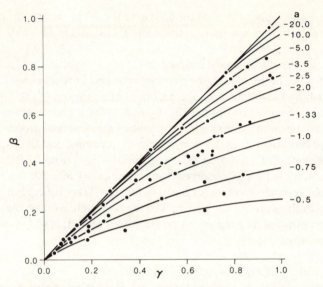

Figure 12.7 Variability in junction-specific slope hydraulic geometry relationships, Dart River. $\gamma = S_1/S_2$ and $\beta = S_0/S_2$.

for each hydraulic geometry variable requires that we have field measurements on each channel. Such detailed information on the morphometry of drainage networks is usually not available. However, the model could be applied using channel slopes measured from topographic maps. If one assumes that channel slope approximates water surface slope then only the minimum power losses model may be tested since it only requires slope measurements. As will be seen, predictions based on these data were unsatisfactory.

Channel slopes were measured from the 1:25 000 map and the junction-specific hydraulic geometry exponents were calculated for the Dart River. Roy and Woldenberg (in press) have developed a simple nomograph that is useful to display the variability in morphometric changes at junctions. The data are reported on the nomograph in Figure 12.7. In this nomograph γ is equal to S_1/S_2 and β to S_0/S_2. Because S_2 (the slope of the minor tributary) is expected to be larger than or equal to S_1 and greater than S_0, the data points should plot in the lower right half triangle of the graph. This area represents the zone where Equation 12.11 has a solution for a. When S_0 lies between S_1 and S_2, then Equation 12.11 has no solution. The nomograph obtained for the Dart River shows that the values of a range from -0.5 to values lower than -20.0. This implies that junction-specific z values are highly variable and range from 0 to -2.0. Furthermore, only 60% of all junctions of the Dart River could be plotted on the nomograph (i.e. a solution existed for a). Substituting these values of z and the symmetry ratios into the model yielded optimal angles for each junction. Another way of showing the agreement between the junction-specific predictions and the observed angles is to plot the observed junction

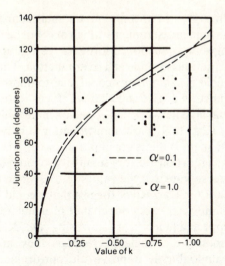

Figure 12.8 Predictions of the junction angle based upon the junction-specific $z = k$, Dart River.

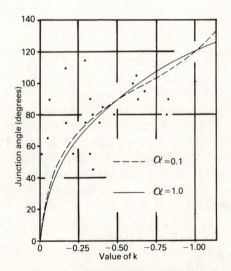

Figure 12.9 Predictions of the junction angle based upon the junction-specific $z = k$, Eastman.

angles as a function of z using the graph on Figure 12.3. ($z = k$ on this graph). As displayed in Figure 12.8 we see that there is a very poor agreement between the observed junction angles and the optimal lines. In general, the observed ψ values do not increase as z declines and the optimal angles overestimate the observed values. The situation is reversed when z is larger than -0.25. This application of the model using junction-specific parameters is clearly not satisfactory.

Several reasons may be invoked to explain this lack of fit between observed and optimal junction angles. In my opinion, the main source of discrepancy is that z is derived from map measurements which may not be accurate enough to evaluate slope changes at a junction. Moreover, cartographic generalization may mask the true variations in slope at a river junction. Thus, accurate values of z obtained from field measurements could lead to improved predictions.

As an example, junction angles and channel slopes were both measured in the field for a small badland network developed on mine tailings near Eastman, Quebec. The results (Fig. 12.9) show a much better fit between observed and optimal angles although the outliers are important when k is larger than -0.30, (i.e. slope changes at a junction are very small). Also junction angles are never smaller than 48 degrees. Note that the values of z never go below -0.8, while map measurements for the Dart River suggest that z could be as low as -2.0. Such low values of z imply a tremendous change in channel slope at a junction. Such variations in slope at a junction are surprising considering the maturity of the landscape of south-west England; perhaps they are spurious. Field measurements will undoubtedly improve the performance of the optimal model and provide for a much better test.

Conclusion

The optimality principle affords a general theoretical approach to the investigation of river branching angles. Based upon a link length minimization procedure, the model states that branching angles are a function of branching symmetry and of the hydraulic geometry relationships used in the definition of costs. The approach proposed herein imbeds as a special case the well-known Hortonian model of branching angles (Horton 1932, 1945, Howard 1971, Roy 1983). Empirical tests of the model suggest that it predicts well the average angles within a drainage network but fails to estimate accurately angles of individual junctions. The performance of the model is improved, however, when field measurements are used to compute the value of k. Map measurements should be used only to derive an exponent k that describes the whole system rather than individual junctions.

The optimal angular geometry models were originally designed for biological trees. Thus, it is surprising to see that the agreement between observed and optimal angles is much better for rivers than for airways (Roy 1982) and arteries. For instance, in the bronchial tree, the models do not predict the angles at all (Roy 1982). Rivers appear to adjust their branching geometry so as to achieve an optimum as defined by the length minimization. This tendency is strongly related to the effect of slope on the organization of river networks. Biological systems may be governed by other optimality rules which override the weighted link length minimization principle.

References

Cohn, D. L. 1954. Optimal systems; I. The vascular system. *Bull. Math. Biol.* **16**, 59–74.

Cohn, D. L. 1955. Optimal systems: II. The vascular system. *Bull. Math. Biol.* **17**, 219–27.

Hess, W. R. 1903. Eine mechanische bedingte Gesetzmässigkeit im Bau des Blutgefässystems. *Archiv. Entwicklungsmech. Organ.* 16, 632.

Horton, R. E. 1932. Drainage basin characteristics. *Trans. Am. Geophys. Union.* **13**, 350–61.

Horton, R. E. 1945. Erosional development of streams and their drainage basins: hydrophysical approach to quantitative morphology. *Geol. Soc. Am. Bull.* **56**, 275–370.

Howard, A. D. 1971. Optimal angles of stream junction: geometric stability to capture, and minimum power criteria. *Water Resources Res.* **7**, 863–73.

Leopold, L. B. and T. J. Maddock 1957. The hydraulic geometry of stream channels and some physiographic implications. *U.S. Geol. Surv. Prof. Pap. 500–A.*

Leopold, L. B., M. G. Wolman and J. P. Miller 1964. *Fluvial processes in geomorphology.* San Francisco: W. H. Freeman.

Lubowe, J. R. 1964. Stream junction angles in the dendritic drainage pattern. *Am. J. Sci.,* **262**, 325–39.

Miehle, W. 1958. Link length minimization in networks. *Oper. Res.* **6**, 232–43.

Murray, C. D. 1926a. The physiological principle of minimum work: I. *Proc. Nat. Acad. Sci.* **12**, 204–14.

Murray, C. D. 1926b. The physiological principle of minimum work applied to the angle of branching of arteries. *J. Gen. Physiol.* **9**, 835–41.

Murray, C. D. 1927. A relationship between circumference and weight in trees and its bearing on branching angles. *J. Gen. Physiol.* **10**, 725–29.

Pieri, D. C. 1979. *Geomorphology of Martian valleys.* Ph.D diss., Cornell Univ. Washington: NASA TM 8179.

Pieri, D. C. in press. Junction angles in drainage networks. *J. Geophys. Res.*

Rashevsky, N. 1960. *Mathematical biophysics-physico-mathematical foundations of biology.* II. New York: Dover.

Rosen, R. 1967. *Optimality principles in biology.* New York: Butterworths.

Roy, A. G. 1982. *Optimality and its relationship to the hydraulic and angular geometry of rivers and lungs.* Ph.D. diss., State Univ. New York, Buffalo. Ann Arbor: Univ. Microfilms.

Roy, A. G. 1983. Optimal angular geometry models of river branching. *Geogr. Anal.* **15**, 87–96.

Roy, A. G. and M. J. Woldenberg 1982. A generalization of the optimal models of arterial branching. *Bull. Math. Biol.* **44**, 349–60.

Roy, A. G. and M. J. Woldenberg (in press). Changes in channel form at a river junction. *J. Geol.*

Weber, A. 1909. *Über den Standort der Industrien.* Tübingen.

Wesolowsky, G. O. 1973. Location in continuous space. *Geogr. Anal.* **5**, 95–112.

Woldenberg, M. J. and K. Horsfield 1983. Finding the optimal lengths for three branches at a junction. *J. Theor. Biol.* **104**, 301–18.

Woldenberg, M. J. and K. Horsfield (in preparation). Relation of branching angles to optimality for four cost principles.

Zamir, M. 1976. Optimality principles in arterial branching. *J. Theor. Biol.* **62**, 227–51.

Zamir, M. 1978. Non symmetrical bifurcations in arterial branching. *J. Gen. Physiol.* **72**, 837–45.

13

Models of fluvial activity on Mars

Victor R. Baker

Introduction

The appearance of sinuous troughs fed by a network of tributaries does not cause great concern to a geologist studying space photographs of the Earth. Pictures of such features dissecting the cratered terrains of Mars (Fig. 13.1) have now been available to the scientific community for over a decade, yet considerable controversy rages concerning the origin of Martian channels and valleys. This paper cannot present an exhaustive review of the subject; indeed such a review is available (Baker 1982). Rather, the purpose here will be to discuss the problem for the insights that it conveys for the scientific endeavor of model-making. For geomorphology in particular the exploration of Mars has proven to be a stimulating source of analogic models (Sharp 1980).

Models may be defined as analogies or abstractions that are drawn in verbal, mathematical, or physical terms for rationalizing the necessary connections between changes in a system (Matalas *et al.* 1982). Models may be derived intuitively, without formal deductive capability. They comprise part of an upward spiral between analogic reasoning and the testing of that reasoning with scientific observations. At some point a model may begin to abstract from raw data the facts that its inventor perceives to be fundamental and controlling, placing these in relation to each other in ways that were not understood before, and thereby generating predictions of surprising new facts (Judson 1980). At this point the model has the qualities of a theory. True theories bind diverse consequences together in such an elegant manner that they compel belief by the scientific community. Geomorphology has little in the way of true theory, but is a fertile ground for the development of models that move from one level of analogic reasoning to another.

When a science is limited in its ability to predict new facts germane to its field or speciality there is a need for new models. Thus geomorphology is immensely improved when various observations, such as drainage network topology or river-channel hydraulic geometry, are fitted to predictive models (Shreve 1979). On the other hand, predictive ability does not necessarily constitute

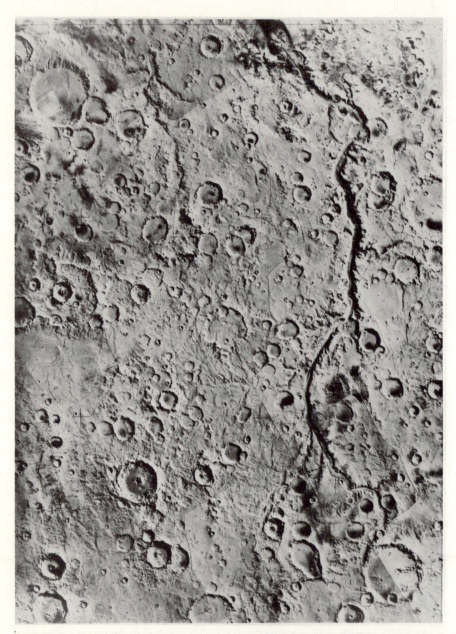

Figure 13.1 Mosaic of Viking Orbiter pictures showing Ma'adim Vallis and the heavily cratered terrain that it dissects. Small tributaries feed the upstream parts of this drainage system (bottom right). The flow was towards the north (top of the picture). The scene is approximately 500 × 750 km.

system understanding (Klemes 1982). The level of science that leads to new theory does not merely arise from intricate model-building with the existing data set. The puzzles generated by new discoveries are the most stimulating sources of new models that expand understanding beyond the limited range of existing theory. The general vitality of geomorphology during the discoveries of new landscapes in the American West, Africa, and Australia is more than coincidence. Similarly the exploration of the ocean basins led directly to the plate tectonic theory that revitalized geology.

Thanks to planetary exploration, geomorphology finally has a new source of elaborate puzzles with which to test the generalities of its existing theory. The upward spiral of model building has occurred at a phenomenally rapid pace, from discovery to overly simplistic initial models to equally simplistic reactionary models to improved models and to general syntheses. The rapid pace of planetary geomorphic model-building and its discussion, predominantly in nongeomorphological journals, has led to incomplete assimilation and to misunderstanding by many geomorphologists.

The explosion of Mars geomorphic studies (Baker 1981a) has occurred in the last decade. Semipopular accounts of Mars in the early part of this century envisioned a life-sustaining planet with an earth-like atmosphere and macro-engineering projects (the famous 'canals'). A more accurate though less exciting view was proposed by Dean B. McLaughlin in the 1950s. McLaughlin proposed that volcanism and wind action were dominant processes on the planet. His predictions subsequently were substantiated by the spacecraft missions of the last decade (Veverka & Sagan 1974).

The first spacecraft pictures of Mars were generated in the 1960s, but these low-resolution images failed to show prominent regions of volcanism, channeling, or valley development. The heavily cratered terrains revealed by these early missions resembled the lunar highlands. Consequently these missions inspired very conservative models of water abundance and atmospheric evolution. Leighton and Murray (1966) developed a computer model of the thin, cold modern Martian atmosphere. Given the small amount of carbon dioxide in that atmosphere and the probable accumulation of dry ice at the polar caps, their model suggested very modest volatile outgassing by the planet. These preliminary results led to a very conservative view about the Martian environment, comparing it more closely to the dry, lifeless Moon rather than to the water-rich Earth. The view became more polarized as a result of debates concerning the possibility of life on Mars. The semipopular, overly enthusiastic legacy of the speculative 'canal school' of Mars science only served to reinforce this conservative view.

The Mariner 9 mission in 1971–72 and the Viking mission of 1976–80 revealed the Martian channels and valleys as well as a whole host of water- and ice-related processes (Baker 1982). Models that allowed only modest outgassing would seem inconsistent with a planet that had obviously experienced such a long history of aqueous processes. Nonetheless a fascinating array of nonfluvial models was proposed to avoid the need to explain valleys and channels by water-related processes. Part of the reason for the persistent attempts to develop nonfluvial models is a scientific reaction to past models that tied hypothesized water on Mars to highly speculative accounts of life on the planet. Another reason for this persistence is the fact that the various models were being proposed just before large amounts of data of increasing resolution arrived with which to test the models.

The channels and valleys of Mars are interpreted from images produced by the sensing devices on spacecraft. The interpreter of landforms on those images relies on analogic reasoning to reconstruct the complex interaction of processes responsible for the observed features (Mutch 1979). The highest quality Viking orbiter images can resolve surface features as small as seven meters per picture element (pixel). This resolution exceeds that of familiar Landsat orbital pictures of Earth (resolution of 80 m per pixel for Landsats 1, 2, and 3). However, many of the early hypotheses for Mars channel genesis were proposed after studies of Mariner 9 imagery with resolutions of 100 to 1000 m per pixel. Even today the immensity and complexity of the Viking data resource (60 000 pictures) has prevented a complete study of the data.

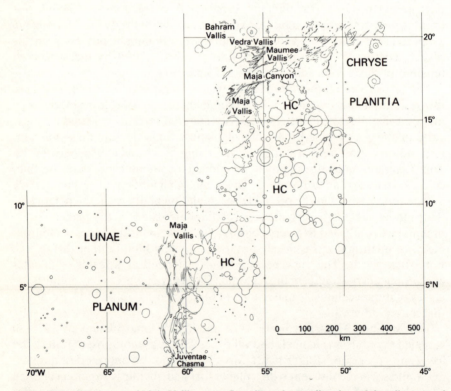

Figure 13.2 Sketch map of Maja Vallis, an outflow channel extending through heavily cratered uplands, HC, from Juventae Chasma (a region of chaotic terrain) to Chryse Planitia.

Models of channel formation

The term *channel* is properly restricted to those Martian troughs which display at least some evidence of large-scale fluid flow on their floors. Outflow channels show evidence of flows emanating from zones of complex collapse known as chaotic terrain. The Viking data have shown that the channels are immense features, as much as 100 km wide and 2000 km in length (Fig. 13.2). The gradients of channel floors range from about 1–2.5 m/km. A suite of bedforms on the channel floors indicate that large-scale fluid flows were involved in channel genesis. Many outflow channels show evidence of headward extension of the fluid-release collapse zones and concomitant erosion of downstream troughs by the channel-forming fluids. Morphological relationships in the channels show that the eroding fluid had a free upper surface, demonstrated by

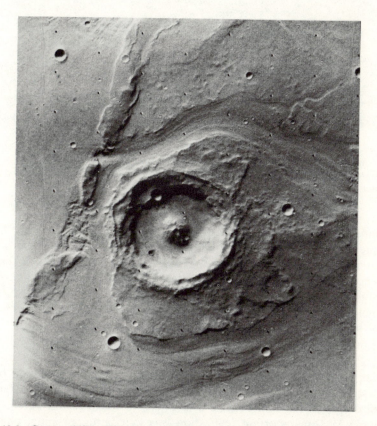

Figure 13.3   Crater at 20°N, 49°W near mouth of Maja Vallis (Fig. 13.2). Note that the ridge to the left (west) of the crater appears to have ponded fluid flows entering from the left. These flows breached low points in the ridge, eroding at constrictions and converging downstream with other flows. The crater is approximately 18 km in diameter.

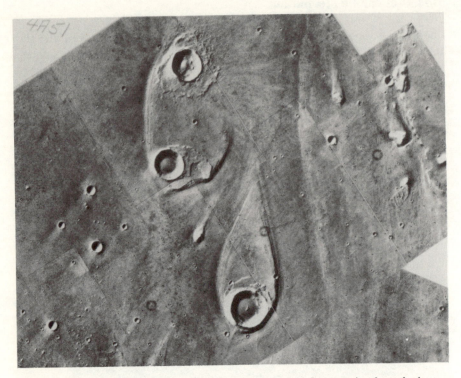

Figure 13.4 Streamlined hills that formed in large-scale fluid flows coming from the bottom (south) and flowing to the top (north). The raised crater rims acted as barriers to the flows, as crater ejecta were eroded on their upstream sides and preserved on the downstream sides. The crater at the bottom, center, is approximately 10 km in diameter.

its ponding upstream of flow constrictions (Fig. 13.3). It eroded scour holes, deposited bar-like sediment accumulations, spilled over low divides, and shaped magnificent streamlined hills (Fig. 13.4). Among cosmically abundant substances water seems best suited to satisfy these and other constraints on the nature of the primary agent for channel genesis (Baker 1978a, 1982, Mars Channel Working Group 1983, Masursky *et al.* 1977, Milton 1973, Sharp & Malin 1975).

The outflow channel geometry and bedforms require aqueous erosion on an immense scale, a scale achieved only in some terrestrial examples of catastrophic flooding (Baker 1973, 1978b, 1978c, 1981b). The evolution of the catastrophic flooding model of channel genesis shows an increase of model complexity as better data became available with which to test earlier ideas (Table 13.1). An important realization was that, after an initial phase of immense fluid flows, the outflow channels experienced extensive modification of their floors and walls by processes that included cratering, ground ice melting, eolian erosion and deposition, landsliding, debris flowage, and rilling.

Table 13.1 Development of the fluvial model for outflow channel formation on Mars.

Phase	Reference	Model Characteristics
discovery	McCauley *et al.* (1972)	recognition of fluvial features
	Masursky (1973)	significance of fluvial features
early models (monogenetic)	Milton (1973)	braided and meandering rivers
	Baker and Milton (1974)	cataclysmic flooding
	Nummedal *et al.* (1976)	braiding; fans; flooding
	Masursky *et al.* (1977)	fluvial history of Mars
	Trevena and Picard (1978)	braiding
reactionary models (fluvial-related alternatives)	Nummedal (1978)	liquefaction; mudflow
	Nummedal (1980); Nummedal and Prior (1981)	debris flow; submarine analogs
	Thompson (1979)	debris flow
	Lucchitta (1982)	glacial sculpture and small floods
revised models (polygenetic)	Sharp and Malin (1975)	endogenetic and exogenetic processes
	Baker (1978a, 1979)	multiple processes of flood erosion
	Baker and Kochel (1979)	post-fluvial modification
	Carr (1979)	relation to aquifers
	Theilig and Greeley (1979)	multiple ages of channeling
synthesis	Baker (1982)	multiple models reviewed
	Mars Channel Working Group (1983)	complex model by consensus of investigators

Despite the current concensus about an aqueous origin for Martian channels (Mars Channel Working Group 1983), it is fascinating that in the scientific literature nearly every conceivable fluid has been invoked to explain the Martian channels (Table 13.2). The list includes low-viscosity turbulent-flow lava, wind, glacial ice, liquefaction of crustal materials, debris flow, and water. Some fluids were proposed without reasonable analogs to their geomorphic effects (e.g. liquid alkanes and liquid CO_2). Some models achieved theoretical elegance (e.g. the eolian hypothesis of Cutts and Blasius 1981) but failed in their detailed geomorphological implications. The important comparison of models to observed Martian landforms can be illustrated by a compatibility matrix (Table 13.3). Note that only the aqueous hypotheses (debris flow, glacier, catastrophic flooding) are consistent with most of the Martian features.

Testing the models

Models for the origin of Martian outflow channels (Table 13.2) provide excellent illustrations of the scientific method. Suppose we assume that Table

Table 13.2 Single-process models for the origin of outflow channels on Mars.

Model	Reference	Model characteristics or terrestrial analog
eolian	Cutts and Blasius (1981)	theoretical
	Cutts et al. (1976)	yardangs
	Whitney (1979b, 1979c)	vortex features
debris flow	Nummedal (1978)	liquefaction and mudflows
	Nummedal and Prior (1981)	submarine debris flows
	Thompson (1979)	theory
glacial	Lucchitta et al. (1981)	sculpture by ice
	Lucchitta (1982)	ice and water flows
lava	Schonfeld (1977)	erosion by turbulent lava flows
	Cutts et al. (1978)	lava erosion
	Carr (1974)	rejection of lava hypothesis
fluvial	Baker and Milton (1974)	Channeled Scabland
	Sharp and Malin (1975)	flooding and seepage
	Masursky et al. (1977)	fluvial history
tectonic (no fluids)	Schumm (1974)	tensional fractures
carbon dioxide	Lambert and Chamberlain (1978)	CO_2 thermodynamics flows of CO_2 slurry
liquid alkanes	Yung and Pinto (1978)	theoretical

13.3 constitutes a true test of various models. Used in this way the models are really elaborate hypotheses. Either a given model (1) is consistent with all the features in Martian channels (barring observational error), or (2) it is not. The second result certainly shows that the model (hypothesis) is not completely satisfactory. However, the first result does not establish model validity. Tests showing a lack of disagreement do not prove hypotheses; they only serve to show failure at disproving them. Indeed in science most models are never proven; they are only shown to be more probable than other models. A model does not enhance its probability by surviving tests; it gains acceptance when tests reduce the probability of its rivals. Thus, as stated by Bayly (1968, p. 119): '... the probability of a hypothesis is established, not by tests, but by the improbability of its rivals.'

In human terms the reality of scientific methodology (Popper 1959) seems an unfortunately destructive way to proceed. To establish a favored hypothesis the investigator cannot simply test the analogies (or models) that relate to that hypothesis; rather, the investigator must test the rival hypotheses. If the hypotheses are too closely identified with the hypothesizer, then this may be viewed as a regrettable, even an offensive, way to proceed. Nevertheless it is

Table 13.3 Compatibility matrix listing the important morphological features of Martian outflow channels (Baker 1982) and ability of various flow systems to explain these features.

Morphological features	Wind	Debris flow Mudflow	Glacier	Lava	Cataclysmic Flood
anastomosis	?	X	X	X	X
streamlined uplands	X	X	X	?	X
longitudinal grooves	X	X	X	?	X
scour marks	X	?	?	?	X
scabland	?	–	?	–	X
inner channels	?	X	?	X	X
lack of solidified fluid of channel mouth	X	–	X	–	X
localized source region	?	X	X	X	X
flow for thousands of kilometers	X	–	X	X	X
bar-like bed forms	?	?	?	?	X
pronounced upper limit to fluid erosion	–	X	X	X	X
consistent downhill fluid flow	?	X	X	X	X
sinuous channels	*	X	*	X	X
high width-depth ratio	X	X	X	–	X
headcuts	–	?	X	–	X
U-shaped valleys	?	?	X	?	X

? questionable;   – incompatible;   X compatible;
* occur as modification of pre-existing fluvial valleys.

the only valid scientific approach. Bayly (1968, p. 120) summarized the reason for this as follows: '. . . science is not the orderly accumulation of facts; it is the orderly accumulation of rejected hypotheses.'

Models may fail tests either because the models are invalid or because the tests are invalid. A fascinating rationale for rejecting the cataclysmic flood model of outflow channel genesis was that it appealed to catastrophic processes. Thus Lambert and Chamberlain (1978) reject water-related processes on Mars with the following statement: 'Any satisfactory H_2O-based theory is necessarily nonuniformitarian or even catastrophic in nature and unreasonable to that extent.' Whitney (1979b, p. 1127) describes the fluvial hypothesis as follows: '. . . a turning away from the principles of uniformitarianism to catastrophism to explain the Martian channels.' Cutts and Blasius (1981, p. 5097) explain their work on the eolian model with the statement: '. . . our approach has been to attempt to explain the channels in terms of processes which are still at work on Mars rather than to invoke catastrophic events for which there is little independent evidence.'

That the above arguments about uniformitarianism can still be used reflects considerable misunderstanding of the concept. Catastrophism and uniformitarianism ceased to be issues when geology separated itself from religious issues in the past century (Baker 1981b). The power of this old idea must be profound when it is invoked for one landform (channels) on a planet that is dominated by another landform (impact craters). On Earth, of course, it is the impact craters that many would consider catastrophic.

Just as fascinating an exercise is the justification of model hypotheses by the weight of the evidence. Whereas this approach has decided numerous litigations, it has not served as a scientific tool. If, like lawyers, scientists emphasized the evidence that supports their models (cf. cases), then science would advance because of an investigator's skill at model presentation, not his skill at model testing. For example, an elegant theoretical model has been developed to explain the origin of Martian outflow channels through sand blasting by wind (Cutts & Blasius 1981). This model is quite consistent with numerous factors in the Martian environment and with many aspects of Martian geologic history. The key point for channel genesis, however, is that the model provides highly improbable explanations for a number of fundamental geomorphic consequences (Nummedal *et al.* 1983), including the following: (1) confined flow in sinuous channels; (2) ponding of flows upstream of constrictions; (3) the ubiquitous nature of flow down regional topographic gradients; (4) the demonstrated low efficacy of eolian processes on Mars for eroding even small impact craters; and (5) inconsistency of channel directions with reasonable global wind patterns. The eolian hypothesis was considered in the first studies of Martian channels and found inconsistent with the observation of sinuous valleys developed by a fluid with a free surface and an uncanny tendency to always flow downhill (Milton 1973).

Models that are reasonable at one scale of operation are not necessarily workable at vastly different scales. An example of this is the interesting work of Whitney (1978, 1979a) on the development of small-scale pits, linear erosion trends, and other micro-sculpturing of ventifacts by suspended particles in wind transport. Whitney's model of wind vorticity and aerodynamic lift works well in the small-scale boundary layer around ventifacts. However, the extension of this wind-erosion process to the planet-wide scale for Mars (Whitney 1979b, 1979c) went so far beyond the bounds of reasonable analogic model testing that few channel investigators will refer to the papers. Whitney (1979b, 1979c) extended her model to obviously faulted terrains in the Valles Marineris, structurally controlled valley networks, ridge spurs, impact craters (termed vortex pits), shield volcanoes with calderas at their summits (considered to be erosional remnants with vortex pits at their summits), and nearly every other type of terrain on the surface of Mars. Whitney (1979c, p. 1129) observed: 'While I concede that meteoritic impact and volcanoes produce craters, to date there seem to be no very good criteria for distinguishing the morphologies of the several wind-formed pits from those of the Martian craters.' This statement verbally dismisses a rich scientific literature, hundreds of papers documenting the details of impact-generated landforms, shield volcano studies in Hawaii, indeed entire books on comparative planetology. The publication of this speculation in the prestigious *Geological Society of*

America Bulletin shows that the pace of model development can race ahead in spite of the availability of excellent data with which to test those models. There was a lack of awareness which extended to the editorial board and reviewers of a major scientific society. Of course models that clearly violate numerous tests are sometimes ignored by the scientific community. (The Whitney Mars papers were never discussed in the journal, presumably to avoid drawing attention to them.) If published in a prominent journal, however, such models may severely mislead scientists who are unfamiliar with the field.

A fine line may exist between the outrageous hypothesis that goes beyond the bounds of existing theory to explain startling new facts, and the outrageous hypothesis that ignores important facts merely to present speculation as a startling new theory. The former can serve as a stimulating source of scientific advancement, as in the famous Spokane Flood Debate over the origin of the Channeled Scabland (Baker 1981b). However, the latter tends to suppress facts that are inconsistent with the favored hypothesis. Thus, Whitney (1979c, p. 1143) states: '. . . there are no features of any so-called impact crater on Mars that could not also have been produced by wind vorticity.'

An example of a reactionary model that usefully advances knowledge is the glacial hypothesis of outflow channel genesis by Lucchitta *et al.* (1981) and Lucchitta (1982). Lucchitta (1982) illustrates striking morphological similarities between some glacially eroded terrains (Antarctica and Alaska) and some features of the Martian outflow channels. In particular the large grooves of glaciated terrains are seemingly well explained, although grooved terrain has been produced by cataclysmic flooding (Baker 1981b). This similarity of certain landforms produced by glaciation and those produced by flooding is a classic example of equifinality (von Bertalanffy 1950, 1952; Chorley 1962) which is the convergence of somewhat dissimilar processes to produce apparently identical landforms.

The problems with the glacial hypothesis are in its unanswered regional and theoretical questions. Most of the terrestrial examples of pronounced glacial erosion come from situations where glacial ice has converged by flow from extensive accumulation zones. The erosive ice streams of Antarctica did not form from accumulation in valleys, as would have to be the case on Mars. Instead ice from the extensive accumulation area of the West Antarctic Ice Sheet converges to flow at high velocity through narrow bedrock gaps. Likewise the Alaskan examples occur where accumulation from numerous tributary valley glaciers converge and modify proglacial sinuous, fluvially eroded valleys. All the Martian examples are elongate troughs without possible ice sheet source areas. To press the glacial analogy for Martian channels would require a paradoxical formation of ice sheets in valleys. The glacial model also raises improbable mass balance considerations for the hypothesized Martian glaciers. Lucchitta (1982) considers atmospheric precipitates and wind drifts to be unlikely sources of mass addition to the glaciers, favoring instead water from springs and small catastrophic outbursts. While a big enough pile of jammed ice would probably flow, the key question is what would maintain the flow.

Glaciers are effective erosive agents, not because of high power, but because of their prolonged application of moderately high power. It seems unlikely that springs or small floods could provide just enough water to freeze and maintain a net positive mass balance for the glacier. It would be more likely that either the ablation would exceed the positive input of small water additions, or that excessive water would dam up behind the ice jams, eventually breaking out from behind them as proglacial floods.

Other difficulties with the glacial hypothesis include (1) the occurrence of scour and depositional features that seem most consistent with a turbulent erosive and sediment-transporting fluid (Baker 1982), and (2) streamlined upland shapes more consistent with fluvial than glacial flow processes (Komar in press). Nevertheless, unlike the eolian model, the glacial model is in reasonable accord with many channel landforms (Table 13.3). We simply know too little about the operations of glaciers to reject this model. Though it seems to me less probable than the fluvial model, it remains a worthy hypothesis for portions of some outflow channels. Of historical interest is the fact that a glacial hypothesis for the origin of the Channeled Scabland was argued for decades on the basis of some of the same evidence used to demonstrate the catastrophic flood origin (Baker 1981b). The question was only resolved through painstaking and thorough field work by J. Harlen Bretz.

In a relatively short time period the scientific debate over the origin of Martian outflow channels has illustrated the classic wisdom of multiple working hypotheses as a systematic method of scientific thinking. This method utilizes several competing hypotheses, rather than a ruling or controlling one, to evaluate alternative explanations of phenomena (Chamberlin 1890, 1897). The alternatives are then tested by crucial experiments or with new data (from the field or laboratory). The whole procedure is then recycled to refine remaining hypotheses. This basic philosophy, honestly and vigorously pursued, contributes to the strong inference that distinguishes a rapidly advancing scientific field from a stagnant one (Platt 1964). By consciously incorporating hypothesis testing, analogic reasoning (Gilbert 1886), and the dispassionate recognition of scientific advancement by disproof (Popper 1959), the method of multiple working hypotheses probably provides the best hope for making competing models converge on an explanation for complicated landforms or landscapes (Twidale 1977).

Models of fluvial erosion and transport

Quantitative physical and/or mathematical models develop in parallel to more qualitative models, such as those that merely compare landforms produced by different processes. Although terrestrial analogs have been immensely useful in evaluating alternative modes of channel genesis, no one analog has fully accounted for the suite of features observed in the Martian examples. This has led to theoretical studies of the basic flow physics and of various flow systems

Table 13.4 Models of fluvial erosion and transport

Process	Reference	Model characteristics
streamlining	Baker and Kochel (1978)	comparison to Channeled Scabland
	Baker (1979)	lemniscate theory
	Komar (in press)	flume experiments
sediment	Komar (1979)	Bagnold's (1966) theory
transport	Komar (1980)	washload theory
macroturbulence	Baker (1979)	theoretical; Channeled Scabland
cavitation	Baker (1979)	theoretical
	Pieri (1980a)	theoretical
river ice	Baker (1979, 1982)	arctic rivers
ice sculpture by accumulated river ice	Lucchitta (1982)	glacial erosion processes

that might induce unique landforms that can be recognized on Mars (Table 13.4).

Of the various erosional-deposition systems studied, the streamlined forms (Fig. 13.4) may be the most important indicators of flow dynamics. Baker and Kochel (1978) suggested that these shapes developed to generate minimum resistance by the residual form to the fluid flow field. By morphometric analysis Baker and Kochel (1978) and Baker (1979) demonstrated the similarity of streamlined forms in the Channeled Scabland to those in several Martian outflow channels. Lucchitta (1982) extended this analytical technique to glacially carved 'islands' associated with the Antarctic ice sheet. The Antarctic glacial forms are less elongate than Martian streamlined islands, but the scabland forms tend to be more elongate than the Martian forms.

Komar (1983, and in press) has completed a rigorous morphometric and theoretical analysis of both glacial and water produced streamlined forms. He found that glacial drumlins exhibit a high variability of shape ranging from ellipsoid to lemniscate (streamlined). However, streamlined islands in rivers, the Channeled Scabland, and Martian outflow channels are all consistent with lemniscate shapes. The islands acquire their shapes by minimizing the drag or resistance to the flowing fluid when they are formed. Only water and eolian forms seem to resemble the kinds of ideal shapes observed in the Martian outflow channels. Since wind-induced streamlining is excluded by evidence of a free surface on the responsible fluid, water emerges as the most probable fluid to explain the streamlining.

Models of fluid release

On Mars the appearance of flowing surface water must be explained in terms of known present atmospheric conditions or postulated ancient conditions. For the outflow channels the required short-duration floods are not severely constrained even by the present Martian atmosphere, but a mechanism must be found to yield the necessary immense quantities of water. The proposed release mechanisms include outbursts of melted ground ice from an ice-rich permafrost heated by volcanism, meteor impacts into the ice-rich permafrost, liquefaction of sensitive subsurface materials on Mars, and sudden release of immense aquifers of very high permeability confined by the ice-rich permafrost (Table 13.5). Although an extensive planet-wide permafrost system appears to have been involved, the precise release mechanism for fluids to form outflow channels remains highly speculative. The major obstacle to testing the various fluid-release models is that the key evidence is in the subsurface. Here then is an important problem to be investigated by future scientific missions to Mars.

Once water finds its way to the surface of Mars in liquid form a mechanism must be found to maintain it there sufficiently long to perform the indicated work of erosion and sediment transport. Models of prolonged surface flow on Mars postulate a warmer, denser ancient atmosphere, ice-covered rivers, freezing point depressants, and other environmental controls (Table 13.6). Calculations by Carr (1982) suggest that water can easily flow on Mars even for the present conditions of extreme cold and low atmospheric pressure. Carr (1982) calculates that flows 0.5 to 1.0 m deep will persist for perhaps 100 km at reasonable gradients. The key problem seems to be that of defining conditions for water to become liquid. Instabilities in an ice-rich permafrost are generally

Table 13.5 Some models of fluid release for Martian floods.

Mechanism	Reference
meteor impact into ground ice	Maxwell *et al.* (1973)
crater-controlled igneous melting of ground ice	Schulz and Glicken (1979)
volcano-ground ice interaction	McCauley *et al.* (1972)
	Masursky *et al.* (1977)
seepage zones	Sharp and Malin (1975)
geothermal heating of water confined beneath clay-rich regolith	Clark (1978)
rupture of confined aquifers	Carr (1979)
liquefaction of sensitive materials in megaregolith	Nummedal (1978)
release of ponded water from lakes	Lucchitta and Ferguson (1983)
phase transformation of carbon dioxide hydrate in Martian permafrost	Milton (1974)
breakout of liquid water from a permafrost zone rich in ground ice	Peale *et al.* (1975) Soderblom and Wenner (1978)
hydrothermal melting of ground ice associated with rejuvenation of ancient impact basin structures	Schultz *et al.* (1982)

Table 13.6 Models for maintaining surface-water flow on Mars.

Hypothesis	Reference
warmer, denser ancient atmosphere	Pollack (1979)
ice-covered rivers	Wallace and Sagan (1979)
freezing point depressants	Ingersoll (1970)
short-duration floods	Baker and Milton (1974)
present conditions are adequate	Carr (1982)

considered most likely for outflow channels, but the valley networks pose a very different set of problems.

Models of valley network development

The term *valley network* applies to Martian trough systems, which appear to form by fluid flow, but which, unlike the outflow channels, lack a suite of bedforms on their floors. The Martian valleys of greatest interest consist of interconnected, digitate networks that dissect extensive areas of heavily cratered uplands on the planet (Fig. 13.5). Although they were recognized on Mariner 6 and 7 images (Schultz & Ingerson 1973), the high-resolution Mariner 9 and Viking pictures revealed the most important network properties (Table 13.7). As with the development of the fluvial model of outflow channel genesis, the fluvial model of the Martian valley network morphogenesis also shows several phases of development (Table 13.8).

Because the valley networks are much smaller than the outflow channels, the lower resolution of the first pictures of these systems was a severe impediment to model development. Based on the Mariner 9 pictures Masursky (1973), Milton (1973), and Sagan *et al.* (1973) concluded that rainfall and overland flow processes probably had occurred on Mars. In an extreme example of extrapolation from empirical relationships, Weihaupt (1974) applied various discharge versus 'channel' dimension equations to the sinuous 'valley' of Nirgal Vallis (Fig. 13.6). The resulting calculation indicated a mean annual discharge similar to that of the Mississippi River. Not only did this calculation apply channel geometry considerations to a valley, it also failed to consider the influence of reduced Martian gravity on the basic hydraulics (Komar 1979).

The steep-walled, theater-like terminations of many Martian valleys (Fig. 13.7) led to the suggestion that basal undermining and wall collapse (sapping) are the important processes of valley formation on Mars (Pieri 1980a, 1980b; Higgins 1982). These details of valley morphology were recognized on high-resolution Viking imagery, so the sapping model arose as a reaction to simple overland flow concepts of network genesis that were suggested by the low-resolution Mariner 9 pictures.

Following earlier studies on terrestrial tributary junction angles (Howard

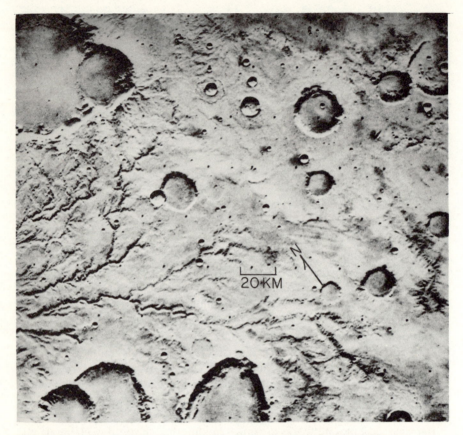

Figure 13.5 Parana Vallis, a valley network in the heavily cratered terrain of Mars.

1971, Lubowe 1964), Pieri (1980a) developed a quantitative model for junction angle statistics. The model shows a close fit to various terrestrial networks, but Martian valleys deviate significantly from it. The junction angles in Martian networks do not show the small tendency of angles to increase downstream as the size of recipient streams increases. This relationship was also interpreted as consistent with network generation by sapping.

The sapping mechanism of network formation has now been modeled in the laboratory (Kochel *et al*. 1985). The model networks developed by Kochel *et al*. show a close correspondence to the tributary networks of the Valles Marineris canyon system on Mars. However, considerably more work is needed to understand the complex morphogenesis of valley networks in the heavily cratered terrains of Mars (Baker *et al*. 1982). The latter show extensive modification by cratering, mass movement, volcanic, thermokarstic, and eolian processes. Some highly degraded networks show such extensive development that overland flow runoff processes may be implied.

Complexity and polygenetic developmental models are also apparent in

Table 13.7 List of typical morphological features for the valley networks on Mars (Baker 1982).

theater-like valley heads of short first-order segments
low drainage density
hanging valleys
structural control of network
steep-sided, crenulated valley walls
valley width does not increase uniformly in downstream direction
tributaries join mainstream at low junction angles
networks are parallel to quasi-parallel, but dendritic
distal termini of networks are closed depressions on obscure terrains
very high crater densities (indicating great age, probably >3.5 billion years)

Table 13.8 Developmental phases of the fluvial model of valley network morphologies on Mars.

Phase	Reference	Model characteristics
recognition	Masursky (1973)	small fluvial networks
	Milton (1973)	gullies
	Sagan et al. (1973)	latitudinal control
	Hartmann (1974a)	arroyos
	Weihaupt (1974)	empirical equations; Mississippi River
early models	Sharp and Malin (1975)	runoff processes
	Mutch et al. (1976)	runoff processes
	Pieri (1976)	regional distribution
	Sharp (1973)	dry sapping
	Pieri (1980a, 1980b)	sapping; junction angles
sapping models	Baker (1980a, 1980b)	sapping developed by Dunne (1980)
	Higgins (1982)	sapping; beach analog
	Laity and Malin (in press)	Colorado Plateau analog
	Baker (1980c)	valleys on volcanoes
	Carr and Clow (1981)	global distribution
revised models	Mouginis-Mark et al. (1982)	valleys on volcanoes
(multigenetic)	Breed et al. (1982)	Gilf Kebir Plateau
	Maxwell (1982)	Gilf Kebir Plateau
	Kochel et al. (1985)	laboratory model
	Baker (1982)	review of multiple models
synthesis	Mars Channel Working Group (1983)	consensus of investigators

studies of other types of valley systems on Mars that do not fit so readily into the usual classification schemes. Thus Breed et al. (1982) argued that the prominent fretted channels of the Martian highlands-plains boundary reflect a climatic morphogenesis from a past climate favoring fluvial processes to a modern hyperarid climate. This morphogenesis might parallel that which occurred in the hyperarid eastern Sahara, where the relatively wet Tertiary

Figure 13.6   The so-called 'meandering reach' of Nirgal Vallis. This scene depicts a region 80 × 85 km.

climate permitted valleys to develop on uplands like the Gilf Kebir Plateau of southwestern Egypt (El-Baz & Maxwell 1982). During the Pleistocene, eolian and mass wasting processes in an arid climate replaced fluvial action as the dominant factor in landscape development.

Whereas outflow channels are easily distinguished by their diagnostic bedforms, the absence of these bedforms in other valleys would not seem to necessitate hypothesizing a mechanism involving flowing water. This problem is especially pronounced for valley development on some of the Martian volcanoes, where lava channels are possible (Carr 1974). Mouginis-Mark *et al.* (1982) mapped the distribution of sinuous valleys on Hecates Tholus, a low shield volcano that rises about 6 km above the surrounding plain and has a diameter of about 160 km. The valleys radiate from the summit area, showing convergent parallel-dendritic patterns. Riemers and Komar (1979) developed a model for explosive eruption that explained these valleys as the products of erosion by volcanic density currents. However, Mouginis-Mark *et al.* (1982) point out that many of the Hecates valleys originate 20–40 km from the potential eruptive source at the summit caldera. Moreover, some valleys cut across craters without exhibiting the infilling that would be expected for various kinds of volcanic flows.

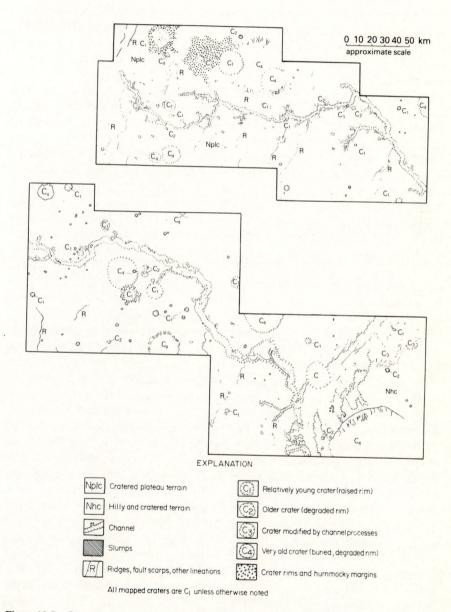

0 10 20 30 40 50 km
approximate scale

EXPLANATION

Nplc	Cratered plateau terrain	C_1	Relatively young crater (raised rim)
Nhc	Hilly and cratered terrain	C_2	Older crater (degraded rim)
	Channel	C_3	Crater modified by channel processes
	Slumps	C_4	Very old crater (buried, degraded rim)
R	Ridges, fault scarps, other lineations		Crater rims and hummocky margins

All mapped craters are C_1 unless otherwise noted

Figure 13.7 Geomorphic map of Nirgal Vallis showing network development of probable sapping origin.

Models of atmospheric evolution

Channels and valleys on Mars are important for the information they convey on the evolution of the atmosphere and the climatic history of the planet. Early attempts to refute fluvial models of Martian landforms were consistent with the atmospheric models that claimed Mars began as a planet impoverished in volatiles (Leighton & Murray 1966). Later attempts to estimate volatile outgassing by Mars yielded greatly variable amounts (Table 13.9). Depending on the initial assumptions the range of water outgassing spans two orders of magnitude.

An inventory of existing water on Mars cannot be made because most of it probably resides in a planet-wide permafrost zone that is hidden from orbital inspection. Estimates for the thickness of this zone range from 1–3 km(Sharp 1973, Soderblom & Wenner 1978, Rossbacher & Judson 1981) to 3.5–8 km (Anderson 1982). Many of the unsolved problems of Martian geomorphology relate to the operation of this permafrost system, including the origin of the outflow channel fluids (Table 13.5). An ingenious global hydrologic cycle for Mars has been proposed (Clifford 1981) in which water is outgassed from equatorial regions and replenished by upward thermal migration of water from an extensive subpermafrost groundwater system. This ground water is recharged by atmospheric transfer of water and precipitation at the poles. Insulation of the polar caps generates basal melting, which then allows water to pass into the planet-wide subpermafrost groundwater system. A hydraulic gradient must be maintained to transfer this ground water from the poles to equatorial regions.

Some of the models that suggested low water outgassing by Mars relied on an assumed scaling of various gas abundances to the noble gases neon, argon, and krypton. Presumably these gases are too heavy to escape to space and too inert to be incorporated into rocks and sediments. Therefore they remain in the planetary atmospheres even when more reactive gases (oxygen, nitrogen, water vapor, and carbon dioxide) have been lost. Mainly because of the argon data for Mars, Anders and Owen (1977) predicted rather modest outgassing. In 1979 the Pioneer Venus spacecraft returned data showing that the Venus atmosphere does not conform to the model of strict scaling of volatile abundances to noble gases abundances on the inner planets. By comparing Earth, Mars, and Venus, Pollack and Black (1979) showed that Mars must have outgassed 10 to 20 times more water than was previously assumed. Thus plenty of water seems to have been available for fluvial processes.

The small valley networks of the Martian heavily cratered terrains are the only fluvial features on Mars that seem to require climatic conditions different from those of today. To form channels by runoff processes a hydrologic cycle must have functioned, if only to recharge aquifers for sapping. Since the small valley networks are extremely ancient, probably dating to the first several hundred million years of planetary history, or at least the first billion years, they imply special atmospheric conditions at that time. A paradox in this is that

Table 13.9 Estimates of Martian volatile outgassing.

	Fanale (1976)	Anders and Owen (1977)	Pollack (1979)	Pollack and Black (1979)
water*	1 km	9.4 m	75 m	80–160 m
CO_2	several bars	140 mb	500 mb	500–1000 mb
Volatile inventory	same as Earth (pre-Viking)	less (low Argon 36)	less (high Argon 40)	scaled to Earth and Venus
efficiency of outgassing	same as Earth	inefficient	more efficient	scaled to Earth and Venus

*Equivalent to depth of water averaged over the planetary surface.

stellar evolutionary theory indicates that the sun probably had a lower luminosity during the early history of the solar system, implying that Martian conditions might have been even colder than at present.

The answer to this dilemma lies in various models of climatic change, some of which are listed in Table 13.10. An attractive idea is that the accretionary process and an initial phase of active tectonism or volcanism leads to very rapid outgassing early in the planet's history. If enough CO_2 were outgassed, it would have created a greenhouse effect, warming the surface above the freezing point of water. The temperatures would probably have fluctuated because of the immense obliquity variations of the Martian spin axis (Ward *et al.* 1979). The high CO_2 atmosphere could not last, however, because when water is present CO_2 is readily precipitated into carbonate rocks, removing it from the atmosphere. When the early Martian atmospheric temperature fell below the freezing point of water, the planetary water had to concentrate in the vast permafrost reservoirs. It was subsequently released only by specialized processes related to the permafrost.

The view that Mars evolved from a volatile-rich state has been linked to probable evolutionary schemes of the Earth and Venus atmospheres by Walker (1978). Walker's evolutionary models seem to hold up well under the testing provided by the Earth's geologic record (Walker 1977). It is intriguing that the problem of planetary atmospheric evolution requires the comparison of all planetary atmospheres to study the origin of any one. Furthermore since little or no geologic record remains of the first billion years of Earth's history, the geomorphology of the Martian valley networks may be a clue to the early climatic conditions of Earth.

Conclusions

This review has not attempted to summarize and evaluate all the various models for fluvial activity on Mars, although Tables, 13.1, 13.2, 13.4, 13.5, 13.6, and 13.8 provide a relatively complete list of important references.

Table 13.10 Models of the primordial Martian atmosphere as related to the origin of valley networks.

Reference	Model characteristics
Leighton and Murray (1966)	Mars dry and cold throughout its history
Sagan et al. (1973)	high pressure CO_2 atmosphere modulated by insolation changes
Mutch et al. (1976)	catastrophic outgassing; dense water-rich atmosphere
Sagan (1977)	reducing greenhouse atmosphere
Pollack (1979	
Ringwood (1978)	Mars was volatile-rich in its primordial state
Walker (1978)	evolution of atmosphere from an early volatile-rich state
Ward et al. (1979)	immense obliquity variations could have generated periodic polar warming
Cess et al. (1980)	CO_2 greenhouse

Rather, the foregoing has been a highly selective discussion of the decade of scientific study devoted to a fascinating geomorphic puzzle. The origin of channels and valleys on Mars is important not merely as a scientific question to be answered but also as a rich source of new geomorphic models. These models address the same fundamental question that led James Hutton and John Playfair to establish a scientific basis for the study of landforms. Only the scope of the question has been expanded: what is the origin of valleys—on Earth and on other planets?

Acknowledgments

My research on Mars geomorphology has been supported through the N.A.S.A. Planetary Geology Program, Grants NSG-7326, NSG-7557, and NAGW-285. This particular study has benefited from my participation in the Mars Channel Working Group, which is also sponsored by the Planetary Geology Program of N.A.S.A. Over the years my close work with G. R. Brakenridge, R. C. Kochel, and P. C. Patton has contributed immensely to ideas developed in this paper. Discussions with P. D. Komar and J. C. G. Walker provided important insights, and M. Woldenberg's review of the manuscript was especially helpful.

References

Anders, E. and T. Owen, 1977. Mars and Earth: origin and abundance of volatiles. *Science* **198**, 453–65.

Anderson, D. M. 1982. *Physical and mechanical properties of permafrost on Mars*. N.A.S.A. Tech. Memo. 85127, 264.

Bagnold, R. A. 1966. An approach to the sediment transport problem from general physics. *U.S. Geol. Surv. Prof. Pap.* 422–I, 1–37.

Baker, V. R. 1973. Paleohydrology and sedimentology of Lake Missoula flooding in eastern Washington. *Geol. Soc. Am. Special Pap.* 144, 1–79.

Baker, V. R. 1978a. A preliminary assessment of the fluid erosional processes that shaped the Martian outflow channels. *Proc. Lunar and Planet. Sci. Conf.* IX, 3205–23.

Baker, V. R. 1978b. The Spokane Flood controversy and the Martian outflow channels. *Science* **202**, 1249–56.

Baker, V. R. 1978c. Large-scale erosional and depositional features of the Channeled Scabland. In *The Channeled Scabland*, V. R. Baker and D. Nummedal (eds.), 81–115. Washington: National Aeronautics and Space Administration.

Baker, V. R. 1979. Erosional processes in channelized water flows on Mars. *J. Geophys. Res.* **84**, 7985–93.

Baker, V. R. 1980a. *Some terrestrial analogs to dry valley systems on Mars.* N.A.S.A. Tech. Memo. 81776, 286–88.

Baker, V. R. 1980b. *Nirgal Vallis.* N.A.S.A. Tech. Memo. 82385, 345–47.

Baker, V. R. 1980c. *Degradation of volcanic landforms on Mars and Earth.* N.A.S.A. Tech. Memo. 82385, 234–35.

Baker, V. R. 1981a. The geomorphology of Mars. *Prog. Phys. Geog.* **5**, 473–513.

Baker, V. R. (ed.) 1981b. *Catastrophic flooding: the origin of the Channeled Scabland.* Strouds-burg, Penn.: Dowden, Hutchinson and Ross.

Baker, V. R. 1982. *The channels of Mars.* Austin, Tex.: Univ. of Texas Press.

Baker, V. R., G. R. Brakenridge, and R. C. Kochel 1982. *Valley networks on Mars: mapping and morphogenesis.* N.A.S.A. Tech. Memo. 85127, 206–8.

Baker, V. R., and R. C. Kochel 1978. Morphometry of streamlined forms in terrestrial and Martian channels. *Proc. Lunar and Planet. Sci. Conf.* IX, 3193–203.

Baker, V. R., and R. C. Kochel 1979. Martian channel morphology: Maja and Kasei Valles. *J. Geophys. Res.* **84**, 7961–83.

Baker, V. R., and D. J. Milton 1974. Erosion by catastrophic floods on Mars and Earth. *Icarus* **23**, 27–41.

Bayly, B. 1968. *Introduction to petrology.* Englewood Cliffs, N.J.: Prentice-Hall.

Breed, C. S., J. F. McCauley, and M. J. Grolier 1982. Relict drainages, conical hills, and the eolian veneer in southwest Egypt—applications to Mars. *J. Geophys. Res.* **87**, 9929–50.

Carr. M. H. 1974. The role of lava erosion in the formation of lunar rilles and Martian channels. *Icarus* **22**, 1–23.

Carr, M. H. 1979. Formation of Martian flood features by release of water from confined aquifers. *J. Geophys. Res.* **84**, 2995–3007.

Carr, M. H. 1982. *Mars valley networks as indicators of former climatic conditions.* N.A.S.A. Tech. Memo. 85127, 203–5.

Carr, M. H., and G. D. Clow 1981. Martian channels and valleys: their characteristics, distribution, and age. *Icarus* **48**, 91–117.

Cess, R. D., V. Ramanathan, and T. Owen 1980. The Martian paleoclimate and enhanced atmospheric carbon dioxide. *Icarus* **41**, 159–65.

Chamberlin, T. C. 1890. The method of multiple working hypotheses. *Science* **15** (old series), 92–6.

Chamberlin, T. C. 1897. Studies for students: the method of multiple working hypotheses. *J. Geol.* **5**, 837–48.

Chorley, R. J. 1962. Geomorphology and general systems theory. *U.S. Geol. Surv. Prof. Pap.* 500–B.

Clark, B. C. 1978. Implications of abundant hygroscopic minerals in the Martian regolith. *Icarus* **34**, 645–55.

Clifford, S. M. 1981. A model for the climatic behavior of water on Mars. In *Third International Colloquium on Mars*, 44–45. Houston: Lunar and Planetary Institute.

Cutts, J. A., and K. R. Blasius 1981. Origin of Martian outflow channels: the eolian hypothesis. *J. Geophys. Res.* **86**, 5075–102.

Cutts, J. A., K. R. Blasius, and K. W. Farrell 1976. Mars: new data on Chryse Basin landforms. *Bull. Am. Astron. Soc.* **8**, 480.

Cutts, J. A., W. J. Roberts, and K. R. Blasius 1978. Martian channels formed by lava erosion. *Lunar and Planet. Sci.* **9**, 209. Houston: Lunar and Planetary Institute.

Dunne, T. 1980. Formation and controls of channel networks. *Prog. Phys. Geog.* **4**, 211–39.

El-Baz, F., and T. A. Maxwell (eds) 1982. *Desert landforms of southwest Egypt: a basis for comparison to Mars*. Washington, D.C.: National Aeronautics and Space Administration.

Fanale, F. P. 1976. Martian volatiles: their degassing history and geochemical fate. *Icarus* **28**, 179–202.

Gilbert, G. K. 1886. The inculcation of scientific method by example. *Am. J. Sci.* **31**, 248–99.

Hartmann, W. K. 1974a. Geological observations of Martian arroyos. *J. Geophys. Res.* **79**, 3951–57.

Higgins, C. G. 1982. Drainage systems developed by sapping on Earth and Mars. *Geology* **10**, 147–52.

Howard, A. D. 1971. Optimal angles of stream junction: geometric stability to capture, and minimum power criteria. *Water Resources Res.* **7**, 863–78.

Ingersoll, A. P. 1970. Mars: occurrence of liquid water. *Icarus* **168**, 972–73.

Judson, H. F. 1980. *The search for solutions*. New York: Holt, Rinehart and Winston.

Klemes, V. 1982. Empirical and causal models in hydrology. In *Scientific Basis of Water-Resource Management*. N.R.C. Geophysics Research Council, 95–104. Washington: National Academy Press.

Kochel, R. C., A. D. Howard, and C. McLane. 1985. Channel networks developed by groundwater sapping in fine-grained sediments: analogs to some Martian valley networks. In *Models in geomorphology*, M. Woldenberg (ed.), 313–41. Boston: Allen & Unwin.

Komar, P. D. 1979. Comparisons of the hydraulics of water flows in Martian outflow channels with flows of similar scale on Earth. *Icarus* **37**, 156–81.

Komar, P. D. 1980. Modes of sediment transport in channelized water flows with ramifications to the erosion of the Martian outflow channels. *Icarus* **43**, 317–29.

Komar, P. D. 1983. Shapes of streamlined islands on Earth and Mars: experiments and analyses of the minimum drag form. *Geology* **11**, 651–54.

Komar, P. D. in press. The lemniscate loop-comparisons with the shapes of streamlined landforms. *J. Geol.*

Laity, J. E., and M. C. Malin in press. Groundwater sapping and the origin of theater-headed canyons of the Colorado Plateau. *Geol. Soc. Am. Bull.*

Lambert, R. St. J., and V. E. Chamberlain 1978. CO_2 permafrost and Martian topography. *Icarus* **34**, 568–80.

Leighton, R. B., and B. C. Murray 1966. Behavior of carbon dioxide and other volatiles on Mars. *Science* **153**, 136–44.

Lubowe, J. K. 1964. Stream junction angles in the dendritic drainage pattern. *Am. J. Sci.* **262**, 325–37.

Lucchitta, B. K. 1982. Ice sculpture in the Martian outflow channels. *J. Geophys. Res.* **87**, 9951–73.

Lucchitta, B. K., D. M. Anderson, and H. Shoji 1981. Did ice streams carve Martian outflow channels? *Nature* **290**, 759–63.

Lucchitta, B. K. and H. M. Ferguson 1983. Chryse Basin channels: low gradients and ponded flows. *J. Geophys. Res.* **88**, A553–86.

Mars Channel Working Group 1983. Channels and valleys on Mars. *Geol. Soc. Am. Bull.* **94**, 1035–54.

Masursky, H. 1973. An overview of geological results from Mariner 9. *J. Geophys. Res.* **78**, 4009–30.

Masursky, H., J. M. Boyce, A. L. Dial, G. G. Schaber, and M. E. Strobell 1977. Classification and time of formation of Martian channels based on Viking data. *J. Geophys. Res.* **82**, 4016–38.

Matalas, N. C., J. M. Landwehr, and M. G. Wolman 1982. Prediction in water management. In *Scientific basis of water-resource management*, 118–27. Washington: National Academy Press.

Maxwell, T. A. 1982. Erosional patterns of the Gilf Kebir Plateau and implications for the origin of Martian canyonlands. In *Desert landforms of southwest Egypt: a basis for comparison to Mars*, F. El-Baz and T. A. Maxwell (eds.), 281–300. Washington, D. C.: National Aeronautics and Space Administration.

Maxwell, T. A., E. P. Otto, M. D. Picard, and R. C. Wilson 1973. Meteorite impact: a suggestion for the origin of some stream channels on Mars. *Geology* 1, 9–16.

McCauley, J. F., M. H. Carr, J. A. Cutts, W. K. Hartmann, H. Masursky, D. J. Milton, R. P. Sharp, and D. E. Wilhelms 1972. Preliminary Mariner 9 report on the geology of Mars. *Icarus* 17, 289–327.

Milton, D. J. 1973. Water and processes of degradation in the Martian landscape. *J. Geophys. Res.* 78, 4037–47.

Milton, D. J. 1974. Carbon dioxide hydrate and floods on Mars. *Science* 183, 654–56.

Mouginis-Mark, P. J., L. Wilson, and J. W. Head, III 1982. Explosive volcanism on Hecates Tholus, Mars: investigation of eruptive conditions. *J. Geophys. Res.* 87, 9890–904.

Mutch, T. A. 1979. Planetary surfaces. *Rev. Geophys. and Space Phys.* 17, 1694–722.

Mutch, T. A., R. E. Arvidson, J. W. Head, K. L. Jones, and R. S. Saunders 1976. *The geology of Mars*. Princeton: Princeton Univ. Press.

Nummedal, D. 1978. *The role of liquefaction in channel development on Mars*. N.A.S.A. Tech. Memo. 79729, 257–59.

Nummedal, D. 1980. *Debris flows and debris avalanches in the large Martian channels*. N.A.S.A. Tech. Memo. 81776, 289–91.

Nummedal, D., J. J. Gonsiewski, and J. C. Boothroyd 1976. Geological significance of large channels on Mars. *Geol. Romana* 15, 407–18.

Nummedal, D., H. Masursky, and M. Mainguet 1983. Comment on 'Origin of Martian outflow channels: the eolian hypothesis' by James A. Cutts and Karl R. Blasius. *J. Geophys. Res.* 88, 1243–44.

Nummedal, D., and D. B. Prior 1981. Generation of Martian chaos and channels by debris flows. *Icarus* 45, 77–86.

Peale, S. J., G. Schubert, and R. E. Lingenfelter 1975. Origin of Martian channels: clathrates and water. *Science* 187, 273–74.

Pieri, D. C. 1976. Martian channels: distribution of small channels in the Martian surface. *Icarus* 27, 25–50.

Pieri, D. C. 1980a. *Geomorphology of Martian valleys*. N.A.S.A. Tech. Memo. 81979, 1–160.

Pieri, D. C. 1980b. Martian valleys: morphology, distribution, age, and origin. *Science* 210, 895–97.

Platt, J. R. 1964. Strong inference. *Science* 146, 347–53.

Pollack, J. B. 1979. Climatic change on the terrestrial planets. *Icarus* 37, 479–553.

Pollack, J. B., and D. C. Black 1979. Implications of the gas compositional measurements of Pioneer Venus for the origin of planetary atmospheres. *Science* 205, 56–59.

Popper, K. R. 1959. *The logic of scientific discovery*. London: Hutchinson and Co.

Reimers, C. E., and P. D. Komar 1979. Evidence for explosive volcanic density currents on center Martian volcanoes. *Icarus* 39, 88–110.

Ringwood, A. E. 1978. Water in the solar system. In *Water, planets, plants, and people*. A. K. McIntyre (ed.), 18–34. Canberra: Australian Acad. Sci.

Rossbacher, L. A., and S. Judson 1981. Ground ice on Mars: inventory, distribution, and resulting landforms. *Icarus* 45, 39–59.

Sagan, C. 1977. Reducing greenhouses and the temperature of Earth and Mars. *Nature* 269, 224–26.

Sagan, C., O. B. Toon, and P. J. Gierasch 1973. Climatic change on Mars. *Science* 181, 1045–49.

Schonfeld, E. 1977. Martian volcanism. *Lunar Science* 8, 843–45. Houston: Lunar Sci. Inst.

Schultz, P. H. and H. Glicken 1979. Impact crater and basin control of igneous processes on Mars. *J. Geophys. Res.* 84, 8033–47.

Schultz, P. H. and F. E. Ingerson 1973. Martian lineaments from Mariner 6 and 7 images. *J. Geophys. Res.* **78**, 8415–27.

Schultz, P. H., R. A. Schultz and J. Rogers 1982. The structure and evolution of ancient impact basins on Mars. *J. Geophys. Res.* **87**, 9803–20.

Schumm, S. A. 1974. Structural origin of large Martian channels. *Icarus* **22**, 371–84.

Sharp, R. P. 1973. Mars: Fretted and chaotic terrains. *J. Geophys. Res.* **78**, 4222–30.

Sharp, R. P. 1980. Geomorphological processes on terrestrial planetary surfaces. *Ann. Rev. Earth and Planet. Sci.* **8**, 231–61.

Sharp, R. P., and M. C. Malin 1975. Channels on Mars. *Geol. Soc. Am. Bull.* **86**, 593–609.

Shreve, R. L. 1979. Models for prediction in fluvial geomorphology. *Math. Geol.* **11**, 165–74.

Soderblom, L. A., and D. B. Wenner 1978. Possible fossil H_2O liquid-ice interfaces in the Martian crust. *Icarus* **34**, 622–37.

Theilig, E., and R. Greeley 1979. Plains and channels in the Lunae Planum-Chryse Planitia Region of Mars. *J. Geophys. Res.* **84**, 7994–8010.

Thompson, D. E. 1979. Origin of longitudinal grooving in Tiu Vallis, Mars: isolation of responsible fluid-types. *Geophys. Res. Lett.* **6**, 735–38.

Trevena, A. S., and M. D. Picard 1978. Morphometric comparison of braided Martian channels and some braided terrestrial features. *Icarus* **35**, 385–94.

Twidale, C. R. 1977. Fragile foundations: some methodological problems in geomorphological research. *Revue Geómorph. Dyn.* **26**, 81–95.

Veverka, T., and C. Sagan 1974. McLaughlin and Mars. *Am. Scientist* **62**, 44–53.

Von Bertalanffy, L. 1950. The theory of open systems in physics and biology. *Science* **111**, 23–29.

Von Bertalanffy, L. 1952. *Problems of life*. London: Watts and Co.

Walker, J. C. G. 1977. *Evolution of the atmosphere*. New York: Macmillan.

Walker, J. C. G. 1978. Atmospheric evolution on the inner planets. In *Comparative planetology*. C. Ponnamperuma (ed.), 141–63. New York: Academic Press.

Wallace, D., and C. Sagan 1979. Evaporation of ice in planetary atmospheres: ice-covered rivers on Mars. *Icarus* **39**, 385–400.

Ward, W. R., J. A. Burns, and O. B. Toon 1979. Past obliquity oscillations of Mars: the role of the Tharsis uplift. *J. Geophys. Res.* **84**, 243–59.

Weihaupt, J. 1974. Possible origin and probable discharges of meandering channels on the planet Mars. *J. Geophys. Res.* **79**, 2073–76.

Whitney, M. I. 1978. The role of vorticity in developing lineation by wind erosion. *Geol. Soc. Am. Bull.* **89**, 1–18.

Whitney, M. I. 1979a. Electron micrography of mineral surfaces subjected to wind blast erosion. *Geol. Soc. Am. Bull.* **90**, 917–34.

Whitney, M. I. 1979b. Aerodynamic and vorticity erosion of Mars: Part I. The formation of channels. *Geol. Soc. Am. Bull.* **90**, 111–27.

Whitney, M. I. 1979c. Aerodynamic and vorticity erosion of Mars: Part II. Vortex features, related systems, and some possible global patterns of erosion. *Geol. Soc. Am. Bull.* **90**, 1128–43.

Yung, Y. L. and J. P. Pinto 1978. Primitive atmosphere and implications for the formation of channels on Mars. *Nature* **273**, 730–32.

14

Channel networks developed by groundwater sapping in fine-grained sediments: analogs to some Martian valleys

R. Craig Kochel, Alan D. Howard, and Charles McLane

Introduction

Mariner 9 and Viking Orbiter imagery uncovered a wide variety of channel forms on the surface of Mars (Masursky 1973, Masursky *et al.* 1977, Baker 1982). Much previous research on Martian channels has focused upon the large outflow channels such as Kasei and Maja Vallis (Baker & Milton 1974, Sharp & Malin 1975, Malin 1976, Masursky *et al.* 1977, Baker & Kochel 1979). The emphasis of recent channel research has shifted toward the smaller-scale valley systems (Pieri & Sagan 1979, Pieri 1980a, b, Malin, Laity, & Pieri 1980, Laity & Saunders 1981, Higgins 1982). These investigators generally agree that the morphology and geometry of Martian valley systems indicate that groundwater sapping processes were probably much more important than surface runoff processes in forming these valley networks (Pieri & Sagan 1979, Baker 1982).

The sections below report on our preliminary studies of formative processes responsible for certain valley systems on Mars, in particular for networks developed along escarpments like Valles Marineris. The focus of our studies is on whether the linear, areal, angular and topologically related parameters of Martian channel networks are more closely aligned with patterns developed by surface runoff or those developed by groundwater sapping processes. Our interpretations are based on quantitative measurements of these systems, but the results are preliminary and meant to serve as guidelines for focusing our upcoming quantitative studies with more elaborate experimental design. First, we report on spatial and geometric parameters measured for 70 channel networks along the Valles Marineris escarpment on Mars from Viking Orbiter imagery. Second, measurements of the same spatial and geometric parameters are reported for 69 channel networks developed in fine-grained sediments by groundwater sapping experiments in a three-dimensional laboratory model.

Finally, we explore the mechanics of groundwater sapping of fine-grained sediments using a two-dimensional experimental model.

Terrestrial drainage networks exhibit an organized spatial and geometric pattern which reflect their adjustment to the hydrologic, lithologic, and geomorphic characteristics of the basin (Schumm 1977). This organization is manifest topologically by a hierarchical system of ordered tributaries and spatially by relationships involving channel length, orientation, basin shape, drainage density, and junction angles. Horton (1945) and many subsequent investigators (Strahler 1964) have demonstrated quantitative regularity in many of these parameters for terrestrial networks. Thus we will use morphometry to compare and contrast terrestrial fluvial, Martian and sapping box networks. We are aware that link-based topologic measures may be more sensitive than Strahler ordered measures to certain types of processes, especially those involving space filling. For reviews of this type of morphometry see Jarvis and Woldenberg (1984) and Abrahams (1984).

Groundwater sapping erodes indurated rocks and unconsolidated sediments. Groundwater sapping is important in forming large, theater-headed valleys in the Colorado Plateau (Laity & Saunders 1981) and was probably a dominant formative mechanism for similar valleys on Mars (Pieri 1980a, Laity & Saunders 1981, Higgins 1982). Emerging groundwater accelerates physical and chemical weathering of rocks exposed at the base of valley walls. Sapping of unconsolidated sediments occurs on tidal beaches (Higgins 1982) as a result of diurnal stage variations along the Colorado River in the Grand Canyon (Howard & Dolan 1981), as an agent of streambank and streamhead erosion (Dunne 1980) and as a cause of dam failure (Boffey 1977).

The processes of groundwater sapping of fine-grained sediments are amenable to theoretical analysis and laboratory experimentation. Because of the similarity in form of sapping valley networks developed in fine sediments and indurated rocks, the process rate laws may be similar so that study of small-scale sapping processes holds promise of general applicability. As an example, Manker and Johnson (1982) summarized a series of low pressure, low temperature laboratory experiments in simulating chaotic terrain and outflow channels on Mars. Their experiments with unconsolidated sediments containing water ice and ground ice were successful in creating sudden collapse and release of water upon heating from a local subsurface source. When the containers were given a regional slope sapping occurred rapidly and caused sudden collapse.

Morphology of Martian valley networks

Baker (1982) identified three major types of valley networks on Mars: (a) small valley networks; (b) longitudinal valley networks; and (c) slope systems. Small valley networks occur in heavily-cratered regions of Mars and exhibit nearly parallel network geometries. Tributaries in small valley networks are abundant, terminate in cirque-like valley heads, and join main valleys at angles

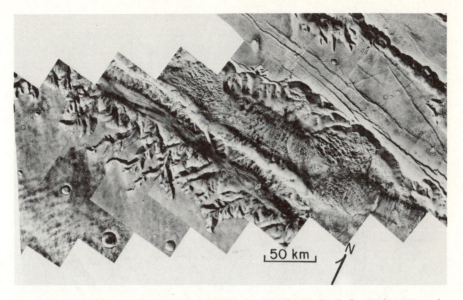

Figure 14.1   Viking Orbiter photomosaic of a portion of Valles Marineris. Image shows examples of the slope networks tributary to the Valles Marineris canyon.

much lower than is typical for terrestrial runoff systems (Pieri & Sagan 1979). The tributaries are usually as wide and deep as the trunk valley they enter. These morphological characteristics and the lack of evidence of competition for drainage between surface divides led Pieri and Sagan (1979) and Pieri (1980a,b) to conclude that surface runoff processes were unimportant in forming these networks. They proposed that wet sapping by fluids from restricted source areas could better explain the spatial and morphologic characteristics observed in the small valleys.

The second type of valley network described by Baker (1982) is the longitudinal network such as Nirgal Vallis. Longitudinal networks are long valleys with poorly-integrated, deeply incised, short tributaries. Tributaries in these networks terminate in cirque-shaped heads and have cross-sections comparable to the trunk channel at their junctions (Pieri 1980b, Baker 1980). Ratios of tributary length to mainstream length are very low in longitudinal networks compared to other valley systems on Mars or terrestrial drainage systems.

The third class of valley networks noted by Baker (1982) are slope systems. Slope networks extend perpendicularly to the strike of large escarpments on Mars such as Valles Marineris. These slope networks generally exhibit a quasi-dendritic pattern (Figs. 14.1 and 14.2). Some tributary valleys narrow upstream and terminate in cirque-like heads. Tributary walls are generally smooth, steep, and probably talus-covered. There is a general transition in valley form with distance from the escarpment (Fig. 14.1). Close to the escarpment adjacent valleys have merged via mass wasting, hence, the spatial relationships of individual networks are complex. In the upper part of

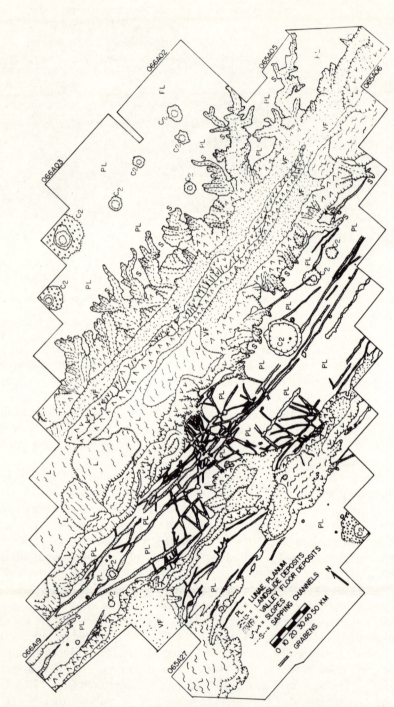

Figure 14.2 Generalized geomorphic map of the Valles Marineris study area.

the networks channels bifurcate and extend headwardly as distinct valley systems.

Slope valleys along Valles Marineris: area and length measures
Our simulation models were constructed as possible analogs to all three types of valley systems on Mars. Pieri (1980a,b) has summarized the morphology and morphometry of small valley networks on Mars. We have selected the slope valleys along Valles Marineris for additional quantitative measurements of Martian valley systems to compare to the valley networks generated in our experimental models. A summary of morphometric parameters for the 70 Valles Marineris networks is given in Table 14.1. Basin areas were calculated using the tributary heads to outline the basin perimeter. The average basin area for the Valles Marineris networks studied is 603 km^2 with a range from 53 km^2 to 3275 km^2. These slope networks are generally elongated with an average length to width ratio of 2:1 (Fig. 14.3). The mean lemniscate k is 2.33 (Chorley 1959). The elongate nature of these basins is also reflected by the high trunk stream length to tributary stream length ratio which averages 3.9:1. Tributaries are numerous, but they are typically short, first-order streams. None of the 70

Table 14.1 Summary of network morphometry, Valles Marineris, Mars.

Parameter	n	Mean	Standard Deviation
basin area (km^2)	70	603.00	627.00
basin shape			
lemniscate k	70	2.33	0.75
length/width	70	2.06	0.64
junction angles	275		
θ_1		42.00	20.00
$\theta_1 + \theta_2$		49.00	21.00
dimensionless			
drainage density[1]	70	12.61	9.53
length main channel/			
length first order			
channels	27	3.89	0.85
bifurcation ratio	55	3.53	1.46
Shreve magnitude (M)	70	5.01	3.85
Strahler order	70	1.74	0.85
channel length ratios			
1:2	54	1.39	0.42
1:3	24	1.69	0.38
2:3	24	1.23	0.29
basin area/channel			
area	70	3.40	1.85
interchannel			
spacing	54	20.50	5.70

[1] Dimensionless drainage density $= \Sigma L^2/A$.
This should have been divided by $(2M-1)$. See Appendix.

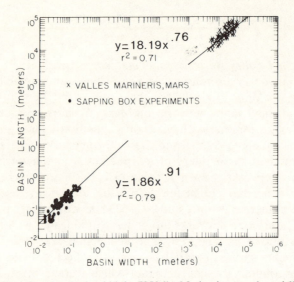

Figure 14.3 Basin length vs basin width for 70 Valles Marineris networks and 69 networks created in the sapping box experiments. Length was taken as the maximum straight-line distance through the basin and width was measured as the longest line perpendicular to the length axis. Note the similarity of the relationships (except for the scale) of the Martian and experimental networks. The slopes are not statistically different. [This figure suggests that for larger networks, length/width decreases. If extrapolated to much larger basins, width would exceed length. Hack (1957) found that length varies as the 0.6 power of basin area for fluvial networks, indicating a higher L/W ratio for larger basins. Perhaps this difference in length–width allometry reflects the different processes involved. *Editor's note*.]

networks measured along the Valles Marineris have a Strahler order greater than three.

The traditional drainage density is a dimensional parameter; therefore, the miniature sapping basin experiments can not be directly compared to Martian and terrestrial river basins. To overcome this scale problem the drainage density was modified by multiplying by the total channel length term. It was thought that this dimensionless drainage density would provide a relative measure of the completeness of the total area of the basin covered by channels. The average dimensionless drainage density of Martian valley networks is very low compared to terrestrial basins. The mean value for 70 Valles Marineris networks is 12.61. We have calculated the dimensionless drainage density for 52 terrestrial basins in varied physiographic regions studied by Patton and Baker (1976). The average for the 52 basins is 706, over an order of magnitude larger than the Martian figure. We have subsequently learned that we erred in not dividing by the number of links (2M–1) which will create a corrected dimensionless drainage density.[1] When this is done the large differences

---

[1] Editor's comment: I have found that the dimensionless drainage density is almost perfectly correlated with the total number of links (2M−1) where M is the magnitude or number of first order streams. Thus the high dimensionless drainage density for terrestrial fluvial systems is an artifact of the magnitude of the systems. For further comments, see the Appendix.

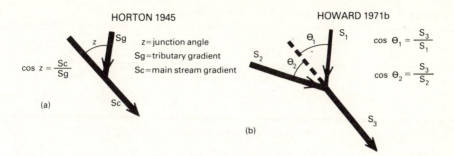

Figure 14.4 Definition of junction angles. (a) Junction angle as defined by Horton (1945). (b) Junction angle as modified by Howard (1971b). We have followed Pieri's (1980a) convention in labeling the smaller tributary with a 1, the larger tributary with a 2 and the mainstream with a 3.

between the Martian and terrestrial networks are greatly reduced. We still need to investigate whether there is any significant difference between the corrected dimensionless drainage densities of the two kinds of networks.

Channel area is generally very insignificant compared to total basin area in terrestrial runoff networks. Measurements of 44 third order runoff networks in central Virginia, west Texas, and southeastern Pennsylvania indicate that the ratio of total basin area to channel area is consistently in excess of 25:1, averaging 36:1 (Kochel 1980). For the Valles Marineris networks this ratio averages 13:5, when channels are defined as the areal extent of a relatively smooth floor. Therefore, channel areas in sapping networks are 2.7 times more areally extensive relative to their respective basin areas compared to runoff systems. This is consistent with observations of the evolution of channel networks in the sapping experiments which indicate that after initial channel extension the dominant geomorphic process is channel widening by slumping.

A fairly uniform spacing of about 20 km exists along the Valles Marineris escarpment between major networks visible in Viking photograhs. Drainage divides are not always identifiable in the Viking photographs and there is little evidence of surface competition for drainage area between adjacent networks. Thus this regularity is probably caused by limitations imposed on the groundwater collecting system by lithologic permeability and subsurface interference of neighboring sapping fields.

Junction angles

Stream junction angles are defined in various ways. Horton originally defined junction angle as shown in Figure 14.4a. Howard (1971b) modified Horton's cosine law by distinguishing two independent angles with respect to the trend of the outlet stream (Fig. 14.4b). Our measurements for Valles Marineris and the sapping box were obtained using the method proposed by Howard (1971b).

Tables 14.1 and 14.2 show that Martian junction angles from our study average 42° for θ_1 and 49° for $\theta_1 + \theta_2$ (for all values of mainstream magnitude).

Table 14.2 Mean junction angles, Valles Marineris, Mars.

| | θ_1 (degrees) where $m = 1$, $M = 1$–6 | | | | All m, M | | | | |
	1,1	1,2	1,3	1,4	1,5	1,6	θ_1	θ_2	$\theta_1 + \theta_2$
$\bar{x}$	38	39	44	50	45	44	42	8	49
σ	19	18	20	25	21	17	20	10	21
n	92	54	35	16	8	9	275	275	275

m = magnitude of the small tributary valley; M = magnitude of the large tributary.

The values for $\theta_1 + \theta_2$ are smaller than those found in terrestrial stream networks (Lubowe 1964, Pieri 1980a).

Tributary junction angles in stream networks on Earth are dependent on the relative slopes of the joining channels (Fig. 14.4, Horton 1945, Howard 1971b). Horton and Howard demonstrated that as the contrast in gradient between the smaller (or larger) tributary stream and the gradient of the mainstream below the junction increases the junction angle θ_1 (or θ_2) will increase. In a temporal sense, this implies that as basins evolve from nearly flat relief (with nearly equal or random gradients throughout) to a well-graded network with consistent downstream decrease in gradient, junction angles should increase and become more regular. If junction angles are a function of relative gradients, the difference in gravity between Earth and Mars should have little or no effect upon junction angles of equivalent maturity.

Terrestrial networks do generally exhibit a regular increase in tributary junction angles in the downstream direction as predicted. Lubowe (1964) found that $\theta_1 + \theta_2$ increases as the order difference increases between the two tributaries. Pieri (1980a) showed that θ_1 increases in the down network direction due to the increased contrast between tributary and mainstream magnitudes and gradients. He also noted that the rate of increase of θ_1 declines with increasing magnitude of the trunk stream.

In contrast to fluvial networks on Earth, analysis of three networks in the Margaritifer Sinus area of Mars showed almost no correlation between junction angles and position within the network (Pieri 1980a p. 110). We have found a slight tendency toward increasing junction angles with increasing magnitude of the trunk stream (about 1.5° for each increment of magnitude) but this small change is masked by the wide scatter of data indicated by a large standard deviation of angle at each increment of magnitude. Larger samples are required to determine with greater statistical confidence if this trend is valid. There is much greater variation in Martian junction angles ($\theta_1 + \theta_2$) than in terrestrial runoff systems studied by Pieri (1980a) and Lubowe (1964).

Figure 14.5 illustrates that in general Martian networks have lower values for θ_1 than terrestrial runoff networks. However, exceptionally low θ_1 values were observed in experimental networks at Colorado State University (Parker, 1977). These low values may be caused by the youth of the networks; their development was terminated prior to attaining a mature, equilibrated state.

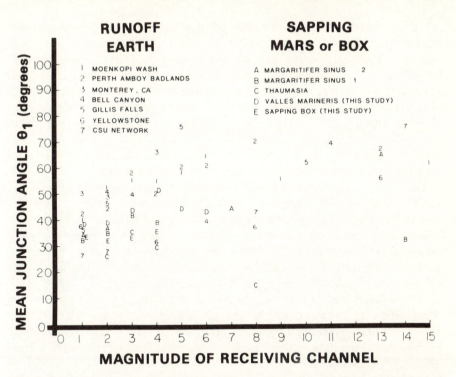

Figure 14.5 Mean junction angles for θ_1 of Martian networks, sapping box networks and terrestrial runoff networks. Data from terrestrial cases are indicated by numbers while data from Martian basins and sapping box networks are shown by letters. No regression lines have been fitted to these data, however one can see that the Martian and sapping networks generally show junction angles below those expressed in terrestrial runoff networks. The Martian and terrestrial networks were taken from Pieri (1980a). CSU data are from Parker (1977).

In summary, Martian junction angles deviate from terrestrial patterns in three respects: (1) individual θ_1 values are smaller for equivalent tributary and mainstream magnitudes; (2) θ_1 values show weaker increase with magnitude contrast; and (3) junction angles exhibit greater variability for a given magnitude contrast. The data of Pieri (1980a) as well as those presented here convincingly show the smaller θ_1 values for Martian networks. The weaker increase of θ_1 values with magnitude and the greater scatter are also exhibited in our sample; greater sample sizes are needed for both terrestrial and Martian networks to make firm conclusions as to the relation of angle to magnitude increase.

Four explanations may be advanced for these contrasts between Martian and terrestrial junction angles: (1) relative youthfulness of the Martian networks; (2) structural controls on network development in presumed Martian sapping networks; (3) differences in equilibrium channel gradient relationships between terrestrial and Martian networks related to process differences; and (4) resolution limits in Martian images which lead to underestimation of tributary and mainstream magnitudes. Most observers of the Martian channels have agreed that the Martian channels present a youthful stage of development and are not presently active. The low junction angles of the Martian networks suggest little systematic downstream change in channel gradients, which is consistent with interrupted development. Similar low junction angles occur in very youthful terrestrial runoff networks, such as on the dissected pediment shown in Howard (1971a, Fig 14.5) and the experimental networks created at Colorado State University (Parker 1977). However, these youthful terrestrial networks generally exhibit a more uniform network pattern between basins and more regular junction angles than the Martian networks.

The greater irregularity of junction angles in Martian networks may be due to structural controls exerted on the sapping processes. Sapping networks should be subject to a greater degree of controls by fractures, faults, and lithologic inhomogeneities than runoff networks, due to the strong influence of groundwater flow by these structural controls and due to the elongation of the networks by differential weathering rather than by scour. Recent surveys of the orientations of probable sapping channels along the margins of Valles Marineris and Kasei Vallis on Mars indicate a strong correspondence to trends of other structural features visible in the region (Kochel & Burgess 1983, Kochel & Capar 1983). Terrestrial networks presumed to be strongly influenced by sapping processes show strong structural influences (Laity 1980, Laity & Pieri 1980, Laity & Saunders 1981).

Evidence of structural influence in these systems observed by Kochel and Capar (1983) includes: (a) the abundance of straight channel segments with sharp bends; (b) the correspondence between orientations of channel segments and grabens; and (c) the observation of several channels that extend headwardly along structural features such as grabens (Fig. 14.6). Sharp (1973) interpreted these slope networks as having formed by headward sapping of ground-ice. Carr and Schaber (1977) noted that the valley cross-sections were V-shaped and probably smooth talus slopes.

If stream junction angles are determined by downstream changes in channel gradient, then junction angles should vary with the characteristic relationship between gradient and drainage area, being greater the more rapid the decrease in gradient with drainage area. Sand-bed channels generally show weak downstream change in gradient and small junction angles, whereas coarse-bed and bedrock-floored channels show more rapid decrease and larger junction angles (Howard 1980). (Weak downstream change in gradient can also be due to youthfulness, as discussed above). Unfortunately no evidence is available regarding the original bed characteristics of Martian channels due to limited image resolution and subsequent modifications by eolian and mass-wasting processes, so that this factor cannot be evaluated.

One possible explanation for the weak change of junction angle with network position in Martian valleys may be that limited photo resolution does not permit recognition of lower magnitude channels and that we are really picking up data further down in the system, where junction angles increase

Figure 14.6 Viking Orbiter photograph (Frame 064A16) showing examples of headward tributary sapping along structural features.

only slightly with mainstream magnitude. However, we believe this is not the case for valleys studied along Valles Marineris because there is sufficient high resolution coverage to suggest that we have observed the complete network.

Simulation of sapping channel networks (3-dimensional)

Our experiments with a three-dimensional sapping chamber were designed to create surface channel networks by groundwater sapping so their morphometry could be compared with the morphometries of terrestrial surface-runoff networks and Martian slope networks along Valles Marineris. These experiments represent the preliminary stages of our simulation studies of channel networks and erosion mechanics related to groundwater sapping. The three-dimensional experiments are a semiquantitative pilot study which we will use to refine our future quantitative studies.

A one cubic meter sapping box was constructed from marine plywood and sealed with roofing tar (Fig. 14.7). The sedimentary medium used was a mixture of 90% 2.25 ϕ fine sand with 10% clay-sized coal fly ash to provide some cohesion in channel walls. A perforated plexiglass plate was used to establish a water reservoir to maintain a constant head throughout the experimental runs. Twenty-three runs were conducted using the same head. These varied in

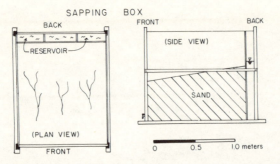

Figure 14.7 Schematic drawing of the three-dimensional experimental sapping chamber.

duration from 90 to 200 min, measured from the onset of basal sapping. Prior to each run, the box was drained for at least 24 hr, the sand surface regraded, and the new slope was recorded. Time zero was taken as the time when sapping was first observed at the surface somewhere along the base of the slope. Sapping was initiated on slopes ranging from 3° to 16°, but most of the runs used slopes between 8° and 14°. Several runs were made with the presence of an initial basal escarpment of 2 to 5 cm while others had no initial escarpment. Temporal and spatial development of sapping networks were recorded by tracing channel patterns at selected time intervals on a transparent frame mounted about 2 cm above the sand surface.

Temporal evolution of sapping networks
Although each of the 23 runs of the sapping box resulted in network evolution that differed in detail, there were two general evolutionary scenarios that described the development of the sapping channel networks, depending upon the initial slope of the sand surface. Groundwater upwelling at the channel heads dominated sediment transport processes near the site of sapping. Within several centimeters downvalley, enough groundwater had been discharged for fluvial processes to become the dominant transporting agent. In all cases after sapping began there followed a period of rapid escarpment retreat and channel extension for 15 to 60 min (Fig. 14.8). After the adjustment period sapping processes continued to cause escarpment retreat and headward extension of channels, but at a much reduced rate (Fig. 14.8).

For runs with initial surface slopes in excess of 12°, the initiation of sapping was contemporaneous with the onset of large-scale slumping and basal escarpment formation. This was followed by rapid headward slumping along the escarpment and eventual stabilization as slopes were reduced to a quasi-stable angle. Once the stable phase was reached, sapping channels developed in the slump debris and began to bifurcate and extend headwardly through the slump debris.

For runs with initial slopes less than 12°, sapping began as a discontinuous seepage zone near the base of the slope. Sapping channels were initiated at

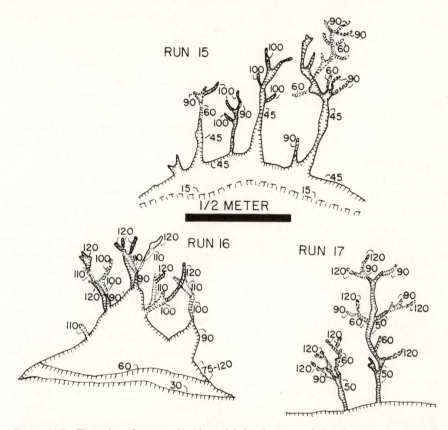

Figure 14.8 Examples of temporal and spatial development of three experimental sapping network runs. Numbers along the channel margins indicate time measured from the onset of basal sapping.

approximately regularly-spaced intervals in the central two-thirds of the slope surface. Rapid headward extension characterized the initial phases of sapping. This was followed by a period of increased rate of bifurcation and slow headward extension of tributaries (Fig. 14.8). During this phase of relatively slow channel extension, channel widening became the dominant geomorphic process. Tributaries tended to form mostly at the headward areas of the network. Very little bifurcation occurred in down-network areas of the basins. Groundwater that would normally be available to down-network areas was apparently being intercepted by extending and bifurcating up-network tributaries. This process of up-basin pirating of groundwater would probably be very effective for situations involving regional groundwater flow systems and should be readily evident by a disparity of drainage densities in the upper versus lower reaches of the channel network. In the Mars basins studied, upper basin drainage densities average $0.14 \, km/km^2$ while whole basin drainage densities average $0.09 \, km/km^2$ (ratio of 1.6:1). In the sapping box basins the average upper basin drainage density was $0.32\,cm/cm^2$ while whole basin drainage density averaged $0.20\,cm/cm^2$ (ratio of 1.6:1). In almost every case the upper basin drainage density was significantly higher than lower basin drainage density for sapping box and Martian networks. Terrestrial runoff networks exhibit considerable variability in upper basin vs lower basin drainage densities, but typically do not show the large decrease in lower basin drainage density seen in these sapping networks.

Schumm (1977) has shown in experimental studies of runoff network evolution that drainage density is temporally variable. He showed that drainage density initially increases rapidly by bifurcation and channel extension near the basin divides even though abstraction occurs near the mouth of the basin. Eventually, drainage density reaches a maximum as a balance is met between addition of basin margin tributaries and lower basin abstraction (Schumm 1977). During the later phases of basin evolution Schumm (1977) observed a decrease in drainage density which was primarily explained by abstraction in the center of the basin. Figure 14.9a shows that after the initial drainage density was established in typical experimental networks, insignificant and irregular changes occurred in drainage density for evolving networks in the sapping box experiments.

Horton (1945) and Howard (1971b) showed that junction angles in terrestrial drainage networks developing on an initially planar surface should

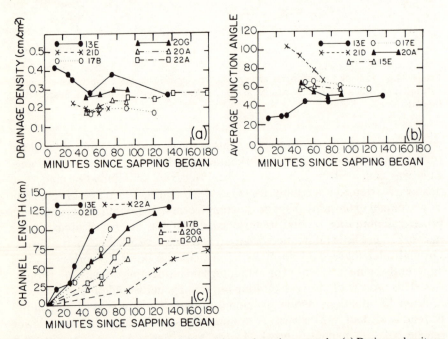

Figure 14.9 Temporal changes in typical experimental sapping networks. (a) Drainage density vs time. (b) Average basin junction angle $(\theta_1 + \theta_2)$ vs time. (c) Channel length vs time. In all cases time is measured in minutes from the onset of basal sapping.

increase from youth to maturity because the ratio of mainstream to tributary gradients becomes smaller (Fig. 14.4). Figure 14.9b depicts the temporal development of junction angles $(\theta_1+\theta_2)$ in typical networks formed in the sapping box. The tendency of increasing junction angle as the networks evolved, as predicted by Horton (1945), did not occur in general. However, Figure 14.9b shows that regardless of early junction angles formed in the sapping box, there was a tendency toward a value between 40° to 50° which remained relatively constant after about 60 minutes from initiation of the sapping channels.

The major similarity between observations of basin evolution in terrestrial or experimental runoff networks and the experimental sapping networks is the steady increase in channel length as the basin evolves. Figure 14.9c shows that sapping-box channels increased slowly in length at first, then went through a period of rapid extension, and finally reached a more or less steady state if the experiment was of long enough duration (in excess of about 120 min) (Fig. 14.9c, Runs 13E, 22A). In many cases (Fig. 14.9c, Runs 21D, 20A, 20G) the experiments were terminated before the network achieved a steady state with regard to channel extension. The nonintegrated nature of surface drainage networks on Mars in heavily cratered terrain (Pieri 1979) and along Valles Marineris suggests that the period of Martian history when channeling was active was brief; therefore, network evolution was interrupted by the changing climatological-hydrological regime prior to maturity.

Final network morphometry
Table 14.3 is a summary of basin morphometric data for the 69 networks developed in the experimental sapping box. Basin areas of the experimental channels were delineated by using the headward ends of tributaries as the basin perimeter, hence, the method of measuring basin area is comparable to that used on Viking photographs. Experimental basin areas averaged 101 cm^2 and ranged from 3 cm^2 to 822 cm^2. Sapping basins were generally elongate, having length to width ratios averaging 2.4:1 (Fig. 14.3) and a mean lemniscate ratio of $k = 2.83$ (Chorley 1959) similar to the elongate networks along Valles Marineris. Tributaries were abundant in the experimental networks, but like the Mars examples, no basins with Strahler orders greater than 3 were observed. Trunk streams were generally 4.5 times longer than tributaries entering them. Drainage density for the sapping networks was low compared to terrestrial runoff channels, averaging 0.47 cm/cm^2, but was somewhat higher than that measured in the Martian Valles Marineris networks. Channels appeared to occur along the sapping escarpment with rather uniform interchannel spacing that averaged about 9 cm. A similar regularity of interchannel spacing occurred along Valles Marineris.

Junction angles in the sapping networks had a mean and standard deviation of 35° and 9° for θ_1, and 44° and 18° for $\theta_1+\theta_2$ (Table 14.4). The means are considerably lower than those observed in terrestrial runoff networks, but very similar to those observed in the Martian networks. Another similarity to

Table 14.3 Summary of network morphometry, sapping box experiments.

Parameter	n	Mean	Standard deviation
basin area (cm^2)	69	101.00	120.00
basin shape			
lemniscate k	69	2.83	1.20
length/width	69	2.43	0.77
junction angles	221		
θ_1		35.00	17.00
$\theta_1+\theta_2$		44.00	18.00
dimensionless			
drainage density	69	9.10	6.61
bifurcation ratio	56	2.93	1.01
length main channel/			
length first order			
tributaries	23	4.50	1.10
Shreve magnitude	69	4.45	3.28
Strahler order	69	1.94	0.89
channel length ratios			
1:2	55	1.09	0.45
1:3	27	1.37	0.53
2:3	27	1.20	0.52
basin area/channel			
area	69	6.70	2.72
interchannel	57	9.20	2.80
spacing			

the Valles Marineris networks is the weak increase in θ_1 as the magnitude of the mainstream increases (Fig. 14.5). It is likely that a systematic downstream relationship with gradient did exist, but because of the qualitative nature of our preliminary experiments our sample grid was not adequate to detect it. In addition, the system may have been terminated in a youthful stage. Finally, the channels could be easily perturbed by anomalous large sand grains.

In terrestrial runoff systems Horton (1945) found excellent correlations between log mean stream length and stream order, and log stream frequency and stream order within a drainage basin. Sapping networks developed in our experiments exhibited the same general relationships, but in about 50% of the cases there was more variation than is normally found in terrestrial streams. Martian slope networks showed similar variation about the general trends. The small differences between the terrestrial runoff relationships and those observed in Martian networks and in the sapping box may be due to the low order of the networks; therefore no firm conclusions should be made based on these preliminary data. However, the tendency for increased variability is consistent with the non-uniform conditions associated with sapping versus fluvial networks.

Morisawa (1959) recognized the power function relationship between mainstream length and area implied by geometric progressions with order of

Table 14.4 Mean junction angles, sapping box experiments.

| | θ_1 (degrees) where $m = 1$, $M = 1\text{--}4$ | | | | All m, M | | |
	1,1	1,2	1,3	1,4	θ_1	θ_2	$\theta_1 + \theta_2$
$\bar{x}$	34	31	34	37	35	9	44
σ	14	15	17	12	17	9	18
n	90	45	27	7	221	221	221

m = magnitude of the small tributary valley; M = magnitude of the large tributary.

length and area. Figures 14.10a and b show the length-area relation by orders for individual basins. Note that again there is considerable variation in about half the cases. Figures 14.10c and d indicate that a well defined relationship exists when all first, second and third order basin lengths and areas are plotted together. (Note that mean segment lengths are cumulated to create an analog to basin length. See Broscoe 1959, Bowden & Wallis 1964).

The low drainage densities of sapping-box networks reflect the relatively poor integration of drainage in these systems. Drainage divides were not readily apparent at the surface and will have to be monitored in future experiments with piezometers. Competition for drainage by neighboring basins was limited and large, undissected divides remained throughout the experiments. On a few occasions interbasin divides or intertributary divides were breached by subsurface piracy and subsequent surface collapse.

Channel morphology
Channels developed in the sapping experiments exhibited many similarities to the presumed sapping channels on Mars and to sapping channels found on tidal beaches (Higgins 1982) and in stage-variable fluvial systems (Howard & Dolan 1981). Channels were steep walled and had relatively flat floors and a notable lack of typical fluvial bedforms such as point bars and sand waves. All of the channels terminated in steep-walled, cirque-like heads at the locations of discharging groundwater. At these localities basal sapping of the sediments was occurring causing subsequent collapse of overlying sediments after basal sediments were removed. Channel cross sections were generally broadly U-shaped. Tributary channel cross sections were typically similar in scale to those of the trunk streams they joined. This phenomenon was also observed in Martian slope networks and in the small valleys on Mars studied by Pieri (1980a).

In terrestrial settings, sapping typically develops in zones of prominent jointing or faulting in permeable bedrock. Correspondence of many of the Valles Marineris channel orientations to structural orientations lends support to this model. Future experiments with laboratory sapping will introduce heterogeneous permeabilities into the sediments to simulate increased permeability in structurally weak zones. Local lithologic heterogeneities cause more efficient piping which causes groundwater flow to converge at the

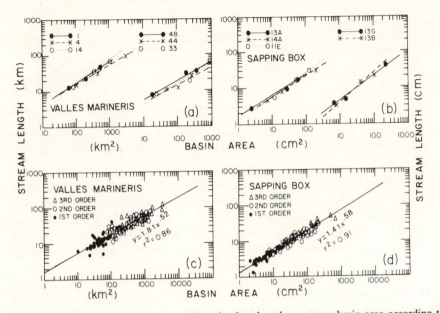

Figure 14.10 Mean cumulative stream length of each order vs mean basin area according to Broscoe (1959) and Bowden and Wallis (1964) for Valles Marineris networks and for typical experimental sapping networks. The examples show there is considerable variation in the relationship between these parameters. For example, good correlation exists for Martian networks 1, 4, and 14 (a) and for experimental networks 13A, 14A, and 11E (b) whose slopes are not statistically different; while poor relationships exist for Martian networks 33, 44, and 48 (a) and for experimental networks 13B and 13G (b), whose slopes are statistically different. (c) and (d) show the entire data set collected for various order networks.

springhead (Dunne 1980). Once initiated, a positive feedback process occurs whereas the converging groundwater flow increases chemical weathering which increases piping and further distorts the flow field. Therefore the sapping channel migrates in a headward direction. Dunne (1980) noted that when other zones of high piping susceptibility are intersected, a tributary will form and enlarge in the same manner. Eventually some optimum drainage density will be achieved when a quasi-equilibrium is established between the demands of neighboring sapping networks competing for available groundwater (Dunne 1980). This optimal drainage is reflected in the experimental channels. It is unclear whether we are observing a similar optimal drainage density in the Valles Marineris networks. The Martian systems probably never achieved maturity because the groundwater reservoir was not replenished as the climate changed, interrupting network development

In future experiments we expect to address the problem of immature network development. Just as Martian networks may be immature, it is possible that the Colorado State networks (Parker 1977) may also be immature. Additional complications arise when analyzing experimental runoff networks because as in natural networks, there may be an important component of subsurface sapping involved in the development of the surface channels. Our future experiments will involve an extensive grid of piezometers along the base of the sapping box to help clarify the relationships between surface and subsurface divides and the respective contributions of these two processes to the formation of channel networks.

Examples of terrestrial networks heavily influenced by groundwater sapping occur in areas underlain by the porous Navajo Sandstone in the Colorado Plateau of southeastern Utah (Laity & Pieri 1980, Laity 1980). Morphological similarities of the Colorado Plateau networks with networks developed in the sapping experiments and on Mars include: (a) steep-walled valleys; (b) angular channel segments in areas of prominent jointing; (c) high ratios of length of master-channel length vs tributary length; (d) headward terminations in cirque-like features; (e) abundant evidence of mass wastage along valley walls; and (f) hanging tributaries.

Mechanics of sapping of fine-grained sediments

Because of the similarity of form of sapping valley networks developed in fine-grained sediments and indurated rocks, the process rate laws may be similar. Therefore these sapping processes are amenable to theoretical analysis and laboratory experimentation. In this section we will discuss our preliminary experiments with a two-dimensional sapping chamber designed to develop theoretical relationships of the mechanics of sapping in fine-grained sediments.

Theoretical development

The threshold of erosion of fine-grained sediments subjected to emergent groundwater flow is more complicated than the threshold of motion of flat-lying sediments beneath a fluid flow because of the added effects of the groundwater flow and, often, because of appreciable surface gradient. This equilibrium can be evaluated either by consideration of the bulk stability of the surface layers of the sediment or by considering the balance of torques applied to a particle lying on the surface. The first approach was taken by Haefeli (1948), Burgi and Karaki (1971), Oldenziel and Brink (1974), and Howard and McLane (1981); the second was used by Kezdi (1979) and in the present analysis. The two methods give equivalent results.

Figure 14.11 illustrates a particle lying on the surface of a sedimentary deposit and subjected to four forces: gravity, F_g, acting vertically; fluid drag, F_w, exerted by surface water flow (assumed here to act downslope and parallel to the sediment surface); and fluid drag, F_s, due to the emergent groundwater flowing at an angle, ψ, to the horizontal; and a cohesive resistance, F_c, opposing the particle motion. The particle is assumed to be at the threshold of motion, so that the cohesive resistance has its maximum value and the four rotational torques about the downstream rotational point are balanced. The average

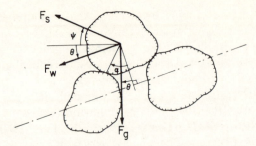

Figure 14.11   Force balance on a particle lying on a sloping surface with seepage and surface flow. F_g is gravity force, F_s is seepage force, and F_w is force due to surface flow.

angle about the particle center of the rotational point (relative to the normal to the surface of the sedimentary deposit) is α, and is approximately equal to the angle of internal friction for the sediment in loose packing. All forces are assumed to act through the particle center, and lift forces due to the external fluid flow are neglected. The soil is assumed to be fully saturated. The balance of torques thus gives:

$$[(F_g d \sin (\alpha-\theta)+F_c d-F_w d \cos (\alpha) - F_s d \cos (\theta+\psi-\alpha)]/2 = 0 \quad (14.1)$$

where θ is the slope angle and d is the particle diameter. The two fluid forces (F_w and F_s) are assumed here to act independently, although this is unlikely to be generally valid, because the emerging water will affect the development of the near-surface boundary layer. The four forces are assumed to be related to the particle dimensions as follows:

$$F_g = (C_1 \pi \gamma_b d^3)/6 \quad (14.2)$$

$$F_s = (C_2 \pi \gamma_w I d^3)/6 \quad (14.3)$$

$$F_w = (C_3 \pi \tau d^2)/4 \quad (14.4)$$

$$F_c = (C_4 \pi d^2)/4 \quad (14.5)$$

where γ_b is the buoyant unit weight of the particle, γ_w is the fluid unit weight, I is the hydraulic gradient, τ is the surface shear stress exerted on the bed by the surface flow, and C_1–C_4 are constants. C_1 and C_2 will be on the order of unity, C_3 is determined by the Sheilds diagram or other threshold-of-motion criterion, and C_4 is a measure of the cohesive strength per unit area of interparticle contact. The hydraulic gradient, I, can be expressed as a unique function of the angles θ and ψ, as illustrated in Figure 14.12, noting that the pressure head is zero at the soil surface:

$$I = \sin (\theta)/\cos (\psi+\theta) \quad (14.6)$$

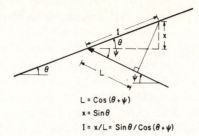

$$L = \cos (\theta + \psi)$$
$$x = \sin \theta$$
$$I = x/L = \sin \theta / \cos (\theta + \psi)$$

Figure 14.12 Relation of hydraulic gradient, I, to flow geometry for emergent seepage.

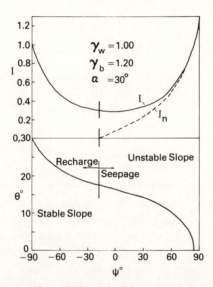

Figure 14.13 Relation of maximum stable slope angle, θ, for seepage emerging at angle ψ relative to the horizontal. The hydraulic gradient, I, and the surface-normal component of the hydraulic gradient, I_n, are also shown as a function of ψ for critical conditions.

Substituting and simplifying:

$$-2C_1 d\gamma_b \sin (\alpha-\theta)/\cos (\alpha)-3C_4/\cos (\alpha)+3 C_3\tau$$
$$+2C_2\gamma_w\sin (\theta)d\{1+\tan (\theta+\psi) \tan (\alpha)\} = 0 \qquad (14.7)$$

In the case where terms involving C_4 and C_2 are negligible (noncohesive sediment with no groundwater flow), Equation 14.7 reduces to the usual threshold-of-motion criterion for surface flow (e.g. Howard 1978).

The importance of the seepage force in determining slope stability can be illustrated by considering the case of noncohesive sediment and negligible surface flow (terms involving C_3 and C_4 omitted). Figure 14.13 illustrates the relationship between the maximum stable slope angle and the angle, ψ, of emergent groundwater flow, as well as the corresponding magnitude of the hydraulic gradient for a particular combination of sediment and fluid properties under conditions of no appreciable surface flow. Note that for a level surface and vertical seepage, erosion and instability occur for a value of I equal to the ratio of the buoyant unit weight of the sediment and the fluid unit weight (e.g. Lambe & Whitman 1969, 353), which is a value of 1.3 for the case illustrated. However, on a sloping surface the required hydraulic gradient is less, reaching a minimum of about 0.3 for the example.

The capacity of the fluid flows to transport sediment is related to the degree by which the applied forces exceed the threshold of motion. The simplest model assumes that the capacity is a power function of the value, T, of the left-hand side of equation 14.7:

$$q_s = KT^\beta \tag{14.8}$$

where k and β are constants, and q_s is the volumetric transport rate (per unit time per unit area of slope surface). The rate of erosion of the slope (E, in units of distance per time) is a function of the downslope (x-direction) change of transport capacity:

$$E = \partial q_s / \partial x \tag{14.9}$$

Experimental program
Experiments have been conducted in a narrow groundwater flow chamber (Fig. 14.14) to quantify the rate of sapping erosion in unconsolidated sediments as a function of flow and sediment parameters. Sediment is placed in the tank. The lateral flow and sapping rate are controlled by establishing the desired water level in the reservoir at the rear of the tank. A screen separates the sediment from the reservoir. The experiments are generally conducted with a fixed head, h_w, (Fig. 14.15) at the rear of the chamber (the head is varied between runs). The total length of the sand-filled portion of the chamber can be varied between runs, as can be the height of the outflow, h_o. Change of outflow height varies the depth of groundwater flow because the tank bottom is an aquiclude. The groundwater flow is monitored by a dense piezometer network on the side of the flow tank.

Experiments to date have been limited to two sizes of noncohesive sands under a variety of flow conditions. The resultant sapping involves an intimate intermixture of erosion by virtue of groundwater flow and overland transport below the sapping face (Fig. 14.15). Near the outflow end of the tank rates of groundwater outflow are minimal and fluvial processes dominate. Near the point of first emergence of groundwater, surface flow rates are small and the transport is driven primarily by the groundwater upwelling. The hydraulic gradient has its maximum value in this zone, called the sapping face. Thus there is a gradual transition from groundwater to fluvial transport proceeding downslope from the sapping face. Figure 14.16 portrays the temporal evolution of one experiment. The rate of backcutting of the sapping face is limited by the

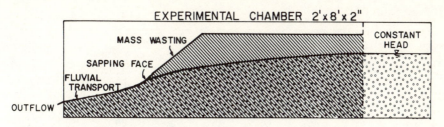

Figure 14.14 Longitudinal section of two-dimensional experimental sapping apparatus. Ruled lines indicate sand, circles show saturated zone.

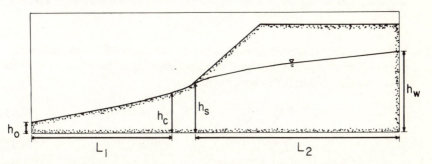

Figure 14.15 Terminology for parameters defining sapping geometry during two-dimensional experiments. h_w is the constant head, h_o is the height of the outflow, h_s is the height at the sapping face, and h_c marks the transition from sedimentary transport dominated by emerging groundwater to surface water.

delivery of sand from the overlying slump face. The slumping occurs intermittently, so that the sapping face geometry and erosion rates vary considerably over a period of a few minutes, but the process progresses smoothly over longer time scales. In experiments using densely packed sand, capillary tension in the sand above the free surface is sometimes sufficient to hold the overlying weight of sand, so that a cave develops which erodes headward more rapidly than where free slumping occurs. It is apparent from Figure 14.16 that most of the erosion occurs at the very head of the sapping face, and that erosion rates in the fluvial zone are relatively small. This implies that although the transport rates (q_s in Eq. 14.8) are high, the rate varies little along most of the sapping face and fluvial zone, so that E in Equation 14.9 is small.

Multiple regression has been used to analyze the results of the sapping experiments in terms of the flow geometry and sediment parameters. The flow geometry is characterized by three dimensionless parameters (see Fig. 14.15):

$$I_a = \text{average hydraulic gradient} = (h_w - h_s)/L_2 \qquad (14.10)$$

$$f = \text{form ratio} = h_s/L_2 \qquad (14.11)$$

$$S = \text{channel gradient} = (h_c - h_o)/L_1 \qquad (14.12)$$

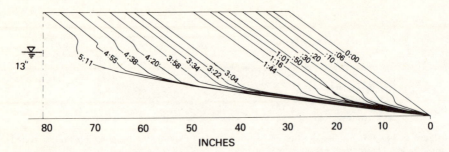

Figure 14.16   Typical temporal progression of the sand surface during sapping erosion in the two-dimensional tank under constant head. Elapsed time since the beginning of the experiment is indicated.

The following interrrelationships characterize the experiments to date, where d is the grain size, c is the sediment concentration, q is the outflow, q_s is the sediment outflow, and u is the water viscosity:

$$S = q^{-0.4}q_s^{0.3}(I_a - 0.11)^{-0.2} \quad (R^2 = 0.80) \tag{14.13}$$

$$S = c^{0.3}I_a^{-0.7} \quad (R^2 = 0.75) \tag{14.14}$$

$$q = I_a^{0.8}f^{0.2}u^{-1.6}d^{1.4} \quad (R^2 = 0.86) \tag{14.15}$$

$$c = (I_a - 0.11)^{0.7}S^{2.6} \quad (R^2 = 0.77) \tag{14.16}$$

$$q_s = (I_a - 0.11)^1 S^{2.4}d^{1.5} \quad (R^2 = 0.62) \tag{14.17}$$

The value of 0.11 which is subtracted from I_a in the first and last two equations is the critical hydraulic gradient for initiation of erosion in the absence of overland flow, which Equation 14.7 suggests should be independent of grain size but dependent upon grain density and shape parameters (as they affect α). The value of the exponents of the grain-size terms are likely to be inaccurate due to the limited range of grain sizes investigated to date.

From these equations and our experiments, it is apparent that the critical hydraulic gradient for initiation of groundwater sapping in cohesionless sediment is somewhat low when expressed in terms of the average hydraulic gradient, I_a. However, as suggested by Figure 14.13 the local hydraulic gradient at the head of the sapping face should be considerably larger, on the order of 0.3. Measurement of the local hydraulic gradients and simulation of the flow net indicate that, in fact, the surface hydraulic gradient reaches a maximum at the head of the sapping face where it has a magnitude of about three times the average hydraulic gradient.

We are conducting additional experiments to further quantify the effects of variations in grain size and sorting upon sapping rates. Eventually we hope to investigate the effects of moderate cohesion upon sapping processes.

Summary and future experiments

Our experiments demonstrated that surface networks can be developed in fine-grained sediments by subsurface sapping processes. The morphometric data gathered from these sapping networks show many spatial and geometric similarities to the Valles Marineris slope networks such as: (a) elongate basins with length to width ratios between 2.1:1 and 2.4:1; (b) abundant short tributaries of the first order; (c) high ratios of channel length to tributary length; (d) perhaps a lower corrected dimensionless drainage density compared to terrestrial basins; (e) large interbasin undissected terrains; (f) regular interbasin spacing along the basal escarpment; (g) higher drainage density in the upper half of the network compared to the overall drainage density; (h) relatively low junction angles with considerable variation compared to terrestrial runoff networks and a lack of regularity in change of junction angle with position in the network; (i) relatively high variability in relationships between stream order with stream frequency, mean stream length, and mean basin area; (j) variability in the relationship between mean stream length and mean basin area; and (k) the low stream order for basin areas of large size. Morphologic similarities between the sapping channels and Martian channels include: (a) steep valley sides with mass wastage features; (b) broad, flattened U-shaped channel cross sections; (c) similarity of tributary and master-channel cross-section areas; (d) abundant straight segments with high angle bends; (e) headward terminations in cirque-like features; and (f) large channel width and area compared to total basin area. These morphometric and morphologic characteristics are inconsistent with observations of terrestrial runoff networks and can be best explained by groundwater sapping processes similar to those observed in the Colorado Plateau (Malin, Laity & Pieri 1980).

Our results serve as an initial progress report for a quantitative program of investigation of sapping processes and resulting network morphologies. These experiments were designed to provide qualitative trends for focusing our future experiments in a more quantitative manner. We intend to integrate our theoretical and experimental investigations by developing a computer model to simulate the temporal evolution of our experiments. The model will combine numerical simulation of the flow net with a theoretical model of transport and erosion based on Equations 14.1 through 14.9. This will allow testing of the assumptions involved in deriving these equations as well as prediction of the empirical relationships of Equations 14.13 through 14.17. Ultimately we plan to develop a three-dimensional computer simulation model that should simulate networks similar to those created in the three-dimensional experiments and illustrate the mechanics of groundwater capture.

We are currently investigating sapping processes and drainage networks in a larger three-dimensional laboratory model. In this model we will use a dense network of small piezometers to monitor groundwater and establish relationships between surface and subsurface drainage divides. This parameter is important in attempts to quantify estimates of drainage density on Viking photographs. In these new experiments we plan to use the following conditions: (a) variable types of sediments; (b) multilayer stratigraphies of variable permeability; (c) artificial structural discontinuities; (d) variable regional slopes; and (e) ground-ice interactions.

We will also continue to search for morphometric criteria to help distinguish between valley networks developed by sapping and those developed by runoff and other processes. The search for suitable parameters is complicated by several factors. First, many morphometric variables require knowledge of drainage-basin areas and shapes. The underground drainage divides cannot be accurately determined for suspected Martian analog valley systems, and are difficult to measure in laboratory experiments. Careful study of the Martian imagery may help delineate contributary areas. Second, the limited resolution of Martian imagery restricts comparative data to coarser features of the valley networks. Finally, neither the laboratory sapping networks nor, presumably, the suspected Martian analogs have approached a steady-state morphology. Morphometric parameters are thus partial functions of the stage of development, as we have discussed.

Acknowledgments

Financial support for our experiments was provided by N.A.S.A. grants NSG 7557 and NSG 7572 and by a grant from the Department of Geology, State University College at Fredonia, New York. Laboratory facilities were provided by the Department of Geology, State University College at Fredonia and by the Department of Environmental Sciences, University of Virginia, Charlottesville. We thank D. Simmons and C. Lis of Fredonia State for their generous lab assistance with the experiments. We also thank T. N. Diggs for his help in preparing some of the figures. We thank V. R. Baker for his support and discussions relating to these experiments. N. Fisher assisted with the manuscript preparation.

Appendix A   *(by Michael Woldenberg)*

Melton (1958) related stream frequency to drainage density:

$$F = c_s D^2 \tag{A.1}$$

where F, the frequency, is the number of Strahler stream segments (composed of one or more links) divided by basin area. D is the drainage density which is defined as the sum of the lengths of all streams divided by the basin area; c_s is a constant.

The reciprocal of c_s is C_s and can be expressed as:

$$C_s = D^2/F \tag{A.2}$$

Smart (1978 p. 155) has drawn an analogy between Melton's C_s and C_l for links:

$$C_l = D\bar{l} = \frac{\bar{l}(2M-1)\bar{l}}{\bar{a}(2M-1)} = \frac{\bar{l}^z}{\bar{a}} = \frac{1}{c_l} \qquad (A.3)$$

where $\bar{l}$ is the mean link length and $\bar{a}$ is the mean link area for the stream network. (No distinction is made in this approximation for interior and exterior links).

The authors of the present paper have erroneously proposed

$$(\Sigma L)^2/A = \text{constant} \qquad (A.4)$$

where ΣL is the total length of all streams in the network. A. Abrahams (pers. comm.) has demonstrated that the equivalent expression

$$\frac{\bar{l}^2(2M-1)^2}{\bar{a}(2M-1)} \neq \text{constant} \qquad (A.5)$$

If, however, Equations A.4 and A.5 were divided by $(2M-1)$ the quotient would indeed be the constant C_l from Equation A.3.

Furthermore, since $(2M-1)$ is more than N (the number of Strahler stream segments) C_l is smaller than C_s. If Melton had used links, rather than Strahler stream segments, he would have derived c_l which is larger than c_s (Eqns A.2 and A.3).

If equation A.4 were divided by N, then the constant would be C_s. N, c_s and C_s depend on M and the bifurcation ratio.

References

Abrahams, A. D. 1984. Channel networks: a geomorphological perspective. *Water Resour. Res.* **20**, 161–88.

Baker, V. R. 1982. *The channels of Mars*. Austin: Univ. of Texas Press.

Baker, V. R., and R. C. Kochel 1979. Martian channel morphology: Maja and Kasei Vallis. *J. Geophys. Res.* **84**, 7961–83.

Baker, V. R. and D. J. Milton 1974. Erosion by catastrophic floods on Mars and Earth. *Icarus* **23**, 27–41.

Boffey, P. M. 1977. Teton Dam verdict: a foul-up by the engineers. *Science,* **195**, 270–72.

Bowden, K. L., and J. R. Wallis 1964. Effect of stream-ordering technique on Horton's laws of drainage composition. *Geol. Soc. Am. Bull.* **75**, 767–74.

Broscoe, A. J. 1959. *Quantitative analysis of longitudinal stream profiles of small watersheds.* O.N.R. Proj. NR 389–042, Tech. Rept. 18. New York: Dept. Geol. Columbia Univ.

Burǵi, A. M., and S. Karaki 1971. Seepage effect upon channel bank stability. *J. Irrig. Drain. Div., Proc. Am. Soc. Civil Engrs.* IR1, 59–72.

Carr, M. H., and G. C. Schaber 1977. Martian permafrost features. *J. Geophys. Res.* **82**, 4039–54.

Chorley, R. J. 1959. The shape of drumlins. *J. Glaciol.* **3**, 339–44.

Dunne, T. 1980. Formation and controls on channel networks. *Prog. In Phys. Geog.* **4**, 211–39.

Hack, J. T. 1957. Studies of longitudinal stream profiles in Virginia and Maryland. *U.S. Geol. Surv. Prof. Paper* 294B.

Haefeli, R. 1948. Stability of slopes acted upon by parallel seepage. *Proc. 2nd Int. Conf. on Soil Mech. and Foundation Engng* **1**, 57–62.

Higgins, C. G. 1982. Drainage systems developed by sapping on Earth and Mars. *Geology* **10**, 147–52.

Horton, R. E., 1945. Erosional development of streams and their drainage basins: hydrophysical approach to quantitative morphology. *Geol. Soc. Am. Bull.* **56**, 275–370.

Howard, A. D. 1971a. Problems of interpretation of simulation models of geologic processes. In *Quantitative geomorphology: some aspects and applications.* M. Morisawa (ed.), 61–82. 2nd Geomorph. Symposium, Binghamton N.Y.: Publications in Geomorphology.

Howard, A. D. 1971b. Optimal angles of stream junctions: geometric, stability to capture, and minimum power criteria. *Water Resources Res.* **7**, 863–73.

Howard, A. D. 1978. Effect of slope on the threshold of motion and its application to orientation of wind ripples. *Geol. Soc. Am. Bull.* **88**, 853–56.

Howard, A. D. 1980. Thresholds in river regime. In *Thresholds in geomorphology*, D. R. Coates and J. D. Vitek, (eds), 227–58. London: George Allen and Unwin.

Howard, A. D. and R. Dolan 1981. Geomorphology of the Colorado River in the Grand Canyon. *J. Geol.* **89**, 269–98.

Howard, A. D., and C. McLane 1981. Groundwater sapping in sediments: theory and experiments. *N.A.S.A. Tech. Mem.* 84211, 283–85.

Jarvis, R. S. and M. J. Woldenberg (eds). 1984. *River networks*. Stroudsburg, PA.: Hutchinson Ross.

Kezdi, A. 1979. *Soil physics: selected topics*. New York: Elsevier.

Kochel, R. C. and C. M. Burgess 1983. Structural control of geomorphic features in the Kasei Vallis region of Mars. *N.A.S.A. Tech. Mem.* 85127, 227–29.

Kochel, R. C. and A. P. Capar 1983. Structural control of sapping valley networks along Valles Marineris, Mars. *N.A.S.A. Tech. Mem.* 85127, 295–97.

Laity, J. E. 1980. Groundwater sapping on the Colorado Plateau. *N.A.S.A. Tech. Mem.* 82385, 358–60.

Laity, J. E. and D. C. Pieri 1980. Sapping processes in tributary valley systems. *N.A.S.A. Tech. Mem.* 81776, 271–73.

Laity, J. E. and R. S. Saunders 1981. Sapping processes and the development of theatre-headed valleys. *N.A.S.A. Tech. Mem.* 84211, 280–82.

Lambe, T. W. and R. V. Whitman 1969. *Soil mechanics*. New York: Wiley.

Lubowe, J. K. 1964. Stream junction angles in the dendritic drainage pattern. *Am. J. Sci.* **262**, 325–39.

Malin, M. C., 1976. Age of Martian channels. *J. Geophys. Res.* **81**, 4825–45.

Malin, M. C., J. E. Laity and D. C. Pieri 1980. Sapping: analog studies on Earth and Mars. *N.A.S.A. Tech. Mem.* 81776, 298–99.

Manker, J. P. and A. P. Johnson 1982. Simulation of Martian chaotic terrain and outflow channels. *Icarus* **51**, 121–32.

Masursky, H., 1973. An overview of geological results from Mariner 9. *J. Geophys. Res.* **78**, 4009–30.

Masursky, H., J. M. Boyce, A. L. Dial, G. C. Schaber and M. E. Strobell 1977. Classification and time of formation of Martian channels based on Viking data. *J. Geophys. Res.* **82**, 4016–38.

Melton, M. A. 1958. Geometric properties of mature drainage systems and their representation in E_4 phase space. *J. Geol.* **66**, 35–54.

Morisawa, M. E. 1959. *Relation of quantitative geomorphology to stream flow in representative watersheds of the Appalachian Plateau Province*. O.N.R. Proj. NR 389–042, Tech. Rept. 20. New York: Dept. Geol., Columbia Univ.

Oldenziel, D. M. and W. E. Brink 1974. Influence of suction and blowing on entrainment of sand particles. *J. Hydraul. Div., Proc. Am. Soc. Civil Engrs.* **HY7**, 935–49.

Parker, R. S. 1977. *Experimental study of drainage basin evolution and its hydrologic implications.* Hydrology Paper No. 90. Fort Collins: Colorado State University.

Patton, P. C., and V. R. Baker 1976. Morphometry and floods in small drainage basins subject to diverse hydrogeomorphic controls. *Water Resources Res.* **12**, 941–52.

Pieri, D. C. 1980a. Geomorphology of Martian valleys. *N.A.S.A. Tech. Mem.* 81979, 1–160.

Pieri, D. C. 1980b. Martian valleys: morphology, distribution, age, and origin. *Science* **210**, 895–97.

Pieri, D. C. and C. Sagan 1979. Origin of Martian valleys. *N.A.S.A. Tech. Mem.* 80339, 349–52.

Schumm, S. A. 1977. *The fluvial system.* New York: Wiley.

Sharp, R. P. 1973. Mars: fretted and chaotic terrains. *J. Geophys. Res.* **78**, 4073–83.

Sharp, R. P. and M. C. Malin 1975. Channels on Mars. *Geol. Soc. Am. Bull.* **86**, 593–609.

Smart, J. S. 1978. The analysis of drainage network composition. *Earth Surf. Proc.* **3**, 129–70.

Strahler, A. N. 1964. Quantitative geomorphology of drainage basins and channel networks. In *Handbook of applied hydrology*, Ven te Chow (ed.), pt. 4, 39–76. New York: McGraw Hill.

15

Ground ice models for the distribution and evolution of curvilinear landforms on Mars

Lisa A. Rossbacher

Introduction

The amount of water available on the surface of Mars has been a subject of debate since Schiaperelli first mapped *canali* cutting the planet's surface in 1877. The small amounts of water available in arid regions on Earth have a significant influence on the geomorphic activity; the presence of water on Mars could profoundly affect the geomorphic processes on that planet.

Estimates of the amount of water on Mars are derived from models for volatile distributions, outgassing history, and atmospheric evolution of the planet. Early estimates concentrated on Mars alone; recent models have also incorporated volatile and noble gas from the Pioneer Venus mission (Pollack & Black 1979). These data have led to a comprehensive model for planetary evolution that fits Martian atmospheric development into an evolutionary scheme for all the inner planets. The 'grain accretion hypothesis' is based on a model with relatively constant temperatures but varying pressures in the primordial nebula. Using this model, Pollack and Black (1979) estimated that Mars outgassed a global equivalent of 80–160 m of water. For comparison, if all the water on Earth were spread out to a uniform depth, the world-ocean would be more than 2.5 km deep (Nace 1967) with a total volume of about $1.3 \times 10^9 \, \mathrm{km^3}$. The volume of outgassed water calculated from Pollack and Black's data for Earth is $1.7 \times 10^9 \, \mathrm{km^3}$.

Estimated volumes of outgassed water on Mars are shown in Table 15.1. The methods of estimating these water volumes are evaluated in detail elsewhere (Rossbacher 1983). Only a small percentage of these estimated volumes can now be observed in the Martian environment (Table 15.2). The most probable storage site for the hypothesized remaining $7.9 \times 10^6 \, \mathrm{km^3}$ of the Martian water is in the subsurface. The zone in which water ice could exist on Mars has been called the cryosphere (Rossbacher & Judson 1981) and is shown in Figure 15.1.

Table 15.1 Assumptions behind estimates of water volume on Mars.

Method of calculation	Water volume (equivalent global depth m)	Reference
$^{36}Ar_{Mars}/^{36}Ar_{degrassed\ ordinary\ chondrite}$	5.9	Rasool *et al.* 1977
$^{36}Ar/^{40}Ar$ (assumes Earth ratio)	9.4	Anders & Owen 1977
$^{36}Ar_{Mars}/^{36}Ar_{Earth}$	10	Owen & Biemann 1976
$^{40}Ar_{Mars}/^{40}Ar_{Earth}$	88.0	Clark & Baird 1979
N_{2Mars}/N_{2Earth}	133	McElroy *et al.* 1977
N/C (grain accretion hypothesis)	80–160	Pollack & Black 1979

Table 15.2 Water inventory for Mars.

Sinks	Estimated volume of water accounted for (km^3)
atmosphere[1]	~1.3
frost[1]	≤1.5
exospheric escape[2]	~4.5×10^5
north polar cap (assuming 2 km thickness)	~1.6×10^6
ground ice[3]	~7.9×10^6
estimated originally outgassed	~10^7

[1] Calculated with data from Farmer and Doms (1979).
[2] Calculated with data from Walker (1977)
[3] Outgassed water minus observed and escaped water.

One way to test this model for the distribution of ground ice is to locate features that may be related to ground ice. Such features that might be expected include rimless depressions or themokarst (Anderson *et al.* 1973, Washburn 1980), lobate debris flows (Squyres 1978, 1979), table mountains (Allen 1979), and polygonally fractured ground (Hunt & Washburn 1966, Morris & Underwood 1978). A survey of orbiter and lander photographs from NASA's Viking mission to Mars indicates that all of these landforms occur on Mars, and they are located in both hemispheres at a range of latitudes (Fig. 15.2).

Based on this geomorphic evidence, there is common agreement that Mars contains subsurface ice. In a review of ice distribution on Mars, a new feature, curvilinear-patterned ground, was identified (Rossbacher & Judson 1981). These features seem to be unique to Mars and may be related to subsurface ice. In the remainder of this paper, I will describe the regional setting of curvilinear ground, summarize its morphometry, map its distribution, and examine models for its genesis.

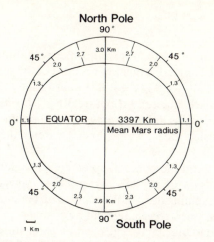

Figure 15.1 Cross-section of the Martian cryosphere, the zone in which the temperature is below the freezing point of water (here assumed to be 273°K) and where ground ice could exist. The thickness of the cryosphere is not to scale with the rest of the planet. The total volume of the cryosphere in this model is $2.5 \times 10^8 \, \text{km}^3$ (Rossbacher & Judson 1981; originally published in *Icarus* and used by permission of Academic Press, Inc.).

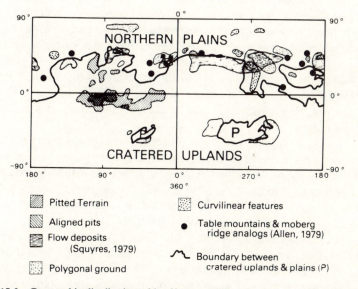

Figure 15.2 Geographic distribution of landforms on Mars that may be related to ice (simple cylindrical projection). The length of the equator is $2.1 \times 10^4 \, \text{km}$. This map is based on observations of over 20 000 Viking orbiter photographs, covering the entire planet, taken under good-to-excellent atmospheric conditions (Rossbacher & Judson 1981; originally published in *Icarus* and used by permission of Academic Press, Inc.).

Regional setting

Mars can be divided into two major terrain types with differing characteristics and ages. The cratered upland, which covers most of the southern hemisphere, has a heavily cratered surface that records the early meteorite bombardment of the planet; these impacts tapered off before 3.5×10^9 years ago (Frey 1980). The smooth northern plains are on average several kilometers lower than the cratered upland. Because the plains are far less cratered they are assumed to be younger.

The transition between the cratered upland and the northern plains can be traced over two-thirds of the planet's circumference; this zone is obscured by volcanics over the remaining third of its length. The boundary between the two terrains varies from a steep escarpment cut by channels to a gradual transition from cratered upland to the plains. The location and nature of this boundary is summarized in Figure 15.3.

Where the transition zone is well exposed, there is a regular gradation from cratered upland to fretted terrain. Fragments of the cratered terrain stand as flat-topped mesas surrounded by debris aprons. Farther from the cratered upland these outliers have eroded into knobby or finely fretted terrain which grades into smooth plains. These relationships suggest that the lower plains may have developed, at least in part, at the expense of the higher cratered terrain.

Soderblom and Wenner (1978) explain this difference in terrain types between the cratered uplands and the northern plains as a variation in composition. Because the elevation difference between the cratered upland and the northern plains is relatively consistent, the lower surface may be a fossil solid/liquid water interface. The material above that elevation contained water ice, they suggest, and below it was liquid water. The presence of liquid water resulted in diagenetic changes in the subsurface material that made the plains more resistant to erosion. Consequently, the northern plains have been stripped to a roughly uniform level (the fossil H_2O solid/liquid interface) below the cratered upland.

Other explanations that have been presented to account for the differentiation between the cratered uplands and the northern plains include (1) the idea that the low plains developed in a region where the crust was extensively fractured and eroded (Saunders 1974) and (2) the proposal that a single-cell convection system in Mars's interior resulted in subcrustal erosion concentrated beneath the northern plains (Wise *et al*. 1979).

The global transition zone on Mars has been described as an escarpment (Sharp 1973), but according to Mariner 9 topographic data the highlands slope gently down to the plains with a regional slope between 0.1° and 3° (Batson *et al*. 1979). The zone is marked by a variety of landforms, including debris flows (Squyres 1979), chaotic terrain (Sharp 1973, Mutch & Saunders 1976), inverted topography (Rhodes 1980), channels (Kochel & Baker 1980), polygonally fractured ground (Carr & Schaber 1977), and the curvilinear features under discussion here (Rossbacher & Judson 1981).

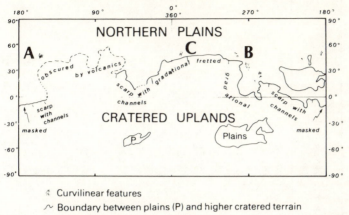

Curvilinear features
~ Boundary between plains (P) and higher cratered terrain
Groups A,B, and C are described in text

Figure 15.3 Nature and location of the boundary separating the ancient cratered terrain and the smoother northern plains on Mars. Groups A, B and C are described in the text.

Over the entire planet, the curvilinear features are associated with the boundary between the northern plains and the cratered upland, but they are not present everywhere along the boundary. The curvilinear features on Mars occur primarily along a zone between approximately 45–55° N latitude, on the plains side of the boundary between the cratered uplands and the northern plains (Fig. 15.3). In the vicinity of 270° W longitude (western Utopia Planitia), however, the features extend south to 15° N, following the plains/upland boundary. The curvilinear features can be assigned to three general groups: centered around 50° N, 180° W in Arcadia Planitia (Group A); 40° N, 270° W in western Utopia Planitia (Group B); and 50° N, 353° W in eastern Acidalia Planitia (Group C). These areas are identified by letter in Figure 15.3.

Morphology of curvilinear features

The Martian curvilinear features have several types of topographic expression, including curvilinear ridges, arcuate troughs, and patterns with no obvious topography at all. These are described in more detail in Rossbacher and Judson (1981), and examples of each are shown in Figure 15.4.

Detailed morphometric data collected on the curvilinear features include the width and length of the arcs, their wavelengths (distance between centers of arcs, perpendicular to long axis), and their height or depth, where measurable. The width, length, and wavelength were all measured along cross-sectional lines on detailed morphometric maps, four of which are presented in this paper to illustrate the variety of curvilinear morphologies. These data are grouped according to the areas where the curvilinear features occur on Mars: Arcadia Planitia (A); Utopia Planitia (B); or Acidalia Planitia (C). The morphometric

Table 15.3 Summary of morphometry of curvilinear features on Mars (number of measurements in sample shown in parentheses). Error reflects accuracy of measurements on the Viking photographs.

Mosaic	Width		Segment length		Wavelength (center to center) (km)	Relief Height or depth (m)	Relief
	Ridge (m)	Depression	Ridge (km)	Depression			
115A11–18 [group A]	360±40 (34)		2.62±.04 (34)		2.25±.04 (30)	80±20 (6)	
115A19–26 [group A]	370±40 (31)	480±40 (2)	3.28±.04 (31)	4.99±.04 (2)	2.28±.04 (29)	60±20 (5)	
608A06–10 [group B]	930±70 (25)		3.68±.07 (25)		4.67±.07 (21)	150±30 (5)	
572A03–09 [group B]	860±120 (27)		5.08±.12 (27)		3.93±.12 (24)	231±60 (3)	
572A30–34 [group B]	830±110 (11)		5.50±.11 (14)		5.01±.11 (12)	30±20 (1)	
57B51–59 [group B]	700±60 (73)	580±60 (21)	5.55±.06 (73)	4.22±.06 (21)	4.71±.06 (87)	90±20 (8)	40±20 (2)
52A39–46 [group C]			3.63±.04 (21)		1.67±.04 (19)		
averages	650±120 (201)	580±60 (23)	4.35±.12 (225)	4.29±.06 (23)	3.73±.12 (222)	110±60 (28)	40±20 (2)

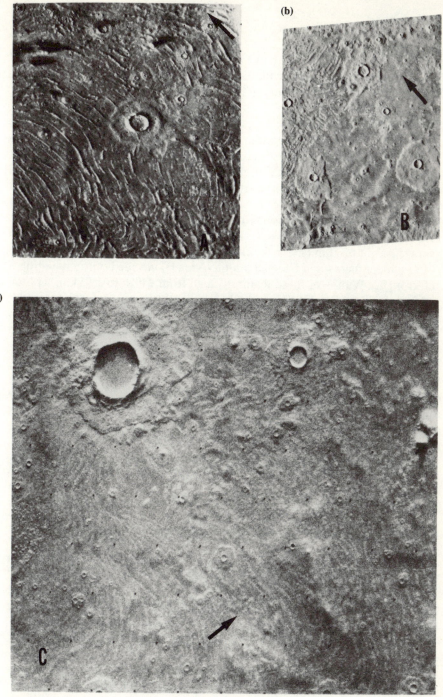

Figure 15.4 Curvilinear features on Mars have a range of topographic expression: (a) alternating positive and negative topography is displayed in western Utopia Planitia. The area shown is about 100 km across (Viking frame 11B03, centered at 49°N, 287°W, covering the same area as frame 57B56); (b) some curvilinear forms are arcuate depressions in a surrounding plateau area, as shown here in Acidalia Planitia. The area shown is about 38 km across (Viking frame 60A52, centered at 49°N, 349°W); and (c) other curvilinear features have no observable topographic expressions, as in this area in eastern Acidalia Planitia, which is about 62 km across (Viking frame 52A43, centered at 47°N, 350°W) (Viking photography supplied by the National Space Science Data Center and the Jet Propulsion Laboratory).

measurements are summarized in Table 15.3. The landform dimensions could be estimated to within 0.1 mm on the 20 cm × 20 cm Viking prints and thus the precision of the measurements is primarily a function of the photographic scale.

In Arcadia Planitia (Area A) almost all of the curvilinear landforms are arcuate ridges. The measured widths of the features range from about 170 m to 720 m. The average width is 360 m in a sample size of 67 (standard deviation of the sample is 106). The lengths of these features, measured perpendicular to the cross-sectional lines, range from 0.56 to 9.65 km, averaging 2.62 km ($N = 67$, s.d. = 1.60). The wavelengths, the distances between the centers of features along the cross-sectional lines, range from 0.55 to 8.51 km, with an average of 2.26 km ($N = 59$, s.d. = 1.71). Eleven curvilinear ridges had an average height of 71 m (s.d. = 0.32).

Curvilinear features in Utopia Planitia (Area B) include both arcuate ridges and troughs. Widths of these features range from 240 m to 1.95 km, with an average of 930 m ($N = 158$, s.d. = 300). The lengths average 5.55 km ($N = 158$, s.d. = 2.93), with individual arcs from 0.96 to 12.18 km. The distances between these arcs range from 0.95 to 23.19 km, and the wavelength averages 4.60 km ($N = 144$, s.d. = 3.59). Measurements of 19 arcuate ridges yielded an average height of 131 m (s.d. = 0.71).

For Group C in Acidalia Planitia, the curvilinear features are mostly arcuate depressions or patterns with no apparent relief. This lack of observable topography may be, in part, because the features are smaller here, but this cannot be demonstrated yet. The width of the curvilinear patterns could not be measured in this area, but the length of the arcs averages 3.63 km ($N = 21$, s.d. = 0.49). The measured wavelengths range from 0.55 to 7.40 km, with an average of 1.67 km ($N = 19$, s.d. = 1.49). Four arcuate depressions had an average depth of 18 m (s.d. = 1.15).

Regional and local geomorphology

The Martian curvilinear features are all located in the northern plains, north or west of the boundary separating the smooth plains and the cratered upland. This section describes the regional relationships among the curvilinear features, the plains/upland boundary, and other landforms that characterize this transitional zone.

Detailed mapping of the Martian curvilinear features reveals geomorphic and topographic relationships among up to four separate photogeologic map units. The local escarpments separating lower units from higher ones are a few tens of meters in height. These units are, from topographically lowest upward:

(a)   a high-albedo layer, *HA*, that appears primarily in isolated patches on the floors of closed depressions in Arcadia Planitia;

(b)   a hummocky, cratered unit *HC*, that is present in all three areas and has an irregular surface and uneven texture;

(c)   a smooth, sparsely cratered plains unit, *PL*, that has a smooth texture, even tone, and fewer craters than the lower HC unit; and

(d)   a unit with low albedo, *DA*, that occurs as small raised patches above the smooth plains unit in Arcadia and Utopia Planitia.

Craters are superposed on all of these units, including partially buried craters with degraded rims, C_1, partially eroded craters that have not been buried, C_2, craters with fresh-appearing rims and ejecta blankets, C_3, and areas that are densely cratered by numerous small impact craters, C_4. The explanation of symbols for the detailed geomorphic maps in this paper are shown in Figure 15.5.

symbol   explanation

SURFACE FEATURES

Craters and associated ejecta blankets
(all craters are type C_3 unless otherwise shown).

C_1 - poorly preserved rim morphology.

C_2 - crater with raised rim, poorly preserved ejecta blanket.

C_3 - crater with distinct rim and relatively fresh morphology.

C_4 - crater swarms forming locally saturated cratered zones.

Curvilinear ridges and depressions (topography indicated by hachures). [Single lines used where features too narrow for width to show on map.]

Valleys with central ridges.

PHOTOGEOLOGIC MAP UNITS

Dark, low-albedo unit.

Smooth-plains unit; smooth surface, even texture, relatively few craters. [This unit should not be confused with the plains unit in the regional geomorphic maps; all of the detailed geomorphic maps cover areas within the regional plains unit.]

Hummocky and cratered unit.

Bright, high-albedo unit.

Figure 15.5   Explanation of symbols used in detailed geomorphic maps.

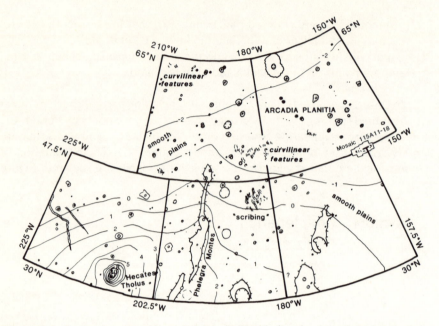

Figure 15.6   Regional geomorphic map of Arcadia Planitia (Lambert's conformal conic projection).

A major uncertainty about these photogeologic units is the elevation of their exposure. If all the occurrences of one unit are at the same elevation, then they might represent a series of horizontal layers. If each unit is exposed at differing elevations, then horizontal layering is unlikely. Similarly, geomorphic and topographic sections across these detailed geomorphic maps also depend on inferences about the subsurface geology. Models for this subsurface geology follow the regional and detailed geomorphic maps in the next section.

Arcadia Planitia (area A)
The curvilinear features in Arcadia Planitia occur primarily in the plains north of Elysium, which is an elliptical uplifted area that according to the Mariner 9 data (Batson *et al.* 1979) stands about 4 km above the surrounding topography. Hecates Tholus, the peak in the lower left of Figure 15.6 is one of the three volcanic shields in the Elysium region. Lava flows have covered some of the surface near Hecates, and many partially filled craters are visible.

Curvilinear features occur in the smooth plains north of 47° N in Arcadia Planitia between 165° and 210° W, near the base of the regional slope. This slope, excluding the volcanic construct of Hecates Tholus, is about 0.1°. The boundary between the cratered terrain and the northern plains is over 2000 km to the south, on the southern side of the volcanics of Elysium Planitia. In map view, the features do not have a uniform orientation; they are concave toward the east, southeast, or west.

In the absence of a global reference level, topography on Mars is described relative to the elevation at which the atmospheric pressure is equal to 6 mb (the triple-point pressure of H_2O); curvilinear features in this area occur between 0 and −2 km relative to the Mars datum.

Another feature that occurs near the curvilinear landforms in this area is finely dissected ground, which is identified in Figure 15.6 as scribing. These features are elongate valleys with central ridges. Some are interconnected in long networks; others are discrete, isolated features. They resemble the shorter and straighter valleys near the boundaries separating two photogeologic map units in areas with curvilinear features. These short valleys that are similar to scribing are described in more detail elsewhere in this paper.

Mosaic 115A11–18

Detailed geomorphic maps of areas containing curvilinear ground are based on mosaics of individual Viking frames (Fig. 15.7). The area covered by this mosaic is shown in Figure 15.6 and the morphometry of the curvilinear features is summarized in Table 15.3.

This mosaic illustrates all four of the photogeologic map units identified in areas with curvilinear features (Fig. 15.7). The scarps separating these units show their topographic relationships. The high-albedo, *HA*, unit (closely spaced stippled pattern) is exposed in the floors of several closed depressions. The hummocky cratered unit, *HC*, is topographically above the high-albedo unit; most of the curvilinear features occur in this HC unit.

The smooth-plains unit is topographically higher than the HC unit. It occurs in patches that suggest the smooth plains may have formed a more extensive

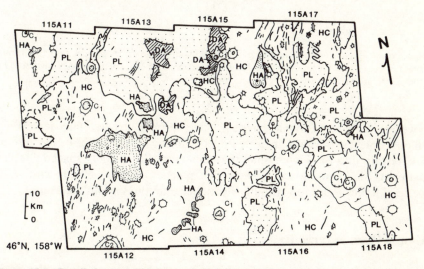

Figure 15.7 Detailed geomorphic map of area with curvilinear features in Arcadia Planitia (Mercator's projection, mosaic of Viking frames 115A11–18). The area covered by this map is outlined in Figure 15.6 47.5°N, 157.5°W.

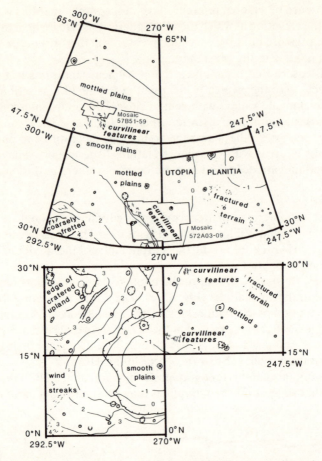

Figure 15.8 Regional geomorphic map of Utopia Planitia. The top map is Lambert's conformal conic projection, and the lower one Mercator's projection. The length of the area shown is 2800 km along 30°N latitude.

covering in the past and may have since been broken up and reduced in extent. In many places, the curvilinear features echo the local boundary between the HC terrain and the higher smooth-plains unit, *PL*, (widely spaced stippled pattern). Surrounding scarps show that patches of a low-albedo unit, *DA*, (striped pattern) are topographically higher than the smooth plains unit.

Utopia Planitia (area B)

Curvilinear features in Utopia Planitia follow the boundary between the plains and the cratered upland at distances between 400 and 800 km from the cratered terrain (Fig. 15.8). The features occur between 260° and 292° W longitude and

15° to 52° N latitude, and at elevations between +1 and −1 km relative to the Mars datum. U.S. Geological Survey photomosaics show a distinct scarp along the western edge of the plains between 5⁰ and 30° N (just west of 270° W), but this scarp only generally corresponds with topographic contours from the Mariner 9 data (Batson *et al.* 1979).

The zone separating the cratered upland and the plains exhibits a transition from the cratered terrain through coarsely fretted and then finely fretted terrain to smooth plains. Locally, the smooth plains grade into mottled and, in one area, fractured terrain. The curvilinear features occur at the base of the 0.1° slope between the cratered upland and the lower plains. The curvilinear features are regionally associated with the polygonally fractured terrain that was discussed by Pechmann (1980).

Mosaic 572A03–09

The high-albedo unit is absent in this area, so the lowest observable layer in this mosaic is the *HC* unit (Fig. 15.9). The HC unit contains numerous curvilinear features and most of them parallel the boundary between the HC unit and the overlying smooth plains. The morphometry of the curvilinear features in this mosaic is summarized in Table 15.3.

The boundary between the *HC* and the *PL* units is cut by several valleys with central ridges. The closed ends of these valleys grade smoothly up to the *PL*

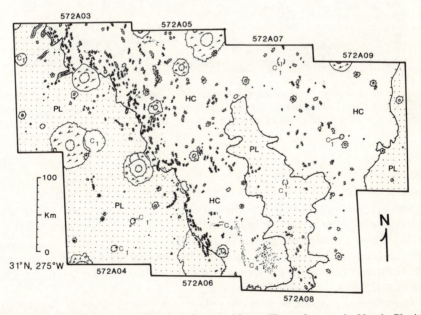

Figure 15.9 Detailed geomorphic map of area with curvilinear features in Utopia Planitia (Mercator's projection, mosaic of Viking frames 572A03–09). The area covered by this map is outlined in Figure 15.8.

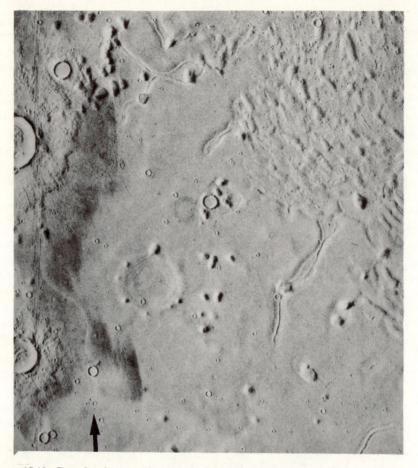

Figure 15.10 Boundary between the smooth-plains, *PL*, and hummocky cratered, *HC*, photo-geologic map units. Note alignment of curvilinear features with edge of *PL* unit and valleys with central ridges in the *PL* unit (Viking photography provided by NSSDC).

unit. Similar features also occur within the *PL* unit at distances of 10 km or more from the *HC/PL* boundary. The ridges that form spines down these valleys are narrower than the curvilinear ridges, and in this area their orientation is generally perpendicular to that of the nearby curvilinear landforms (Fig. 15.10). Thus, although the central ridges in these valleys appear to have some correlation with the curvilinear features, the precise relationship is unclear. The features resemble the interconnected valleys with central ridges in Arcadia Planitia, described previously as scribing.

The origin and significance of the valleys with central ridges is speculative. One possible explanation is that the features were created by a process similar to one on Earth in which flow toward the center of a valley by water-rich, clayey material creates central ridges (Hollingworth *et al.* 1944). The Martian regolith is believed to contain clays (Baird *et al.* 1976), and such an origin seems possible.

Fresh, C_3, and partially buried, C_1, craters are found in both terrain units in this area. A higher percentage of buried craters occurs in the *PL* units (17%) than in the HC unit (5% buried craters, excluding the C_4 type), but partially buried craters do occur in both terrain types. The material filling these craters may be wind-blown dust or thin lava deposits.

In several places in this area the *HC* unit has been pitted by numerous small craters (diameters less than 1 km), sometimes to the extent of saturating an area with impact craters, C_4. These crater concentrations are probably the result of secondary crater swarms or a shower of impacting bodies from the breakup of a large meteorite.

Mosaic 57B51–59
This area is the major exception to the generalization that curvilinear features occur in the hummocky cratered unit (Fig. 15.11). The curvilinear features here occur on a smooth substrate with a reasonably uniform texture, consistent with the smooth-plains unit. These features are at an elevation between 0 and 1 km relative to the Mars datum.

The curvilinear features of this area include both ridges and depressions. Some arcs include both, with ridges changing directly into troughs. In some places the curvilinear ridges are expressed through an ejecta blanket (Fig. 15.4a). Either the ejecta blanket is very thin or the curvilinear landform has developed beneath the crater since ejecta emplacement.

This area also includes good examples of partially buried craters, C_1, and degraded crater rims, C_2. The area contains no craters larger than 10 km in diameter, although several craters in this mosaic approach that limit. Most of the craters are between 0.1 and 1.0 km. These data suggest that the surface was probably not exposed to much, if any, of the heavy bombardment early in Mars's history. More detailed study of the size and abundance of impact craters is needed before the age of the surfaces can be understood.

Acidalia Planitia (area C)

One small patch of curvilinear ground is located in Acidalia Planitia (Fig. 15.12) between 345–360° W and 48–51° N. The features occur in the northern plains at elevations between 0 and −1 km relative to the Mars datum. The curvilinear features are on the lower part of the regional slopes, northwest of the outflow channels of the Chryse Basin and approximately 150–200 km north of the cratered upland. The regional slope is about 0.1°.

The cratered terrain is cut in several places by valleys, including Mamers Valles, Auqakah Vallis, and Huo Hsing Vallis. The cratered terrain and the lower plains are separated by fretted terrain (Sharp 1973) consisting of isolated blocks of the cratered upland surface, separated by crevasses, with debris aprons surrounding many of these mesas.

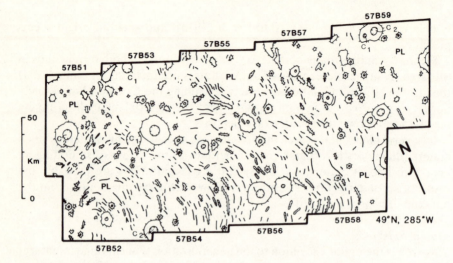

Figure 15.11 Detailed geomorphic map of area with curvilinear features in Utopia Planitia (Mercator's projection, mosaic of Viking frames 57B51–59). The area covered by this map is outlined in Figure 15.8. This area is all within the smooth-plains photogeologic map unit; the fine stippled pattern is omitted in this map for clarity.

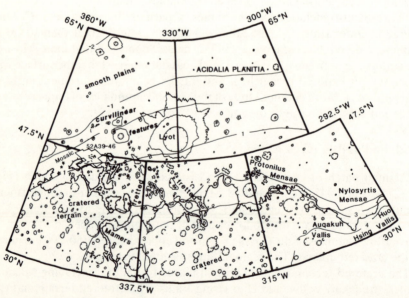

Figure 15.12 Regional geomorphic map of Acidalia Planitia (Lambert's conformal conic projection). The distance across the map base along 30°N latitude is approximately 3300 km; the distance along 65°N latitude is about 1500 km.

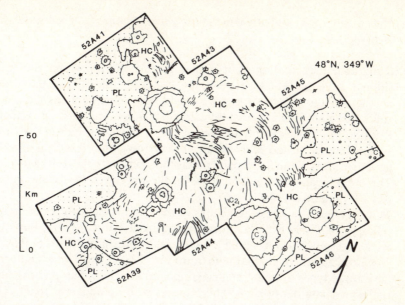

Figure 15.13 Detailed geomorphic map of area with curvilinear features in Acidalia Planitia (Mercator's projection, mosaic of Viking frames 52A39–45). The area covered by this map is outlined in Figure 15.12.

Mosaic 52A39–45

Only the *PL* and *HC* photogeologic map units are observed in this area (Fig. 15.13) and they are clearly separated by a scarp. The curvilinear features are limited to the hummocky, cratered unit. Curvilinear features in this area are generally too poorly defined for their topography (if any) to be observed, and the photographic resolution is too low to measure the width of the landforms. Therefore, most of the curvilinear features are depicted on the geomorphic map for this mosaic as arcuate lines. In the morphometric measurements the wavelength and length of the curvilinear features could be determined, but the widths could not be (Table 15.3). The curvilinear features do seem to parallel the local *HC/PL* boundary in this area.

With the exception of a few degraded craters in the PL unit on the far eastern side of this mosaic, all the craters appear relatively fresh. Crater diameters range between 0.1 and 10 km, but 96% are less than 1 km in diameter; if the area studied is a valid sample, the smooth plains unit in this mosaic may not have been exposed to extensive meteorite bombardment.

Summary of data on Martian curvilinear features

To provide an interim summary, this section will review what is known about the curvilinear features on Mars. Based on this information, models for the subsurface stratigraphy and origin of the curvilinear features can be formulated and discussed.

(a) Geographic distribution. Curvilinear ground occurs within the transitional zone on Mars between the cratered upland and the lower northern plains. The features are found only in three areas along the zone: Arcadia Planitia, western Utopia Planitia, and eastern Acidalia Planitia. The curvilinear landforms occur at elevations between +1 and −2 km relative to the Mars datum, below the cratered terrain and the upper surfaces of the flat-topped mesas that constitute the fretted terrain. The curvilinear features tend to be in the lower reaches of the gentle (0.1°–3°) regional slope.

(b) Morphology. The sizes and shapes of the curvilinear features are fairly uniform. Widths of the landforms range from 170 to 1950 m, but variations are much lower within a local area. Lengths range from 0.56 to 18.63 km, and wavelengths range from 0.55 to 23.19 km. These landforms have three types of topographic expression: arcuate ridges and troughs: rimless arcuate depressions; and features with no apparent relief. The ridges range up to 300 m in height, and the troughs have a maximum measurable depth of about 25 m. Ridges are far more abundant than depressions, but the two seem to be closely associated and in some places the ridges pass directly into troughs.

(c) Associations. In several areas the curvilinear features occur near other features that have been suggested as being related to ground ice, particularly polygonally fractured ground and lobate debris flows. However, this interpretation also depends on a qualitative model for the origin of these features. The associations between the curvilinear features and exhumed landforms are more regional. Both types of feature are associated with the cratered upland/northern plains boundary. Acidalia Planitia is the only area in which inverted craters are found near the curvilinear features.

The curvilinear landforms are also associated, on a local scale, with a group of photogeologic map units: a high-albedo unit; a hummocky cratered unit which contains most of the curvilinear features; a smooth-plains unit that is sparsely cratered; and a low-albedo unit. Based on the adjacent scarps, these are listed in the general order from lowest to highest topography, but the absolute elevations cannot be determined with available data. Some possible interpretations will be offered in the next section. All of these photogeologic map units occur within the northern plains of Mars.

In all three areas where they occur, although not in every specific instance, the curvilinear features parallel the boundary between two adjacent photogeologic map units. This relationship is seen most often where the curvilinear features occur on the hummocky, cratered unit, *HC*, near an escarpment rising to the higher smooth plains, *PL*. This orientation suggests scarp retreat on a local scale, rather than the retreat of the cratered uplands from the northern plains.

Valleys with central ridges occur in the smooth plains unit, near the escarpment separating it from the hummocky, cratered unit above. Most of these valleys grade smoothly to the HC unit, but they also occur in the smooth-plains unit 10 km or more away from the scarp. In every instance these valleys with central ridges are oriented perpendicular to the scarp separating the PL and HC photogeologic map units. In Arcadia Planitia a network of valleys with central ridges, scribing, displays an interconnected pattern that has no clear relationship to local scarps separating photogeologic map units. The depressions suggest some type of solutional or degradational activity, but the origin of these valleys is unknown.

(d) Age. The age of the curvilinear features cannot be determined accurately by crater-counting statistics. However, the surrounding terrain provides some evidence for an intermediate age. The scarcity of craters larger than 10 km in diameter indicates that these surfaces were not affected by the heavy meteorite bombardment early in Mars's history, but numerous small craters (less than 4 km diameters) suggest that the surfaces are not recent. Partially buried craters are found in both the HC and PL units; both areas must have undergone some surface debris deposition.

(e) Analogs. Although numerous terrestrial analogs have been proposed for the Martian curvilinear features, none has provided an adequate explanation for the features. The proposed analogs are based on overall resemblances of morphology, including gilgai (Rossbacher 1978), back-wasting scarps (Carr & Schaber 1977), solifluction lobes (Rossbacher & Judson 1981), ice-cored ridges (Judson & Rossbacher 1979), and glacial moraines (Lucchitta 1981). None of the terrestrial analogs explains the larger size of the Martian features, although Lucchitta (1981) has suggested that glacial moraines are the closest in scale. The analogs all have a single unifying thread: they involve the role of water or water ice as part of their mechanism.

With these observations in mind, we will now turn to models that may explain the subsurface stratigraphy and the origin of the curvilinear features.

Models for subsurface stratigraphy

The topographic relationships among the four photogeologic units shown in the detailed geomorphic maps can be observed from the scarps that separated the units, but the subsurface stratigraphy can only be inferred. A fundamental question is whether the four photogeologic map units represent four discrete stratigraphic layers (Model I in Fig. 15.14), compositional variations in the substrate (Model II), or a buried erosional surface that has been at least partially exhumed (Model III). Models for the subsurface geology that explain

Models for subsurface stratigraphy

I. Layer-cake stratigraphy

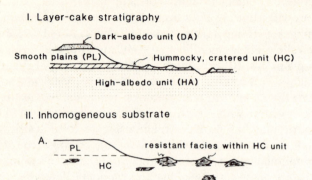

II. Inhomogeneous substrate

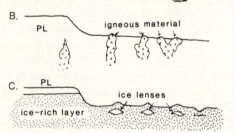

III. Exhumed topography

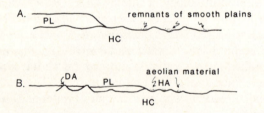

Figure 15.14 Models for the subsurface stratigraphy beneath the Martian curvilinear ground.

the stratigraphic relationships among the photogeologic map units and the curvilinear features are illustrated in the generalized profiles in Figure 15.14 and discussed below. Models I, II, and III do not explain the origin of the curvilinear features; models for the origin are discussed in the following section.

Layer-cake stratigraphy model

The areas covered by the detailed geomorphic maps appear to have up to four discrete photogeologic map units, other than craters and curvilinear features. If each unit corresponds to a separate geologic stratum then the simplest stratigraphy is four superposed horizontal layers, deposited in the order:

high-albedo unit (oldest and lowest); hummocky cratered unit; smooth-plains unit; and low-albedo unit (highest). Nearly all of the models shown in Figure 15.14 exhibit some layering, but Model I has a separate stratigraphic layer corresponding to each photogeologic map unit.

This simple stratigraphy requires a complex history to explain it. Any explanation of the currently observed topography would have to account for the changes in composition or depositional processes that resulted in differences among the strata. Erosional processes would also have to have been sufficiently effective or to have operated over a period of time long enough to remove as many as three complete strata in locations where the high-albedo unit is exposed. Scarps separating the photogeologic map units show the local topographic relationships between the two units, but the information is not detailed enough to allow topographic correlation over any distance, even within one area of detailed mapping.

The layer-cake stratigraphy model seems to explain the relationships among the photogeologic units, but the model requires a complex depositional and erosional history. A key piece of evidence, the elevation of exposure for the various map units, is not available. Without this information, the layer-cake model remains incomplete and impossible to verify.

Inhomogeneous substrate model

The subsurface material around the curvilinear features may not be composed of homogeneous layers but rather of one or more strata of inhomogeneous material. Three possible compositions of an inhomogeneous substrate are shown in Figure 15.14II: a depositional facies that is relatively resistant to erosion; igneous material; and segregated ice lenses.

An erosion-resistant facies within the hummocky cratered unit, which contains most of the curvilinear features, would also account for the topography observed in the detailed mapping (Model II.A), but it would also demand a complex geologic history. The more resistant facies would need to have been emplaced contemporaneously with the rest of the substrate or differentially cemented after its emplacement. A probable mechanism for original emplacement of a more resistant facies is concentration of pyroclastic material by eolian activity and then solidification after deposition. Such eolian concentration of pyroclastic material on Mars is possible but unlikely as an explanation for these features. The curvilinear features are narrower in width and have a much broader arc than barchan dunes or other wind forms. Later differential cementation would probably require local mineralization or formation of ice.

The presence of intrusive igneous material would require magmatic activity at depth on Mars (Model II.B). To date, all products of igneous activity observed on the planet reflect extrusive volcanism. Isolated arcuate intrusions are known on Earth in ring-dike complexes, but the curvilinear patterns on Mars cannot be explained by this mechanism. The model is, at best, improbable.

If the substrate beneath the hummocky cratered unit is rich in ice, then both the ridges and depressions may reflect ice activity (Model II.C). The ridges may be the result of uplift or thrusting of near-surface material by expansion of subsurface ice and growth of segregated ground ice in lenses. The depressions might form where these subsurface ice lenses have become degraded. If the homogeneous substrate does contain partially degraded ground ice, ridges and troughs could occur along a single arc as is observed in several locations. According to this model the proportion of troughs, which is currently low compared with the total number of curvilinear features, should increase as the segregated ground ice lenses continue to degrade. This model is also supported by the presence of craters with fluidized ejecta blankets, which may reflect an ice-rich substrate (Johansen 1978, Mouginis-Mark 1979).

Exhumed topography model
The topographically lower photogeologic map units, especially the hummocky cratered unit, may not be the result of relatively recent erosion of subsurface material. The topography may be a buried erosional surface that has since been exhumed.

In Figure 15.14 Model III.A. illustrates a covermass that has been stripped to expose the buried topography; Model III.B presents a configuration that could account for all four photogeologic map units. If the buried surface has relief greater than the thickness of the blanketing layer, exposures of the dark, low-albedo unit may be the high points of the buried surface. Most erosional surfaces on Earth have 30–150 m of relief, but 300 m of relief on buried surfaces has been reported (Martin 1968). The light, high-albedo patches may be eolian material trapped in closed depressions.

Areas on Mars that may exhibit exhumed topography include the south-polar plains (Sharp 1973) and some areas near the plains/upland boundary, especially in the Mangala region (Rhodes 1980). The presence of buried surfaces suggests a complex history involving formation of the original topography, a change in environmental conditions resulting in burial, another change in environment or base level, and erosion that exhumes and exposes the older topography (Rhodes 1981).

These models address only the question of subsurface stratigraphy. They must be considered, however, as part of the environmental framework for the models that follow, which explain the genesis of the curvilinear landforms.

Models for origin of curvilinear features

Models proposed in the past for the Martian curvilinear landforms have depended on analogs with terrestrial landforms. Several of these analogs, notably gilgai, are created by processes that are poorly understood. However, the processes that now seem likely to have formed these features are relatively limited.

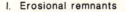

Models for origin of curvilinear ground

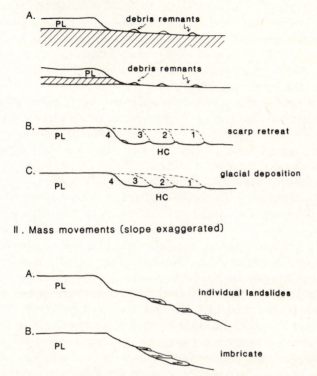

Figure 15.15   Models for the origin of the Martian curvilinear features.

The origin of the curvilinear landforms may be a result of differential erosion, controlled by the subsurface geology described in the models in Figure 15.14. The features might also be residual deposits from other erosional processes, particularly backwasting of scarps or various types of mass movement (Fig. 15.15). These models are presented and briefly discussed here, followed by a consideration of some of the broader implications for Martian geomorphology.

Erosional-remnant model

The local topographic relationships between the smooth-plains unit and the lower hummocky, cratered unit, especially in Utopia Planitia, suggest that the curvilinear features may be erosional remnants of scarp retreat (Fig. 15.15I). This model was originally proposed by Carr and Schaber (1977). The erosion creates the HC terrain at the expense of the smooth plains; the processes involved may range from simple disaggregation and eolian removal of material from the overlying unit to melting of interstitial ice. The curvilinear features may represent accumulations of material at the base of the scarp during a temporary still stand in its retreat.

In parts of Utopia Planitia, the orientation of the curvilinear features roughly parallels that of the nearby scarp. In western Utopia Planitia, however, the curvilinear features occur in the plains and are nearly 1000 km from the cratered upland. These features might reflect former locations of scarps or remnants of higher smooth plains that have since eroded completely, but no evidence remains to support this theory.

Creating the hummocky cratered unit and curvilinear features at the expense of the smooth plains requires disaggregation of particles and their removal. This may have occurred through some combination of erosional processes, possibly including further mass wasting, wind, and perhaps fluvial activity.

Particle disaggregation could be facilitated by ice cement between particles. The mechanism for breakdown would be melting or sublimation of the ice cement. The percentage of ice contained determines whether the substrate exposed in the scarp should be considered to be an ice-cemented sediment (Model I.B), a rock glacier, or a water-ice glacier carrying a load of debris (Model I.C). The amount of ice would also affect the rate of scarp retreat and the quantity of debris that would accumulate at the base of the retreating slope.

The particles freed by disaggregation also need to be considered. On a global scale, an available sink for material eroded from the cratered upland has been a persistent and fundamental question. Assuming the plains cover a third of the planet's surface and average 2 km lower than the cratered upland, the volume of material removed from the northern plain of Mars is almost 10^8 km^3. If this volume of solid material had been spread over the remaining surface of Mars, the cratered uplands would have been covered to a depth of about 1 km. If as much as 50–60% of this volume was water ice, then the problem of where the material went is reduced although not completely resolved. Some of the material could have been transported by wind and deposited elsewhere on the planet, including the polar area. Another partial explanation for the fate of the eroded material is that the major topographic lowlands on Mars may have existed from early in the planet's history, created by subcrustal scour (Carr 1973) or some other process; this would reduce the volume of debris to be deposited elsewhere.

Mass-movement model
The curvilinear features may be the product of mass movement involving some quantity of water (Fig. 15.15 II). The patterns resemble terrestrial solifluction flow deposits in the Canadian Arctic (Mollard n.d.), but the arcuate features in the terrestrial analogs are concave upslope. The curvilinear features on Mars do not show consistent relationships with the regional slope, and detailed information about the local topography is not available. If the features are a result of mass movement, the details of local slope angle and orientation might also indicate the type of mass wasting processes.

On Earth, solifluction can occur on slopes of 2° or less (Andersson 1906). This process may also have been active on Mars (Fig. 15.15 Models II.A. & B), but the details of slope angles and regolith moisture conditions on Mars are unknown. Mass movement of individual or imbricate dry slumps on Mars would require considerably steeper slopes than those needed for solifluction, and the cold temperatures and lower gravity on Mars might also require steeper slopes for soil flow. The regional slopes in areas with curvilinear ground are generally shallow (around 0.1°); local slopes are probably also gentle, but this cannot be determined. If the curvilinear features on Mars are solifluction features, then a high percentage of water would probably have been present in the surficial material when they formed. If they are the result of dry slumping, then steep slopes would be required. Thus, the precise mechanism for mass movement would depend on both the amount of incorporated volatiles and the slope on which these curvilinear features occur.

Arguments for a 'most reasonable' model for origin

Several converging lines of evidence point to an important role for water ice in the formation of the Martian curvilinear features.

(a) The curvilinear patterns on Mars crudely resemble several terrestrial analogs that contain substantial amounts of water as a liquid or solid.

(b) Differential erosion of subsurface ice could explain the combination of positive and negative topography associated with many of the curvilinear features.

(c) Ice as a cementing agent facilitates slope retreat, especially in areas currently near the equator where water ice is unstable and sublimates on exposure to the Martian air.

(d) Mass movements on slopes as low as 1–2°, which seem to be abundant in this area, would be greatly aided by inclusion of even small amounts of volatiles.

(e) Subsurface ice might partially account for differences among photogeologic map units; differing percentages of ice might help to explain the apparent differences in texture and tone on the Viking photographs.

(f) A high percentage of water ice in the substrate reduces the problem of finding a sink for solid material eroded from the northern plains.

(g) Since the cessation of late heavy bombardment at least two periods in Martian geologic history may have had conditions favorable for liquid water to exist on the planet's surface. Fluvial activity carved the dendritic Martian channels, probably around 3×10^9 years ago (Pieri 1980), and volcanism may have melted ground ice and created outflow channels much more recently (Wise *et al.* 1979)

The precise role of water ice in the formation of curvilinear features on Mars is unclear, but a reasonable guess is that it operates through some combination of growth and decay of segregated ground ice and mass movements. These two processes may operate independently or in concert. During mass movements lubricated by water, for example, the water may accumulate in depressions between the ridges and then be buried by eolian deposits. Alternatively, ice might be incorporated in the flow as discrete particles and then concentrated in the ridges of the flow lobes. The persistence or degradation of these concentrations of ground ice would then be expressed as curvilinear ridges or troughs.

Additional evidence supports the idea that extensive amounts of water ice may have been present in the Martian subsurface. The distribution of landforms that may be related to ground ice (Fig. 15.2) suggests that subsurface ice may be—or has been in the past—widespread on the planet. Estimates of the volume of water initially outgassed on Mars (Table 15.1) also support this conclusion.

The proposed correlation between type of crater-ejecta blankets and latitude may also reflect the volume of ice in the substrate. If the northern plains have been created by the retreat of an escarpment composed of mixed debris and water ice, then the cratered upland may still contain substantial volumes of ice. Inferred 'oases' in the southern hemisphere support this idea (Huguenin & Clifford 1980).

Not all of the evidence supports the significance of water ice in the formation of the Martian curvilinear features. Curvilinear features in the northern plains that occur at some distance from the cratered upland do not exhibit any consistent orientation relative to the cratered upland or the nearby smooth-plains photogeologic map unit. Detailed information on the topography might resolve this, however.

Relationship to the plains/upland boundary
The curvilinear features are geographically related to the cratered upland/ northern plains boundary on Mars, and several pieces of evidence suggest that the curvilinear features may also be genetically related to this transition. These landforms do occur consistently in the northern plains near the base of the regional slope. The age of the terrain containing curvilinear features suggests a maximum age for the surface to be after the end of heavy bombardment early in Mars's history, but the plains may have been partially eroded. The evidence that points toward a significant role for water ice in the formation of curvilinear ground cannot shed light on the tectonic theories for the plains/upland dichotomy; the evidence does, however, offer general support for the fossil-permafrost theory of Soderblom and Wenner (1978). Earth-based and space-craft observations have not determined whether ice exists in the Martian substrate in any volume, but the interpretation that water ice has been significant in Mars's history is consistent with the observations discussed here.

The geomorphic fate of water on Mars
The volume of water initially outgassed on Mars can be modelled as equivalent to a global layer around 100 m thick. This entire volume of outgassed water

(about $10^7 km^3$) could be stored in the Martian cryosphere (Rossbacher & Judson 1981) with an available pore space of 4%. However, the pitted features that may have formed by ice degradation suggest that the subsurface ice content could locally exceed 50%. After outgassing, cooling could have condensed this water from the atmosphere and driven it into the subsurface, where it was stored as ground ice (Carr 1979, Pollack 1979).

As subsequent erosion exposed ground ice to the thin Martian atmosphere, the ice probably sublimated rapidly. This would happen under current conditions equatorward of about 40° latitude (Smoluchowski 1968). Once the water entered the atmosphere, it had one of several fates: (a) it could remain in the atmosphere and then be depsoited as night-time surface frost (Jones *et al.* 1979) or reenter the subsurface; (b) it could be transported to the polar regions and be deposited on the polar cap (Cutts *et al.* 1979, Clifford 1980); (c) a small amount could escape from the top of the atmosphere (Walker 1977); or (d) water could be adsorbed on to particles in the regolith (Fanale & Cannon 1978).

Water is a powerful geomorphic agent on Earth, even in arid climates, and the evidence discussed in this chapter suggests that water has been an important agent on Mars as well. The effects of flowing water have been inferred from Martian channels; the curvilinear features suggest other geomorphic processes associated with water, both liquid and solid, are also important on Mars.

Summary and conclusions

The Martian curvilinear ground shows a fairly consistent relationship with both regional and local geomorphology. The landforms occur on the northern plains near the base of the regional slope from the cratered upland down to the northern plains. The curvilinear features are one of a suite of landforms that occur in this transitional zone. The northern plains can be subdivided into as many as four photogeologic map units, with scarps separating adjacent units. The curvilinear features occur primarily, although not exclusively, on the hummocky cratered unit, which appears covered in some places by a smooth-plains unit.

Models for the origin of the curvilinear features suggest that their formation probably incorporates several processes, including scarp retreat by back-wasting, partially through the sublimation of ice from the substrate which then allows the particles to be removed by gravity (mass movement) and wind. Several lines of evidence support the conclusion that curvilinear features are created in part by the activity of water ice, as either interstitial ice or segregated lenses.

On the global scale, the curvilinear features are associated with the transition zone between the northern plains and the ancient cratered terrain. The curvilinear features cannot be used to evaluate the tectonic theories for the dichotomy between the two terrain types, but the significance of water ice in this zone, as presented in this study, indicates that subsurface ice might also have been active in the formation of these large-scale geomorphic features on Mars.

Acknowledgments

Sheldon Judson and Robert Hargraves made valuable suggestions during the course of this study. Helpful comments on drafts of this manuscript were offered by Dr. Judson, Dallas D. Rhodes, and Stephen M. Clifford. Working space and Viking mission images were provided through R. Steven Saunders, Douglas Nash, Nancy Evans, Sandy Dueck Winterhalter, and Leslie Pieri, all of the Jet Propulsion Laboratory. The National Space Science Data Center also provided Viking photographs. This work was done in partial fulfillment of the requirements for the degree of Doctor of Philosophy at Princeton University; financial support was provided by NASA grant NSG-7568 and NASA's Planetary Geology Intern Program.

References

Allen, C. C. 1979. Volcano-ice interaction on Mars. *J. Geophys. Res.* **84**, 8048–59.

Anders, E. and T. Owen 1977. Mars and Earth: Origin and abundance of volatiles. *Science* **198**, 453–65.

Anderson, D. M., L. W. Gatto and F. Ugolini 1973. An examination of Mariner 6 and 7 imagery for evidence of permafrost terrain on Mars. In *Permafrost: The North American contribution to the second international conference*, 499–508. Washington, DC: Nat. Acad. Sci.

Andersson, J. G. 1906. Solifluction, a component of subaerial denudation. *J. Geol.* **14**, 91–112.

Baird, A. K., P. Toulmin, B. C. Clark, H. J. Rose, K. Keil, R. P. Christian and J. L. Gooding 1976. Mineralogic and petrographic implications for Viking geochemical results from Mars. *Science* **194**, 1288–93.

Batson, R. M., P. M. Bridges and J. L. Inge 1979. *Atlas of Mars: The 1:5 000 000 map series.* Washington, D.C.: NASA.

Carr, M. H. 1973. Volcanism on Mars. *J. Geophys. Res.* **78**, 4049–62.

Carr, M. H. 1979. Formation of Martian flood features by release of water from confined aquifers. *J. Geophys. Res.* **84**, 2995–3007.

Carr, M. H. and G. G. Schaber 1977. Martian permafrost features. *J. Geophys. Res.* **82**, 4039–54.

Clark, B. C. and A. K. Baird 1979. Volatiles in the Martian regolith. *Geophys. Res. Lett.* **6**, 811–4.

Clifford, S. M. 1980. Mars: Ground ice replenishment for a subpermafrost ground water system. In *Proc. Third Colloq. on Planetary Water*, 68–72. Washington, DC: NASA.

Cutts, J. A., K. R. Blasius and W. J. Roberts 1979. Evolution of Martian polar landscapes: Interplay of long-term variations in perennial ice cover and dust storm intensity. *J. Geophys. Res.* **84**, 2975–94.

Fanale, F. P. and W. A. Cannon 1978. Mars: The role of the regolith in determining atmospheric pressure and the atmosphere's response to insolation changes. *J. Geophys. Res.* **82**, 2321–5.

Farmer, C. B. and P. E. Doms 1979. Global seasonal variations of water vapor on Mars and the implications for permafrost. *J. Geophys. Res.* **84**, 2881–8.

Frey, H. 1980. Crustal evolution of the early Earth: The role of major impacts. *Precamb. Res.* **10**, 195–216.

Hollingworth, S. E., J. H. Taylor and G. A. Kellaway 1944. Large-scale superficial structures in the Northampton Ironstone Field. *Quaternary J. Geol. Soc. Lond.* C, 1–44.

Huguenin, R. L. and S. M. Clifford 1980. Additional remote sensing evidence for oases on Mars. In *Repts. of Planetary Geol. Prog., 1979–1980* (NASA Tech. Mem. 81776), 153–5. Washington, DC: NASA.

Hunt, C. B. and A. L. Washburn 1966. Patterned ground. In *U.S. Geol. Surv. Prof. Paper* 494-B, B104–33.

Johansen, L. A. 1978. Martian splash cratering and its relation to water. In *Proc. Second Colloq. on Planetary Water and Polar Processes*, 109–10. Washington, DC: NASA.

Jones, K. L., R. E. Arvidson, E. A. Guinness, S. L. Bragg, S. D. Wall, C. E. Carlston and D. G. Pidek 1979. One Mars year: Viking lander imaging observations of sediment transport and H_2O-condensates. *Science* **204**, 799–806.

Judson, S. and L. Rossbacher 1979. Geomorphic role of ground ice on Mars. In *Repts. of Planetary Geol. Prog., 1978–1979* (NASA Tech. Mem. 80339), 229–231. Washington, DC: NASA.

Kochel, R. C. and V. R. Baker 1980. Backwasting and debris accumulation along Martian escarpments. in *Proc. Third Colloq. on Planetary Water*, 53–58. Washington, DC: NASA.

Lucchitta, B. K. 1981. Mars and Earth: Comparisons of cold-climate features. *Icarus* **45**, 264–303.

Martin, R. 1968. Paleogeomorphology. In *The encyclopedia of geomorphology*, R. W. Fairbridge (ed.), 804–12. New York: Reinhold Cook.

McElroy, M. B., T. Y. Kong and Y. L. Yung 1977. Photochemistry and evolution of Mars's atmosphere: A Viking perspective. *J. Geophys. Res.* **82**, 4379–88.

Mollard, J. D. no date. *Landforms and surface materials of Canada: A stereoscopic airphoto atlas and glossary*. Regina, Saskatchewan: Airphoto Interpretation.

Morris, E. C. and J. R. Underwood 1978. Polygonal fractures of the Martian plains. In *Repts. of Planetary Geol. Prog.*, 1977–1978 (NASA Tech. Mem. 79729), 97–9. Washington, DC: NASA.

Mouginis-Mark, P. 1979. Martian fluidized crater morphology: Variations with crater size, latitude, altitude, and target material. *J. Geophys. Res.* **84**, 8011–22.

Mutch, T. A. and R. S. Saunders 1976. The geologic development of Mars: A review. *Space Sci. Rev.* **19**, 3–37.

Nace, R. L. 1967. Water resources: A global problem with local roots. *Environ. Sci. Technol.* **1**, 550–60.

Owen, T. and K. Biemann 1976. Composition of the atmosphere at the surface of Mars: Detection of argon-36 and preliminary analysis. *Science* **193**, 801–3.

Pechmann, J. C. 1980. The origin of polygonal troughs on the Northern Plains of Mars. *Icarus* **42**, 185–210.

Pieri, D. C. 1980. Martian valleys: Morphology, distribution, age, and origin. *Science* **210**, 895–7.

Pollack, J. B. 1979. Climatic change on the terrestrial planets. *Icarus* **37**, 479–553.

Pollack, J. B. and D. C. Black 1979. Implications of the gas compositional measurements of Pioneer Venus for the origin of planetary atmospheres. *Science* **205**, 56–9.

Rasool, S. I., D. M. Hunten and W. M. Kaula 1977. What the exploration of Mars tells us about Earth. *Physics Today* **30**, (July), 23–32.

Rhodes, D. D. 1980. Exhumed topography—a case study of the Stanislaus Table Mountain, California. In *Repts. of Planetary Geol. Prog.—1980* (NASA Tech. Mem. 82385), 397–9. Washington, DC: NASA.

Rhodes, D. D. 1981. Exhumed topography—a review of some principles. In *Repts. of Planetary Geol. Prog.—1981* (NASA Tech. Mem. 84211), 316–8. Washington, DC: NASA.

Rossbacher, L. A. 1978. Terrestrial analogs for Martian striped ground. In *Proc. Second Colloq. on Planetary Water and Polar Processes*, 137–41. Washington, DC: NASA.

Rossbacher, L. A. 1983. *Geomorphic studies of Mars*. Ph.D. thesis, Princeton University. Ann Arbor: Univ. Microfilms.

Rossbacher, L. A. and S. Judson 1981. Ground ice on Mars: Inventory, distribution, and resulting landforms. *Icarus* **45**, 39–59.

Saunders, R. S. 1974. Mars crustal structure. *Bull. Am. Astron. Soc.* **6**, 372.

Sharp, R. S. 1973. Mars: Fretted and chaotic terrain. *J. Geophys. Res.* **78**, 4073–83.

Smoluchowski, R. 1968. Mars: Retention of ice. *Science* **159**, 1348–50.

Soderblom, L. A. and D. B. Wenner 1978. Possible fossil H_2O liquid-ice interfaces in the Martian crust. *Icarus* **34**, 622–37.

Squyres, S. W. 1978. Martian fretted terrain: Flow of erosional debris. *Icarus* **34**, 600–13.

Squyres, S. W. 1979. The distribution of lobate debris aprons and similar flows on Mars. *J. Geophys. Res.* **84**, 8087–96.

Walker, J. C. G. 1977. *Evolution of the atmosphere*. New York: Macmillan.

Washburn, A. L. 1980. *Geocryology*. New York: Wiley.

Wise, D. U., M. P. Golombek and G. E. McGill 1979. Tectonic evolution of Mars. *J. Geophys. Res.* **84**, 7934–9.

16

Wind abrasion on Earth and Mars

Ronald Greeley, Steven H. Williams, Bruce R. White,
James B. Pollack and John R. Marshall

Introduction

'Under the stimulus of aridity, wind scour becomes an erosive agent more constant than the rain, more potent than streams, and more persistent than the sea.' So wrote A. K. Lobeck (1939, p. 378) in his introductory textbook on geomorphology. On earth, eolian processes are particularly important in deserts, along many coastlines, and in glacial outwash plains. Mars experiences frequent dust storms and lacks the inhibiting effects of liquid water and vegetation. Thus, eolian processes play an important role in the evolution of the surfaces of both Earth and Mars.

The objective of our investigation was to determine rates of wind abrasion for a wide range of conditions on Earth and Mars. The study, however, was restricted to considerations of rock abrasion and does not include deflation or the erosion of landforms.

In order to determine rates of eolian abrasion, knowledge of three factors is required: (a) wind characteristics; (b) particle characteristics; and (c) target characteristics. Wind characteristics include wind strengths, durations, and directions. Particle characteristics include particle size, shape, and composition, and the velocities and trajectories of windblown grains; the latter two are functions of wind speeds and heights above the ground. Target characteristics include the rock shape and a measure of the resistance of the rock to abrasion, termed the *susceptibility to abrasion*, S_a (Fig. 16.1). The rate of abrasion may then be calculated at

$$\text{rate} = (S_a)(q)(f) \tag{16.1}$$

where q is particle flux, and f is wind frequency (see Table 16.1).

Although many previous studies have addressed wind abrasion, most analyses have lacked various critical elements and are therefore difficult to use in determining rates of wind abrasion. For example, engineering studies of

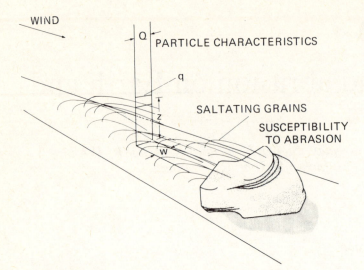

Figure 16.1 Diagram showing the three principal parameters involved in determining rates of eolian abrasion: (1) wind factors; (2) particle characteristics as functions of wind speed and height above the ground; and (3) the susceptibility to abrasion. For explanation, see text. (From Greeley *et al.* 1982). Used by permission of the American Geophysical Union).

abrasion are generally well documented and quantitative in approach, but do not involve geological materials such as rocks. On the other hand, studies of ventifacts in field environments generally lack sufficient data on wind and particle flux to quantify the results.

In this study, we have attempted to gain sufficient information on the characteristics of windblown particles, wind frequencies, and material properties appropriate for estimating rates of wind abrasion on Earth and Mars.

The Martian dilemma

Surface markings on Mars have been observed telescopically for several hundred years. One of the earliest observations that some markings varied seasonally was made in 1884 by the French astronomer E. Trouvelot who attributed the variations to seasonal changes of vegetation (Veverka & Sagan 1974). Speculations that the markings were related to vegetation, water, or even Martian civilizations persisted well into the twentieth century. None of these speculations has been shown to be valid. The first consideration of the markings being related to eolian processes was apparently made by Dean McLaughlin (1954), who published a series of papers on potential Martian wind activity.

In 1971 the Mariner 9 orbiter and the Soviet spacecraft Mars 2 and 3 arrived at Mars during a major global dust storm and verified the predictions of eolian activity. In addition to providing information on the dynamics of dust clouds,

Table 16.1 Nomenclature.

a_m	maximum impact contact radius
A	particle cross-sectional area
C_D	coefficient of drag
C_L	coefficient of lift
D	drag force
D_H	mean crack diameter
D_P	particle diameter
E	subscript denoting Earth
f	wind frequency
g	acceleration due to gravity
G_o	number of saltating grains landing in a unit area per unit time
H	maximum height above surface of saltation trajectory
KE	kinetic energy
L	saltation path length
L_F	lift force
m_p	particle mass
M	subscript denoting Mars
Δ_p	pressure difference
P_a	atmospheric pressure
q	saltation area flux gm/cm^2·sec
Q	saltation lane flux gm/cm·sec
S_a	susceptibility to abrasion
t	time
u	wind speed
u_i	initial particle speed
u_f	final particle speed
u_*	wind friction speed
u_{*t}	threshold wind friction speed
u_∞	freestream wind speed
v	wind speed
v_p	particle speed
v_r	relative speed between particle and wind
w_i	initial particle upward velocity
$\bar{w}_i$	average initial particle upward velocity
w_f	final particle downward velocity
$\dot{x}$	velocity in x-direction
$\ddot{x}$	acceleration in x-direction
$\dot{y}$	velocity in y-direction
$\ddot{y}$	acceleration in y-direction
z	height above surface
z_o	surface roughness
α	angle between saltation trajectory and horizontal surface at impact
θ	angle subtended by crack 'horns'
λ	average saltation path length
ρ_a	density of atmosphere
ρ_p	density of particle
τ_o	total surface shear stress
τ_s	surface shear due to particle impact
τ_t	surface wind shear maximum value (at threshold)
τ_w	surface wind shear
$C_E, C_M, K_1, K_2, K_3, \xi$	constants

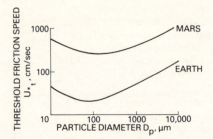

Figure 16.2 Threshold curves for Earth and Mars showing threshold friction speeds required for particles of different sizes. Threshold speeds on Mars are about ten times higher than on Earth because of the lower density atmosphere on Mars. The particle size most easily moved (lowest friction speed) is about 80 μm on Earth and about 100 μm on Mars. (From Greeley *et al.* 1980. Used by permission of the American Geophysical Union).

the Mariner 9 cameras revealed abundant surface features attributed to eolian processes (McCauley 1973). These included yardangs, dune fields (Cutts & Smith 1973), and various bright and dark patterns which changed their size, shape, and position and that were correctly interpreted as representing eolian erosion and deposition (Sagan *et al.* 1972).

The atmosphere was found to be composed predominantly of carbon dioxide with a density very low compared to that of Earth. In this low atmospheric density, threshold wind speeds are more than an order of magnitude higher on Mars than on Earth (Fig. 16.2). Thus it was reasoned that windblown particles would also travel much faster on Mars than on Earth and would be very effective agents of abrasion. This consideration, coupled with the frequent dust storms, led to predictions of extremely high rates of eolian erosion (Sagan 1973, Henry 1975).

A simple qualitative experiment was run in our wind tunnel (Greeley *et al.* 1981) to compare wind erosion on Earth and Mars (Fig. 16.3). In the Earth case a small mirror was positioned downwind from a patch of sand. The tunnel wind velocity was increased to achieve particle movement threshold speed and was maintained until all of the sand was blown away. The experiment was duplicated with a fresh mirror in a simulated Martian atmospheric environment; again, wind velocity was just above threshold speed, but because of the low atmospheric density, the threshold speed was about 16 times higher than in the Earth case. The mirror that was abraded in the Martian environment was much more severely damaged than the one abraded in the Earth environment, a result of the greater kinetic energies of the windblown sand grains on Mars. Thus, at least qualitatively, higher rates of wind abrasion might be expected on Mars if eolian materials and wind frequencies above threshold are comparable to Earth.

The Viking mission to Mars (1976–1981) returned additional evidence of extensive dust storms (Briggs *et al.* 1979) and other eolian activity, including images of vast dune fields (Tsoar *et al.* 1979) and views of the Martian surface

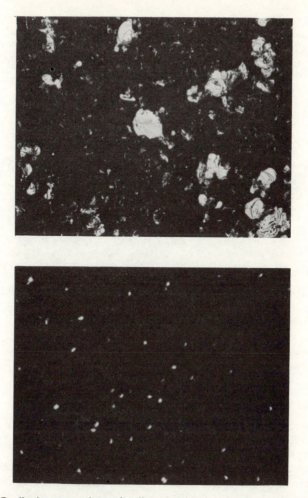

Figure 16.3 Qualitative comparison of eolian abrasion by quartz particles on mirrors for simulated atmospheres on Mars (a) and Earth (b). The mirror in the Martian case is more severely abraded than the Earth case. The lower atmospheric density on Mars requires higher velocity threshold wind speeds resulting in more energetic impacts. Area shown in both cases is about 500 μm by 750 μm.

(Fig. 16.4) showing drifts of eolian sediments (Mutch *et al.* 1976). However, the Viking results have also caused a reassessment of eolian abrasion on Mars (Williams 1981, Greeley *et al.* 1982). Lander images show numerous surface rocks, none showing extensive eolian modification. Orbiter pictures reveal surfaces that have abundant small (~10 m) impact craters. These craters indicate surface ages that are millions or even hundred of millions of years, but have been little modified by erosion of any type (Arvidson *et al.* 1979). If the predicted rates of wind erosion were correct, then these rocks and craters should have been obliterated long ago. Thus there is a conflict between the

Figure 16.4 The surface of Mars as viewed from Viking Lander 1 in the Chryse Planitia region, showing a 2 m boulder, nick-named 'Big Joe,' numerous smaller rocks, and light-toned drifts of material interpreted to be eolian deposits modified by wind erosion.

predicted high rate of eolian abrasion on Mars and the apparent constraints posed by the Viking results.

Approach

Ideally, all of the parameters involved in wind erosion would be observed and measured *in situ*. However, even if a field experiment were designed to obtain complete wind and particle flux data, it would be difficult to separate the effects of the wind from other erosive agents. Our approach was to conduct laboratory simulations, supplemented by field experiments and theoretical considerations, in order to derive empirical results and calculated values for the factors required in estimating eolian abrasion rates. Laboratory experiments have the advantage that conditions can be controlled and the parameters, such as particle velocity, can be isolated and studied separately. Furthermore, the atmospheric environment can be simulated to derive data appropriate for Mars.

Organization

We present our results in seven sections. The introduction is followed by a section describing the rocks, materials, and windblown particles used in our experiments and giving the rationale for their selection. The next three sections present results on the dynamics of windblown particles, susceptibilities to abrasion of the test materials, and wind frequency data. The last sections give estimates of eolian abrasion rates on Earth and Mars and discuss the results.

Experimental materials and rationale for their selection

In this section we discuss the rocks and other materials used to determine susceptibilities to abrasion appropriate for Earth and Mars. We also consider common windblown particles and describe their characteristics in terms of abrasion.

Target materials

Eight materials (Table 16.2) were selected to represent a wide range of potential targets on Earth and Mars. Although there is relatively little information on Martian rock types, the samples within the suite selected are probably appropriate for Mars. Data from the Viking lander inorganic chemistry experiments (Toulmin *et al.* 1977) suggest mafic to ultramafic compositions, probably involving basaltic rocks and their weathering products such as various montmorillonitic clays. Spectral data from remote sensing also suggest that large parts of the Martian surface involve mafic and ultramafic materials (Singer *et al.* 1979). These data on the composition, combined with photogeological mapping, suggest that more than half of the surface of Mars is of volcanic origin (Greeley & Spudis 1981) and provide the basis for selecting a large suite of volcanic rocks for the abrasion tests.

In addition to volcanic materials, parts of Mars also appear to be mantled with sediments (Scott & Carr 1978) that appear to range in thickness from less than 1 m to perhaps hundreds of meters. The degree of induration appears to be variable, including very friable duricrust sediments observed at the Viking landing sites and thick, apparently highly indurated, deposits observed in canyon walls (Carr 1981). Possible origins of the various mantling deposits include windblown sediments, fluvial deposits, glacial deposits, and others. Thus, there may be fine-grained to glassy volcanic rocks on Mars, plus fine-grained pyroclastic materials and possible sedimentary clastic deposits ranging from breccias (generated partly from impact cratering) and conglomerates to clays.

Windblown particles

Most wind-transported particles on Earth range in size from several micrometers (dust) to several millimeters in diameter (Bagnold 1941). Given the similar shape in the threshold curve (Fig. 16.2) it is likely that windblown particles on Mars would be of comparable size. Furthermore, most rock abrasion on Earth is apparently accomplished by sand-size ($\sim$60 to 2000 µm in diameter) grains transported in saltation, although finer material (Whitney & Dietrich 1973) and perhaps even wind alone can be responsible for shaping ventifacts (Whitney 1979). Sand-size particles result from primary sources such as volcanism and from various processes including weathering and fragmentation from impact cratering (Pettijohn *et al.* 1972). On Earth most sand is quartz derived from the weathering of granite. Moreover, most eolian sand has gone through some part of the hydrologic cycle. Because quartz is dominant in eolian processes on Earth, it was selected as one of the primary agents of abrasion in our experiments.

Granite appears to be absent or rare on Mars and consequently there may be very little quartz sand present (Smalley & Krinsley 1979). Furthermore under the present Martian climate, and with the apparent lack of active volcanism and tectonism (Carr 1981), the generation of sand-size particles on Mars is probably suppressed in comparison to that on Earth. Nevertheless, the presence of sand dunes and other eolian features suggest both the presence of sand-size particles and winds capable of transporting them. Remote sensing data for areas of eolian activity on Mars and analyses of thermal inertia measurements also suggest the presence of sand-size particles (Peterfreund 1981) as do analyses of materials at the Viking Landers (Moore *et al.* 1977). Most Martian dunes are very dark and it has been suggested that they are composed of basalt particles derived from volcanic plains (Tsoar *et al.* 1979). Thus sand-size grains of basaltic composition were used in many of our abrasion experiments in order to simulate Mars.

Because of the high wind speeds needed for sand entrainment on Mars, it has been suggested that sand-size particles would be very short-lived; the term *kamikaze* grains was coined by Sagan *et al.* (1977) to describe particles that smashed against rocks and each other, forming fine-grained dust. Estimated particle sizes in the Martian dust clouds is a few μm and smaller and the kamikaze effect may be the principal mechanism to produce the dust. In addition, salt weathering has been shown to be important in the production of fine grains on Earth (Goudie *et al.* 1979) and may be an important process on Mars (Malin 1974).

Laboratory experiments simulating the comminution of windblown sand on Mars give support to the idea of the kamikaze effect (Greeley 1979). Experiments also show that windblown sediments generate electrical charges, a result confirmed by field observations on Earth (Mills 1977). The experiments also showed that the dust generated from the breakup of the kamikaze grains clumped together, forming aggregates that grew to diameters greater than several millimeters. Because the electrical effects would be enhanced on Mars, Greeley and Leach (1978) and Greeley (1979) suggested that aggregates may be an important part of the eolian regime on Mars.

Windblown aggregates also occur on Earth. Some are generated during volcanic eruptions (Krinsley *et al.* 1980) and apparently are bonded by electrostatic charges. Most eolian aggregates, however, are associated with playa deposits. Silt and clay playa deposits are often slightly indurated and/or cemented with various salts. During wind storms, sedimentary accumulations that are fragmented by dessication are picked up by the wind, rounded into sand-size grains and deposited in typical sand bedforms, including dunes (Huffman & Price 1949; and others). Because sand-size aggregates of fine grains may be important on Mars and are known to occur on Earth aggregates were used in our abrasion experiments for comparison with results using crystalline quartz and basalt grains.

Table 16.2 Materials used in determining susceptibilities.

Materials	Source	Description
basalt	Amboy lava flow, California	Typical basalt in thin section. Contains euhedral plagioclase crystals that show some zoning. Crystal size is 0.5–1 mm. Glassy groundmass with some alteration.
obsidian	Lookout Mountain, Long Valley Caldera, Mono Lake, California	In thin section the obsidian is almost totally glass. Contains some very small crystals and shows flow texture.
tuff, welded	Bishop Tuff, Owens River Gorge near Bishop, California	The matrix shows both flow and compaction features. Contains elongated and somewhat devitrified pumice fragments. Has glass shards in matrix. Euhedral sanidine crystals of 1.5–2 mm diameter. Contains a small amount of quartz.
tuff	Bishop Tuff, Owens River Gorge near Bishop, California	Contains some large pumice fragments. Matrix has glass shards but no evidence of flow or compaction. Euhedral sanidine crystals of 1.5–2 mm diameter.
rhyolite	San Carlos Indian Reservation, Arizona	Granular texture with all crystals smaller than 1 mm. Contains microcline phenocrysts with a few quartz crystals and some accessory biotite and garnet.
granite	Unknown	Coarse grained (up to 5 mm) granite. Primarily quartz and microcline, with some plagioclase and thin biotite crystals and accessory zircon.
brick	Common red clay	In hand specimen, the brick is primarily red clay with some quartz and feldspar crystals, ranging in size up to approximately 2 mm.
hydrocal	Common gypsum cement building material	In hand specimen, the hydrocal is uniform chalky mass, with no visible crystals or inhomogeneities, apart from an occasional vesicle.

Summary
Windblown particles and rock targets of a wide range were selected for simulations of eolian abrasion on Earth and Mars. Potential rock targets include igneous and sedimentary rocks which are common on Earth. Experiments involving basaltic rocks were emphasized for Mars, as basalts appear to be common on the red planet. Three types of particles, quartz, basalt, and aggregates of fine grains, were used as the agents of abrasion.

Dynamics of windblown particles
Of the three parameters required to determine the rates of wind abrasion, the one dealing with the dynamic characteristics of windblown particles is the most complicated and difficult to assess. For any given wind the saltation flux and

trajectories of windblown particles vary with surface roughness, height above the surface, and atmospheric density. In order to consider wind erosion on Earth and Mars for times in the past when the climate and atmosphere might have been different, these characteristics must also be known for a range of atmospheric densities.

Methodology
Our approach was to run wind tunnel experiments to determine as many of the particle characteristics as possible for Earth conditions, then to repeat the experiments under simulated Martian conditions. Some of the results for the terrestrial cases were compared with results from field experiments in order to verify the wind tunnel simulations and to provide a basis for extrapolation to Mars. Because not all conditions for Earth or Mars could be satisfied in laboratory simulation, numerical models were derived based on a combination of theory and experiments to allow calculations of flux and saltation trajectories for a wide range of conditions.

Wind tunnel experiments Experiments were performed in an open-circuit wind tunnel using particle collectors (Fig. 16.5) to determine particle flux (Q, q) and an instrument (Fig. 16.6) to determine particle velocity. An experimental matrix was developed combining atmospheric pressure, particle diameter, freestream wind velocity, and height above the surface. Atmospheric pressures of 1 bar (Earth case) and an average of ~7 mb (Mars case) were used. Three particle sizes were selected for analyses: 715 μm, representing coarse sand; 350 μm, for typical dune sand; and 92 μm, the size most easily moved by lowest strength winds (Fig. 16.2).

For both the Earth and Mars experiments a wide range of freestream wind velocities was selected, including high speeds to simulate storm conditions

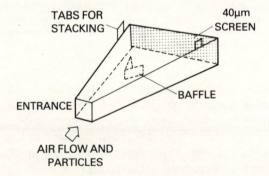

Figure 16.5 Diagram of particle collector used to determine flux. Collector is made of plexiglass; wind tunnel smoke tests were carried out to determine optimum angle for the wedge which would allow maximum efficiency for particle collection with minimum interference to air flow. Baffle prevents particles from bounding back out the front or damaging the screen. Tabs on the side enable collectors to be stacked. (From Greeley *et al.* 1980. Used by permission of the American Geophysical Union).

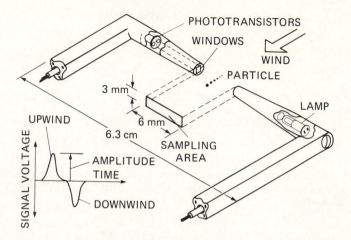

Figure 16.6 Schematic diagram of particle velocimeter; device consists of a light source and two phototransistors; as a particle passes through the beam, the upwind phototransistor registers a positive voltage, then the second (downwind) phototransistor registers a negative voltage; the difference between the two signals yields time (hence velocity); the amplitide is a function of grain size. (From Greeley *et al.* 1982. Used by permission of the American Geophysical Union).

when substantial eolian abrasion probably occurs. Crushed and sieved walnut shells were used in experiments to simulate Mars, as discussed below. Quartz sand was used in terrestrial experiments.

Particle flux as a function of height above the surface was determined as the mass of particles caught per unit area of the collector per unit time. Although the collectors could be stacked to sample the entire cross section of the wind tunnel from the floor to the ceiling, tests showed erratic particle distribution at heights above 65 cm, evidently as a result of particles bouncing from the ceiling. Thus collection in most experiments was made from the surface to 65 cm above the surface. Total particle flux, Q, was determined by summing the masses caught in the individual collectors.

High speed 16 mm motion pictures were taken for some of the Earth-case wind tunnel experiments to determine saltation trajectories and to validate the velocities determined with the velocimeter (Greeley *et al.* 1982, 1983). The camera was positioned perpendicular to the particle flow and recorded the saltation cloud at 2000 to 10 000 frames per second.

Field experiment A field experiment was conducted to provide a calibration of the wind tunnel results under natural eolian conditions using the same techniques and instruments as employed in the laboratory. The field site was located at Waddell Creek State Beach, about 100 km south of San Francisco. In November 1981 a Hycam 16 mm motion picture camera was installed to record saltating particles at framing rates up to 8000 frames per second. Concurrent measurements of wind velocity were made 1 m above the surface. Particle

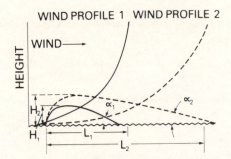

Figure 16.7 Diagram showing two typical saltation trajectories; solid path corresponds to low wind speed (profile 1) and has low (H_1) and short path (L_1) in comparison to path of particle driven by high wind speed (profile 2).

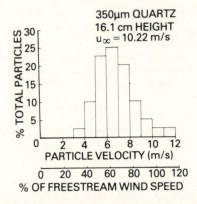

Figure 16.8 Distribution of particle velocities at a height of 16.1 cm above the surface for 350 μm diameter quartz grains subjected to a free-stream wind speed of ~10.22 m/s in a simulation of Earth ($u_* \approx 0.6$ m/s). The velocity distribution is typical and can be expressed as a mean value with an associated one standard deviation range about the mean, as shown. Figure 16.9 will present only the mean and range. (From Greeley *et al.* 1983. Used by permission of Elsevier Press).

collectors were placed in the wind stream to determine particle flux. Seven successful experiments were conducted for wind speeds ranging from about 4.5 to 7.0 m/sec.

Particle velocity

The path that a saltating grain follows is called the saltation trajectory (Fig. 16.7). The velocity of the grain generally increases throughout its trajectory, although by the time it has reached maximum height above the ground most grains will have achieved about 50% of their maximum velocity. Because meteorological data provide only wind speeds (not particle velocities) it is necessary to know particle velocity as a function of wind speed in order to calculate rates of abrasion. Furthermore, at any given instant and at any given place within the saltation cloud there is a complex spectrum of particle paths and speeds. Thus it is also necessary to know the velocity distributions of particles as functions of grain size, wind speeds, and heights above the ground.

Figures 16.8 and 16.9 show typical velocity distribution within saltation clouds measured at a single height above the floor of the wind tunnel. Velocities are given in m/s and as percentages of the freestream wind velocity. In general, particles traveling at greater heights are also traveling at higher speeds. Evidently this is a reflection of the increase in wind speed with height above the ground, through the turbulent boundary layer. In addition, because particles at greater heights tend to have longer saltation trajectories, there is a longer interval of time for them to be accelerated by the wind. Note also that

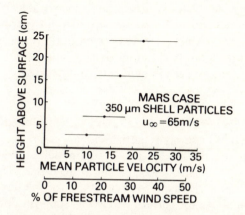

Figure 16.9 Particle velocities for four different heights above the surface for 350 μm walnut shell particles subjected to a freestream wind speed of 65 m/s ($u_* = 3.4$ m/s) in a low atmospheric pressure (6.6 mb) to simulate the Martian environment (From Greeley *et al.* 1983. Used by permission of Elsevier Press).

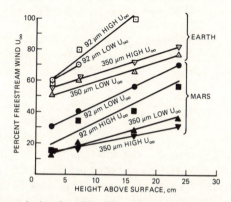

Figure 16.10 Average velocity of particles 350 and 92 μm in diameter as a function of height above surface at 10^3 mb pressure (Earth case) and 6.5 mb pressure (Mars case); both cases are run at freestream wind speeds just above threshold ('low' u_∞) and for storm conditions ('high' u_∞).

some grains travel at velocities exceeding the wind speed, which probably results from elastic rebound of grains in saltation.

Figure 16.10 compares velocity data for particles on Earth with those of Mars under similar dynamic conditions for both low and high wind speeds. *Low* refers to speeds about 25% greater than threshold; *high* refers to storm conditions. In all cases particles travel at a higher percentage of the freestream wind speed on Earth than on Mars. This suggests a more efficient coupling of the particles with the wind in the denser terrestrial atmosphere. Because wind speeds are about an order of magnitude higher on Mars for particle motion, the grains would have to be accelerated to a much greater speed on Mars in comparison to Earth in order to reach the same percentage of the freestream wind speed. Even though the saltation trajectory is longer on Mars, they may not be in flight long enough for an acceleration comparable to Earth. However, the lower percentage could also result partly from experimental error. The saltation path length on Mars is longer than on Earth (discussed below) and in the wind tunnel the saltation cloud in Martian simulations may not become fully developed because of the limited wind-tunnel length.

Saltation model

A computational model of particle saltation was developed in order to determine velocities and trajectories of particles for a wider range of conditions than could be simulated in the laboratory. In the model, a one-dimensional flow regime is assumed for calculations of the trajectories. The velocity in the vertical, y, direction is zero and the velocity, u, in the flux, x, direction is a function of height above the surface. A viscous sublayer develops in the wind flow when the wind speed is below threshold (White *et al*. 1976). However, when minimum threshold speeds, u_{*t}, are reached, the optimum-size particles (Fig. 16.2) begin to saltate, which causes the viscous sublayer to degenerate and form a fully turbulent boundary layer.

Our computational model of saltation trajectories is based on the equations of motion of a single particle (White & Schulz 1977; Fig. 16.11):

$$m_p\ddot{x} = L\frac{\dot{y}}{v_r} - D\frac{\dot{x}-u}{v_r} \qquad (16.2)$$

$$m_p\ddot{y} = -L\frac{\dot{x}-u}{v_r} - D\frac{\dot{y}}{v_r} - m_pg \qquad (16.3)$$

where m_p is the particle mass, u is the wind velocity, $\dot{x}$, $\dot{y}$ are the particle's horizontal and vertical velocity, $\ddot{x}$, $\ddot{y}$ are the particle's horizontal and vertical accelerations, and v_r is the velocity of the particle relative to the velocity of the wind and is equal to $[(\dot{x}-u)^2+\dot{y}^2]^{1/2}$. Aerodynamic force on a particle is proportional to its cross-sectional area, A, times the difference in pressure, Δp, across the particle. Bernoulli's equation states

$$\Delta p = (1/2\rho_a v^2) \tag{16.4}$$

where ρ_a is the atmospheric density and v is the air velocity. Hence

$$L_F \propto 1/2A\rho_a v_r^2 \tag{16.5}$$

$$D \propto 1/2A\rho_a v_r^2 \tag{16.6}$$

Assuming the particles are spherical and of uniform density, we can write

$$L_F = 1/8C_L\rho_a\pi D_p^2 v_r^2 \tag{16.7}$$

$$D = 1/8C_D\rho_a\pi D_p^2 v_r^2 \tag{16.8}$$

where C_L is the coefficient of lift, C_D is the coefficient of drag, and D_p is the particle diameter. We can now substitute Equations 16.7 and 16.8 into 16.2 and 16.3 and rearrange

$$\ddot{x} = -3/4 \frac{\rho_a}{\rho_p} \frac{v_r}{D_p} [C_D(\dot{x}-u)-C_L\dot{y}] \tag{16.9}$$

$$\ddot{y} = -3/4 \frac{\rho_a}{\rho_p} \frac{v_r}{D_p} [(C_D\dot{y}+C_L(\dot{x}-u)]-g \tag{16.10}$$

The drag coefficient of a sphere is strongly dependent upon the Reynolds number. The empirical formulas of Morsi and Alexander (1972) are used for its determination in our numerical calculations. The initial conditions are the same as those used by White *et al.* (1976), in which the particle is at rest on the surface and, as the wind speed is increased above threshold, particles begin to move. For calculations in the Martian atmosphere it is necessary to include the effects of slip flow caused by the low density atmosphere, although the slip flow

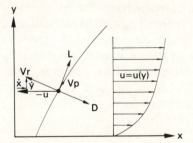

Figure 16.11 Force diagram for a saltating particle. L denotes lift force; D, drag force; v_p is the instantaneous particle velocity and v_r is the velocity of the particle relative to the wind velocity. (Modified from White and Schulz 1977. Used by permission of Cambridge Univ. Press).

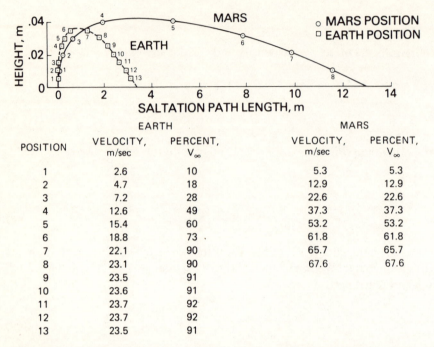

POSITION	EARTH		MARS	
	VELOCITY, m/sec	PERCENT, V_∞	VELOCITY, m/sec	PERCENT, V_∞
1	2.6	10	5.3	5.3
2	4.7	18	12.9	12.9
3	7.2	28	22.6	22.6
4	12.6	49	37.3	37.3
5	15.4	60	53.2	53.2
6	18.8	73	61.8	61.8
7	22.1	90	65.7	65.7
8	23.1	90	67.6	67.6
9	23.5	91		
10	23.6	91		
11	23.7	92		
12	23.7	92		
13	23.5	91		

Figure 16.12 Calculated trajectory for grains in saltation on Earth and Mars for wind speeds about 25% above threshold, showing much longer trajectory for Mars case. Positions along the trajectory show the instantaneous velocity (shown in both m/s and as a percentage of freestream wind speed) of the particle. Note that vertical scale is greatly exaggerated. (From Greeley *et al.* 1982. Used by permission of the American Geophysical Union).

factor becomes significant only for the smaller-size particles in the calculations (White 1979).

Figure 16.12 displays typical computed particle trajectories for Earth and Mars when lift forces due to the Magnus effect (lift force produced by spin) are taken into account (White & Schulz 1977). The percentage values at various positions along the trajectory are the ratio of particle speed to the freestream wind speed. In general the percentage of freestream wind speed that a particle reaches is about the same for all calculated particles at similar positions in their trajectories. Particles accelerate to nearly half of the freestream wind speed in the initial lifting phase of their trajectories, which corresponds to about one quarter of their path length, L. Note, however, that particles more nearly approach freestream wind speed on Earth than they do on Mars, similar to the results determined experimentally.

Particle flux
Numerous investigators have analyzed mass transport, both from experimental and theoretical approaches. Bagnold (1941), for example, carried out experiments to determine flux in terms of momentum loss of the air due to sand in

saltation. Sharp (1964) collected grains within saltation clouds at various heights over long periods of time, then determined the total masses and particle size distributions; however, concurrent wind data are not available. Thus, in order to obtain data appropriate for both Earth and Mars, experiments (Fig. 16.13) were carried out and a numerical model was developed.

Figure 16.14 shows q as a function of height for 92 μm grains at three wind speeds in the wind tunnel compared to results obtained in the field experiment at Waddell Creek where the average grain size was about 300 μm. In general flux decreases with distance from the surface and increases with higher wind velocities. Figure 16.15 shows experimental results for Mars to determine q for 92 μm particles versus height for wind speeds ranging from 58 to 123 m/sec. In these and other Martian simulations particles composed of walnut shells approximately one third of the density of grains expected on Mars were used to compensate partly for the lower Martian gravity (Greeley, White *et al.* 1977). Although this technique is valid for threshold tests and for determining flux in terms of *numbers* of particles, the experimental data for flux *mass* must be

Figure 16.13 View of stacked particle collectors (1) used to determine flux of saltating grains. Samples are taken in intervals of 2 cm above the floor; trap door device allows precise timing of collection during fully-developed saltation cloud. Also shown is particle velocimeter (2) and electrometer probe (3) used to measure electrical current generated by windblown grains. Air flow is from right to left. (From Greeley *et al.* 1982. Used by permission of the American Geophysical Union).

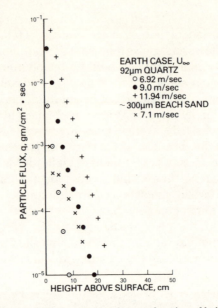

Figure 16.14 Particle flux, q, of 92 μm quartz grains as a function of height for three freestream winds (Earth case), showing general increase in flux with wind speed, and flux for 300 μm particles as determined from field experiment.

adjusted (Greeley *et al.* 1982). Thus in Figure 16.15 the gravity-adjusted data must be shifted upwards by a factor of 2.6 (the density difference between walnut shells and particles expected on Mars, such as basalt). As in the terrestrial cases, there is a general decrease in particle flux with height above the surface. In addition there is an abrupt break in the flux at about 30 cm above the surface. This break may represent the transition from grains that are transported primarily in saltation near the surface to grains moving mostly in suspension at levels above 30 cm.

Total mass flux, Q, was determined by summing incremental flux, q, for the total column sampled. Figure 16.16 shows total flux for freestream wind tunnel speeds ranging from ~6 to 12.0 m/s for 92 μm quartz particles, simulating terrestrial conditions. Martian simulations are shown in Figure 16.17 which give total flux for freestream speeds ranging from about 55 to 90 m/s for 92 μm; both experimental and gravity adjusted data are shown. In general there is an increase in flux with increasing wind velocity. Thus flux is higher on Mars than on Earth because of the higher wind speeds required for particle entrainment under dynamically similar conditions.

Model for calculation of flux   An estimate of the total amount of surface material transported by the wind may be made by examining the physics of the process. The key parameter in estimating flux is the characteristic path length of saltating particles. Several investigators have developed methods to calcu-

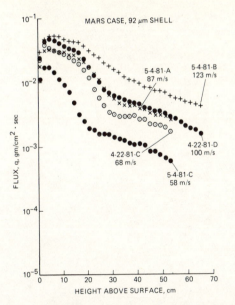

Figure 16.15 Particle flux, q, for 92 µm particles as a function of height for 5 wind speeds in the Martian wind tunnel at ~6 mb atmospheric pressure. To correct for low density of walnut shells (experimental material) the data must be shifted upwards by a ratio of 2.6, the density difference between walnut shells and basalt. See text for explanation.

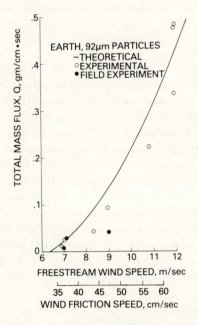

Figure 16.16 Particle flux, Q, as a function of freestream wind speed, u_∞, for 92 µm quartz particles in Earth wind tunnel simulations compared with flux predicted by Equation 16.33.

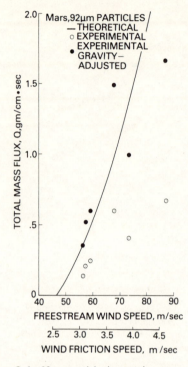

Figure 16.17 Particle flux, Q, for 92 μm particles in experiments, as adjusted for Mars gravity, and compared with flux predicted by Equation 16.33. (From Greeley *et al.* 1982. Used by permission of the American Geophysical Union).

late path lengths (Bagnold 1941, Kawamura 1951, Tsuchiya 1969). Tsuchiya presents a theoretical development based on successive saltating leaps. However, it is necessary to know the initial or maximum vertical component of velocity at lift-off for the particle in order to calculate its path length. This component is generally a function of the flow conditions. A formulation of this type, with the path length a function of the vertical velocity, does not lend itself well to calculations of surface material movement. For our application it is useful to express the path length in relation to the threshold shear stress exerted on the surface by the wind, as presented by Kawamura (1951). The surface shear stress τ_o consists of two elements. The first is the ordinary wind shear, τ_w, caused by the motion of the wind flow over the surface. The second stress, τ_s, comes from the impact of particles colliding with the surface.

Kawamura found that τ_w increases with increasing wind friction speed, u_*, until u_{*t} is reached, then it remains approximately constant at a value of τ_t. Hence

$$\tau_o = \tau_s + \tau_w \qquad (16.11)$$

or

$$\tau_s = \tau_o - \tau_w \tag{16.12}$$

Bagnold (1941) and Kawamura state that the stress on the surface caused by the impact of saltating grains is equal to the momentum lost by the grains:

$$\tau_s = G_o(u_f - u_i) \tag{16.13}$$

where G_o is the mass of grains landing per unit area per unit time, u_f is the final grain speed at surface impact, and u_i is the initial grain speed at particle lift off.

It can be assumed (Kawamura 1951) that there is a linear relationship between the particle momentum lost occurring in the horizontal and vertical directions

$$G_o \mid \overline{u_f - u_i} \mid = \xi G_o \mid \overline{w_f - w_i} \mid \tag{16.14}$$

where ξ is a constant, and u and w are the horizontal and vertical components, respectively, of the particle velocity, v.

After the collision with the surface a bouncing particle traveling vertically with speed w_i will again return to the surface with the same speed $w_f = -w_i$ on the average. Hence

$$\mid \overline{(w_f - w_i)} \mid = 2w_i \tag{16.15}$$

Substituting Equation 16.15 into 16.14 and then 16.14 into 16.13 yields

$$\tau_s = 2\xi G_o \bar{w}_i \tag{16.16}$$

Substituting into Equation 16.12

$$2\xi G_o \bar{w}_i = \tau_o - \tau_t \tag{16.17}$$

From the definition of surface friction speed

$$u_* \equiv [\tau_o/\rho_a]^{1/2} \tag{16.18}$$

where ρ_a is the atmospheric density. Equation 16.16 can be rewritten as

$$2\xi G_o \bar{w}_i = \rho_a(u_*^2 - u_{*t}^2) \tag{16.19}$$

From field studies of saltation Kawamura (1951) found that

$$G_o = K_1 \rho_a(u_* - u_{*t}) \tag{16.20}$$

which also agreed with numerical solutions of particle motion (White 1979). Substituting Equation 16.20 into 16.19 yields

$$2\xi K_1\rho_a(u_*-u_{*t})\bar{w}_i = \rho_a(u_*^2-u_{*t}^2) \tag{16.21}$$

We can solve Equation 16.20 for $\bar{w}_i$, by rearranging and collecting the constants

$$\bar{w}_i = K_2(u_*^2-u_{*t}^2)/(u_*-u_{*t}) = \bar{w}_i = K_2(u_*+u_{*t}) \tag{16.22}$$

The average saltation path length, λ, can be calculated from this result in the following manner. From elementary physics we know that the time aloft for a particle with initial upward velocity, w_i, is

$$t = 2w_i/g \tag{16.23}$$

where t is time aloft and g is the acceleration due to gravity. During time, t, the particle will be moving downstream with a velocity u_i. Applying basic physics equations of distance traveled, the saltation path length, L, is

$$L = u_i t \tag{16.24}$$

Substituting Equation 16.23 into 16.24:

$$L = 2\frac{u_i w_i}{g} \tag{16.25}$$

Because $u_i \propto w_i$ (from Equation 16.14)

$$\lambda = 2\ K_3\frac{\bar{w}_i^2}{g} \tag{16.26}$$

Substituting Equation 16.22 into 16.26 and collecting constants

$$\lambda = K_3\frac{(u_*+u_{*t})^2}{g} \tag{16.27}$$

We now have the necessary information to calculate a total saltation lane flux, Q. If we consider a zone with length, λ, downwind of a unit width lane, no particle can possibly land in that zone without first passing over that lane width; also no particle can pass over the lane width and fail to land in the zone. Hence

$$Q = G_o\lambda \tag{16.28}$$

Thus the amount of material passing over the lane width is equal to the amount of material landing on the surface per unit area times the length of the zone. We can substitute Equation 16.20 and 16.27 into 16.28.

$$Q = K_1\rho_a(u_*-u_{*t})K_3\ \frac{(u_*+u_{*t})^2}{g} \tag{16.29}$$

The final form of the theoretical total flux is

$$Q = K_4 \frac{\rho_a}{g} \ (u_* - u_{*t})(u_* + u_{*t})^2 \tag{16.30}$$

On Mars there appear to be different interparticle forces as well as effects from changes in Reynolds number. These changes will alter the saltation characteristics by affecting the threshold friction speed. The flow field around individual particles (Knudsen and Reynolds numbers effects), shearing rate within the boundary layer, and the existence of a comparatively large (to Earth) viscous sublayer for nonsaltating flow will cause change in threshold values.

Using Equation 16.30, a comparison between Earth and Mars can be made

$$\frac{Q_E}{Q_M} = \frac{\rho_E (u_* - u_{*t})_M g_M (u_* + u_{*t})^2_E C_E}{\rho_M (u_* - u_{*t})_E g_E (u_* + u_{*t})^2_M C_M} \tag{16.31}$$

where the Cs are constants of proportionality. Assuming that the ratio of friction speed to threshold friction speed, U_*/U_{*t}, is the same for both Earth and Mars yields

$$\frac{Q_E}{Q_M} = \frac{\rho_E g_M (u_{*t})^3_E C_E}{\rho_M g_E (u_{*t})^3_M C_M} \tag{16.32}$$

White (1979) has shown experimentally that under Martian conditions substantially less stress is needed to move material than on Earth. This result supports the numerical results that show similar increases in particle path lengths on Mars and that the path length is almost directly related to the mass flux. Thus, it would be useful if Earth and Mars cases could be adequately described by a single universal equation. According to the analytical deriviation, the only differences between the relations describing the two cases are the empirical constants C_E and C_M. If the constants of proportionality C_E and C_M were the same for both cases, this then would indicate that the physics between the two tests remained unchanged, and would imply a universality to the mass flux process. As one would expect, the process of movement .of surface material on Mars should be nearly the same as the process on Earth. Hence the constants of proportionality between the two test cases should be nearly the same. From the wind tunnel data of both cases an empirical constant with a value of 2.61 was determined (White 1979).

Figure 16.18 displays the mass flux versus a function of friction speeds. Here, both high- and low-pressure data collapse to a single line in support of the idea of a universal function describing the movement process. Hence the following equation can be used in estimating mass flux on either Earth or Mars once the friction speed and threshold friction speed are known:

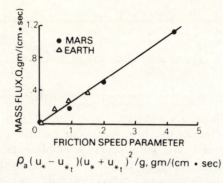

Figure 16.18 Particle flux, Q, as a function of friction speed parameter; one bar (Earth) and low pressure (Mars) data collapse to a single curve which is the straight line on the plot. Note the origin for both cases are data points, as they are used in determining the threshold friction speeds. (From White, 1979. Used by permission of the American Geophysical Union).

$$Q = 2.61 \frac{\rho_a}{g} (u_* - u_{*t})(u_* + u_{*t})^2 \qquad (16.33)$$

The flux of material on Mars is consistently higher than it is on Earth, in spite of the lower Martian atmospheric density, due to: (a) the higher u_{*t} required on Mars; (b) the lower Martian gravity; and (c) the typically longer saltation path lengths on Mars (discussed above). From Equation 16.32 the difference in Q between the Earth and Mars may be expressed as

$$Q_M = \frac{(u_{*t})^3_M Q_E}{(u_{*t})^3_E 13.5} \qquad (16.34)$$

For example, the relationship between the material movement on Earth and on Mars, for equal ratios of u_*/u_{*t} of 1.47, would be $Q_M/Q_E = 6$, or 6 times as great on Mars for 200 μm particles. A direct estimate of flux rates on Mars can be made by employing the results shown in Figure 16.18. To obtain a numerical value, the threshold friction speed must be known. The value will vary with size of particles but, once known, a direct estimate of the mass flux can be made.

Summary
In general, increases in wind speed result in increases in the trajectory height and length for grains in saltation, causing them to achieve higher velocities and to have higher rates of mass transport. Because of the higher wind speeds required to achieve the threshold of particle movement on Mars, both particle velocities and fluxes are higher than they are on Earth. However, particle velocities seldom are greater than about 50% of freestream velocity on Mars, in contrast to Earth where particle velocities commonly meet and sometimes exceed freestream velocity.

Susceptibility of rocks to eolian abrasion

Rocks and minerals of different compositions and textures should have different resistances to abrasion by windblown particles. In this section we review previous studies of wind abrasion, discuss the mechanics of abrasion, and present the results of laboratory experiments that were conducted under controlled conditions in order to derive values that can be applied to Earth and Mars.

Previous investigations

Except for the work of Kuenen (1960), few studies have been made to quantify wind erosion of natural materials, although ventifacts have been long recognized and described. Abrasion of metals, glasses, and ceramics is important in manufacturing and industrial processes and related research has been done for engineering applications. The effects of target composition, impactor size and composition, and impact angle have been investigated experimentally (Neilson & Gilchrist 1968, Sheldon & Finnie 1966b). Studies of the mechanisms of material removal for brittle and ductile materials were conducted by Adler (1976a, 1976b), Smeltzer et al. (1970), Sheldon and Finnie (1966a), Sheldon (1970), Sheldon and Kanhere (1972) and Finnie (1960). The mechanism of abrasion of brittle fused silica was investigated by Alder (1974). The behavior of impacting particles during abrasion was studied by Tilly and Sage (1970). The initial and steady-state erosion characteristics of a target in response to particle impact was noted by Tilly (1969) and discussed by Marshall (1979). Suzuki and Takahashi (1981) have conducted experiments of abrasion in a geological context.

The mechanics of abrasion

The mechanisms by which eolian particles disrupt and remove material in the abrasion process is not well understood. Discussion here is limited to eolian abrasion of brittle materials, appropriate for rocks and minerals, and is based primarily on interpretations of SEM photographs of materials abraded under laboratory conditions. Not all particle impacts lead directly to erosion; some may damage the surface but not necessarily remove material. Under most conditions material removal occurs where propagating fractures intersect.

Fracture patterns produced by rounded particles Well-rounded particles such as sand grains impacting the flat surface of an elastic, brittle surface become flattened at the contact and the target surface is bent downwards. The area of contact increases with particle radius and applied load. If the load exceeds a critical limit a circular crack, called a Hertzian fracture, develops around the identation site, generally at a radius 10 to 30% greater than the contact radius (Johnson et al. 1973). In its ideal form the crack extends a short distance vertically into the specimen before turning outward and propagating into a conical frustrum, the depth of which increases with increasing load.

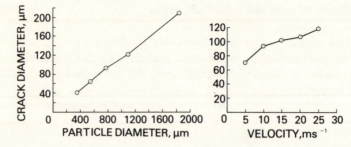

Figure 16.19 Effect of particle diameter and velocity on Hertzian-fracture diameter; each point represents the mean of 100 measurements of quartz grains impacting on quartz plates. (After Marshall 1979).

Andrews (1931), Knight *et al*. (1977), and others have shown that the surface diameter of a Hertzian fracture, D_H, is constant and independent of the impacting grain velocity, a result in agreement with theory for perfectly spherical impactors. However, for well-rounded, but irregularly-shaped particles, the mean diameter of a population of cracks increases with velocity (Marshall 1979, Fig. 16.19). Because the impacting grains were not perfect spheres, the targets were subject to indentation with a wide range of radii and at low impact velocities only the smaller radii (which concentrate the impact stresses) were capable of producing Hertzian fractures. At high impact velocities, v_p, both large and small radii were effective and mean crack diameter, D_H, increased with velocity. The relationship $D_H \propto v_p^{0.47}$ was derived by Finnie (1960), which in comparison to the relationship between maximum contact radius, a_m, and velocity, $a_m \propto v_p^{0.4}$, implies that the cracks formed at about the time of pressure release. Johnson *et al*. (1973) observed cracks smaller than maximum contact size formed during unloading and attributed the behavior to an elastic mismatch between indentor and specimen which affects the stress field via frictional forces at the contact interface.

Figure 16.19 also shows the effect of particle size, D_p, on the surface diameter of Hertzian fractures for impact of irregular particles. Perfectly spherical indenters give rise to a power function relationship of the form $D_H \propto D_p^{2/3}$ when $D_p < 7.0$ cm and a linear relationship when $D_p > 7.0$ cm (Tillett 1956).

At impact angles less than 90° the tangential component of force introduced into the stress system inhibits cracking in front of the grain and the result is an arc-shaped or apparently incomplete Hertzian fracture. As the angle decreases the opening angle θ increases (θ = the angle subtended by the horns of the crack at an imaginary center of curvature of the arc), the diameter decreases, and the preferred orientation of the individual crack increases. (Orientation is defined as the direction pointed to by a line bisecting θ.)

An initially smooth flat surface erodes as a result of continuous bombardment with rounded particles; for quartz (and probably most brittle materials) abrasion proceeds through the sequence shown in Figure 16.20 (Marshall 1979).

Figure 16.20 Stages in erosion of quartz plate, impacted by quartz particles 710–840 μm in diameter at speeds of 20 to 25 m/s, (a) *embryonic stage*, showing triangular, crescentic, and polygonal depressions, (b) *youthful stage*, showing surface represented by cone tops, (c) *advanced stage*, showing curved escarpments, conchoidal fractures, irregular and cleavage-controlled fractures, and (d) *mature stage*, showing blocky, cleavage-controlled structure. Frame widths are: (a) 420 μm, (b) 420 μm, (c) 455 μm, and (d) 150 μm.

Fracture patterns produced by angular particles Angular particles do not abrade in the same manner as rounded particles. A great diversity of erosion patterns can arise from only minor variations in particle shape and from the manner in which the particle strikes the surface. Lawn and Wilshaw (1975) consider the fracture propagation shown in Figure 16.21 to be a basic sequence during normal loading of a surface with a sharp point. Initial contact induces a zone of inelastic deformation (crushing and/or plastic deformation) and at some threshold stress initiates a vertical crack (or median vent) that propagates downward with increasing load. Upon unloading the median vent closes (but does not heal) and lateral vents develop which continue to propagate after

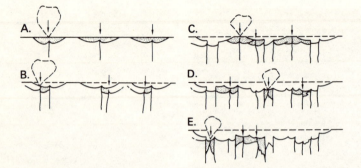

Figure 16.21 Fracturing and chipping below a pointed impacting particle.

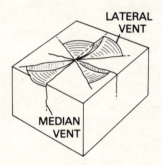

Figure 16.22 Surface pattern produced by intersection of lateral and median vents.

removal of the impacting grain. Chipping can result if the lateral vents reach the surface of the indented specimen. Often the median vent will break through to the surface of the specimen and give rise to a radial fracture trace. The length of the surface trace is an indication of the depth of the vent. For isotropic materials the number of median vents is limited only by the mutual stress-relieving influence of neighbors; several intersecting the surface produce star-shaped patterns (Fig. 16.22). In anisotropic systems the vents tend to follow cleavage directions.

The influence of fracture patterns on abrasion rates Surface textures of ventifacts reflect the rate and manner of abrasion. During the embryonic phase of fracture network development, abrasion is inhibited (Fig. 16.23). At some critical spatial density of conical frustra, material becomes easily removed but abrasion diminishes thereafter because the isolated frustra are difficult to remove. Impact on the top of a cone apparently only deepens the structure, whereas cone spacing prevents contact of rounded particles with surfaces between the cones. Abrasion on the final surface should be somewhat less rapid than during cone removal because surface roughness prohibits cone formation. Such fractures probably have deeper penetration than cleavage-controlled chipping.

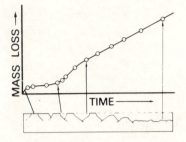

Figure 16.23 Mass lost in eolian erosion as a function of time; profile in lower part of diagram is deduced from SEM observations of Marshall (1979), compared with laboratory experiments of Adler (1976b).

Abrasion experiments

The susceptibility to abrasion, S_a, for different materials is expressed as the ratio of the material mass eroded to either the mass of impacting particles or the number of impacting particles. S_a varies with parameters such as impacting particle velocity, v_p, and angle of impact. Angle of impact is defined as the angle between the surface and the impacting particle and is the complement of the incidence angle.

In order to determine S_a values for various rocks under a wide range of conditions, an apparatus (Fig. 16.24) was fabricated that allows impact velocity, impact angle, impacting particle size and type, sample type, and atmospheric density and composition to be controlled. Marshall (1979) found that anomalous weight losses occur in samples that were not pre-abraded and that have not attained a steady state in eolian abrasion. Thus all targets were pre-abraded prior to experiments to establish fracture patterns in their surfaces. Each target was weighed to ±0.1 mg before and after each run. Dust clinging to the target was removed by compressed air. The fraction of the total mass of particles impacting any target was determined from the total mass expended in the experiment by calculating the angle subtended by the target from the geometry of the apparatus and dividing by 360°. A check on the calculated impact mass was made by placing a particle catcher in the place of one of the sample stations.

Susceptibilities to abrasion were assessed in terms of five parameters: particle size, particle velocity, angle of particle impact, atmospheric density, and impacting particle composition.

S_a versus particle diameter   Figure 16.25 shows S_a as a function of diameter of impacting particles ($D_p = 75$, 105, 140, and 160 µm) for basalt, obsidian, and hydrocal. Particles impacted at an angle of 90° at a velocity of 20 m/s in a CO_2 atmosphere of 5 mb pressure, conditions comparable to a Martian environment. The effect of particle size in the range analyzed is fairly well defined by a slope of ~3 on a log-log plot giving the relationship

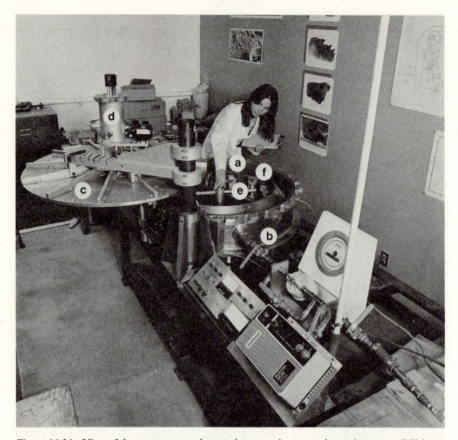

Figure 16.24 View of the apparatus used to conduct experiments to determine susceptibilities to erosion (S_a). Samples to be abraded are placed in holders (a) positioned within cylindrical chamber (b); cover for chamber (c) swings into position such that abrading particles (quartz sand, etc.) in hopper (d) are fed into rotating arm (e) where they are slung against the samples; speed of arm controls the impact velocity; sample holders (a) can be adjusted to control impact angle; 15 samples can be abraded per experiment; one sample station (f) is modified as an open port to collect the same amount of material as strikes each sample to enable an accurate determination of particle flux. Control panel is shown in foreground; chamber can be evacuated to low atmospheric densities and carbon dioxide can be introduced for Martian simulations. (From Greeley *et al.* 1982. Used by permission of the American Geophysical Union).

$$S_a \propto D_p{}^3 \qquad\qquad (16.35)$$

S_a versus particle velocity Figure 16.26 shows the effect of impact velocity on S_a for a wide range of rocks and materials; particle velocities ranged from 5 to 40 m/s. Particles were 125 to 180 μm quartz grains, impacting normal to the target in a CO_2 atmosphere at 5 mb pressure. Although the effect of particle velocity, v_p, is poorly defined for volcanic tuff (both welded and nonwelded) and rhyolite, the graphs for most materials show a slope of 2 on a log–log plot, giving the relationship

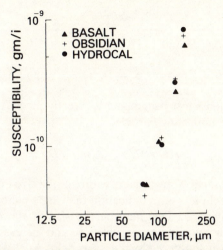

Figure 16.25 Susceptibilities to abrasion of basalt, obsidian, and hydrocal as a function of particle diameter; susceptibility (S_a) is expressed as the ratio of target mass eroded (gm) to the number of impacting particles (i). (From Greeley *et al.* 1982. Used by permission of the American Geophysical Union).

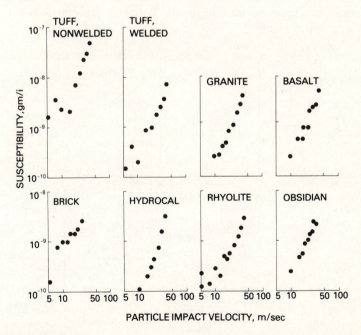

Figure 16.26 Susceptibilities to abrasion of test materials as functions of particle impact velocities. (From Greeley *et al.* 1982. Used by permission of the American Geophysical Union).

$$S_a \propto v_p{}^2 \qquad\qquad (16.36)$$

Combining Equations 23 and 24 gives

$$S_a \propto D_p{}^3 v_p{}^2 \qquad\qquad (16.37)$$

which shows that mass lost per impact is directly proportional to the kinetic energy, *KE*, of the impacting particle.

$$KE = 1/2 m_p v_p{}^2,\, m_p \propto D_p{}^3 \qquad\qquad (16.38)$$

The very slight concave-upwards curvature of the size-effect plots and many of the velocity-effect plots is an indication that some factor other than kinetic energy may be involved. This may be the size of the contact area which increases with both velocity and particle size and increases in the number of pre-existing cracks encountered by the stress field. Apparently the behavior of brick is due to the development of a sintered crust at low and medium velocities that chipped off in large pieces at high velocities.

Sₐ versus particle impact angle   Figure 16.27 shows the effect of particle impact angle on S_a for the test materials; for comparison Figure 16.28 shows relative erosion as a function of impact angle of ideally ductile and brittle materials (after Oh *et al.* 1972). Most of the rocks and materials tested appear to mimic a combination of typical brittle and ductile behavior, except obsidian which behaves as a brittle material.

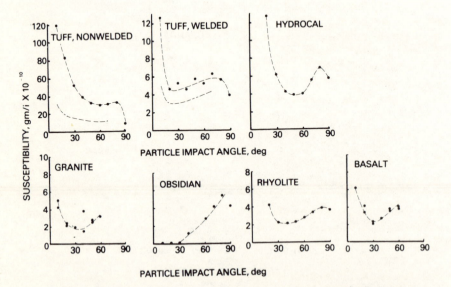

Figure 16.27   Susceptibilities to abrasion of test materials as functions of particle impact angle. (From Greeley *et al.* 1982. Used by permission of the American Geophysical Union).

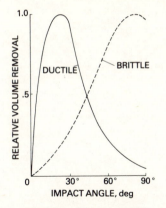

Figure 16.28 Relative erosion as a function of impact angle for typical brittle materials compared with ductile materials. For explanation see text. (After Oh *et al.*, 1972).

The high susceptibility at low impact angles for ductile materials is interpreted as chipping and gouging of microblocks and the efficiency with which they are removed would increase as the angle declines from 90 to about 25 degrees. However fractures propagate to shallower depths as the vertical component of force diminishes. The high susceptibility at high angles for brittle materials suggests an apparent high ratio between indenter and target hardness which could lead to crushing and powdering of the surface at high impact angles. This may provide some degree of protection for the surface at normal (90°) impact and could account for the reduction in susceptibility as angle increases past 85 degrees. The downturn could also be due to impacting particles rebounding from the target and interfering with incoming particles, but this possibility has not been assessed fully.

S_a *versus atmospheric density* Tests to determine the effect of atmospheric pressure on abrasion (Fig. 16.29) were inconclusive. Generally at 1 bar there was slightly more erosion for the three materials tested than at Martian pressure (5 mb). However, this may be related in some way to air turbulence in the confined abrasion chamber.

S_a *versus particle composition* Figure 16.30 shows S_a for basalt targets as a function of impact velocity for quartz, basalt, and basaltic ash particles 125 to 175 µm in diameter, impacting at 90°. There appears to be little difference in abrasion by quartz and basalt; ash appears to be less effective as an agent of abrasion. Note, however, that the data for basalt and ash particles are limited.

Experiments were also conducted using aggregates as the agents of abrasion. Because a different apparatus than MRED was used the results are not entirely compatible for comparisons. However, the critical result was that for impact velocities below about 20 m/s, the aggregates were plastered against the target,

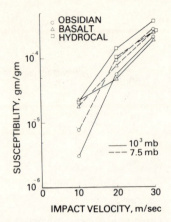

Figure 16.29 Susceptibility to abrasion for three materials as a function of impacting particles velocity in a 10^3 mb (Earth) pressure environment and 7.5 mb (Mars) pressure environment; although the data are limited, there is no apparent difference due to atmospheric density. S_a expressed as ratio of target mass loss to particle mass impacting the target.

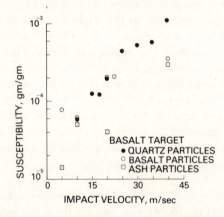

Figure 16.30 Susceptibility to abrasion of basalt as a function of impact velocity by 125 to 175 µm particles of quartz, basalt and basaltic ash. (From Greeley *et al.* 1982).

causing a net gain in mass of the target and forming a protective coating. At higher impact velocities the coating sloughed off but there was no measurable loss in target mass.

Summary
Susceptibility to abrasion is highly dependent on the texture of the rock, composition, velocity, and size of the impacting particles, and the angle of impact. On a per impact basis, glassy materials will erode very quickly for surfaces perpendicular to the wind; crystalline materials such as granite and basalt erode more quickly when surfaces are subparallel to the wind. Quartz and basalt grains appear to be of equal efficiency as agents of abrasion; basaltic

ash is less efficient. Sand-size aggregates of fine grains do not abrade targets at low velocities, but form a veneer on the surface which shields the target from abrasion.

Wind frequency data

The third parameter required for estimating rates of eolian abrasion involves various wind data, including wind directions, strengths, and durations. Ideally, wind data should also include velocity profiles so that surface shear stresses could be derived and utilized for particle threshold and particle flux analyses. In addition, wind data should be available for a sufficiently long period of time to characterize the local eolian regime. Unfortunately, even minimal wind data are seldom available for sites of geologic interest and various techniques must be utilized to extrapolate from other areas.

Terrestrial environment

Wind data are usually obtained from meteorological stations, such as those associated with airports. Commonly, wind velocities with direction are determined for a single height (by convention 10 m above the ground) and reduced in the form of wind rose diagrams that show the frequency, velocities, and directions of winds for a given interval of time. Profiles of the wind velocity as functions of height are seldom obtained, yet this is the critical value for determining u_*.

In principle, it may be possible to extrapolate wind data from nearby recording stations to sites of interest. In practice such an extrapolation is difficult to achieve. One site where this was attempted is the Garnet Hill area in southern California, a classic ventifact locality (Blake 1855). It is a site of frequent strong winds and active windblown sand. Sharp (1964, 1980) established long-term experiments to monitor the erosion of various rocks and man-made materials and to determine mass transport of windblown grains as a function of height above the ground. Particles were collected for intervals as long as four months, which integrated periods of wind activity with periods of quiescence. Unfortunately, on-site wind data were not obtained during the experiment, although Sharp recognized the desirability of such data.

The nearest meteorological station to Sharp's experimental site was the Palm Springs airport, about 9 km southeast of the site. The station has recorded winds for many years, including the interval covering Sharp's experiment. As part of a general study of ventifacts and other eolian features at Garnet Hill, Hunter (1979) obtained wind data near Sharp's site at Whitewater Wash in the fall and winter of 1977. Hunter's results confirmed Sharp's earlier speculations that the Palm Springs airport may not be in a favorable location for providing wind data appropriate for the experimental plot and demonstrates the difficulties in extrapolating wind data from one area to another.

Martian environment

During the past decade intensive observations of Mars have been made from Earth and from spacecraft which provide a variety of data that can be used to assess wind frequencies. Viking landers provided data on the wind velocity, wind direction, and atmospheric temperature and pressure (Hess *et al.* 1976, 1977, Ryan et al. 1978). Although Viking Lander 2 ceased operation in the late 1970s, Viking Lander 1, designated the Mutch Memorial Station, continued to monitor the winds and to obtain images until late 1982.

Observations from orbit provide additional clues to the frequency of threshold-strength winds. Data include measurements of global and local dust storms and of albedo changes on the surface. Global dust storms apparently begin as local dust storms which originate at low or mid latitudes in the southern hemisphere during the summer. During the first Martian year of the Viking mission, two global dust storms were observed, whereas only one such storm took place during the second Martian year. Thus threshold wind speeds are exceeded in the source regions of global dust storms with a frequency of about 3×10^{-2}, averaged over a Martian year.

Strictly speaking, the term *global dust storm* refers only to suspended dust in the atmosphere. However, because supersonic winds would be required to place micron-sized dust directly into suspension (Pollack *et al.* 1976), undoubtedly saltation of $\sim 100\,\mu m$ sand grains also occurs whenever dust is raised into the atmosphere with the saltating grains aiding the infection of fine material into suspension through impact. Because much lower wind speeds are required for saltation to occur, the above estimate of the frequency of raising dust from the surface can also be considered to apply to the frequency of saltation.

Although knowledge of global dust storms is still at a relatively rudimentary stage and strong differences of opinion exist, it seems likely that much of the Martian surface serves as a source region for dust (Pollack *et al.* 1979, Leovy & Zurek 1979). This viewpoint is also supported by the lack of a buildup of a superficial covering of fine dust from the decay of the global storms over much of the surface (Thomas & Veverka 1979). However, the absence of appreciable albedo changes at the two lander sites during the global dust storms of the first Martian year of the Viking mission (Guiness *et al.* 1979) indicates that dust was not raised over all the surface during the genesis of those storms. More generally, sand movement probably occurred to a greater degree in the southern hemisphere than the northern hemisphere (the two Viking landers are in the northern hemisphere).

In addition to grain movement that occurs during global dust storms, movement occurs on local scales, as revealed by numerous local dust storms and by albedo changes of the surface. During the first Martian year of the Viking mission, about 20 local dust storms were observed (Peterfreund & Kieffer 1979). Allowing for incompleteness in coverage perhaps 100 local dust storms actually took place during this time. A typical local dust storm covers an area between about 0.01 and 0.1% of the total Martian surface and lasts for a few days.

Dark wind streaks tend to form behind topographic obstacles such as hills and crater rims (Thomas & Veverka 1979). They are considered to form by wind erosion (Greeley *et al.* 1974) and thus serve to mark localities where saltation may occur more frequently than elsewhere. These streaks tend to be concentrated between 25° and 40° south latitude, although they may be found at all latitudes. Thus saltation occurs more frequently in the lee of craters and ridges and in the upwind portion of crater interiors.

Most places on Mars are therefore subject to saltation at some time during each Martian year, with the southern hemisphere experiencing greater activity. The considerations outlined here lead us to suggest that saltation takes place with a frequency ranging between about 1×10^{-2} and 1×10^{-1}, depending on the region.

Summary

Very little direct information on wind frequencies is available for sites of geological interest on either Earth or Mars. Data which are available seldom cover a sufficiently long period of time to be meaningful in the geological context. For the most part, wind data on Earth must be extrapolated from meteorological stations to sites of interest. Because of the distances and/or varying terrain, such extrapolations allow, at best, only crude estimates of wind frequencies.

On Mars wind frequency data can be obtained directly for two sites from measurements on the surface from the Viking landers, or indirectly through observations from orbit. However, poor knowledge of the surface roughness on Mars hampers translation of wind frequency data into values of threshold and particle flux.

Results and discussion

The previous three sections described the parameters involved in calculating rates of eolian abrasion and present values for each parameter derived from experiments, field studies, and analytical models. These values are now used to estimate rates of abrasion on Earth and Mars.

Rate of eolian abrasion on Earth

Although ventifacts have long been recognized and described (Blake 1855, King 1936, Sharp 1949, and others), there have been relatively few attempts to determine rates of wind abrasion. Rates of abrasion can be derived from data provided from five studies (Table 16.3), but lack of data on wind frequency and particle flux prevent a quantitative assessment.

Sharp (1949) assessed ventifacts in the Big Horn Mountains and recognized at least four separate periods of wind cutting. The inferred rate is $\sim 10^{-3}$ cm/yr based on the estimated minimum abrasion averaged over the entire time since the initiation of wind cutting in the area. Hickox (1959) studied ventifacts

Table 16.3 Rates of eolian abrasion on Earth derived from ventifacts.

Locality	Rate	Material	Reference
Big Horn Mountains, Wyoming	$\sim 10^{-3}$ cm/yr	gneiss	Sharp (1949)
Annapolis Valley, Nova Scotia	$\sim 10^{-1}$ cm/yr	quartzite, sandstone	Hickox (1959)
Western Egypt	1 to 2×10^{-5} cm/yr		McCauley *et al.* (1979)
Mono Craters, California	$\sim 10^{-3}$ cm/yr	dacite	Williams (1981)
Whitewater Wash, California	$\sim 6 \times 10^{-1}$ cm/yr $\sim 5 \times 10^{-1}$ cm/yr	brick hydrocal	Sharp (1964)

composed of vein quartz, quartzite, and indurated sandstone in the Annapolis Valley of Nova Scotia and considered the ventifact faces to have formed in less than 10 years of exposure. McCauley *et al.* (1979) discovered sand-abraded hearth stones in Egypt considered to be of Upper Paleolithic age (200 000 years). The stones are abraded to a depth of 2 to 4 cm, yielding a rate of abrasion of 1 to 2×10^{-5} cm/yr. From examination of a wind-abraded dacite boulder on one of the volcanic domes at Mono craters, California, Williams (1981) obtained a rate of abrasion of $\sim 10^{-3}$ cm/yr.

Sharp (1964) measured abrasion rates of relatively soft materials at Garnet Hill, California, over an eleven year period. He found an abrasion rate of $\sim 6 \times 10^{-1}$ cm/yr for common red brick and $\sim 5 \times 10^{-1}$ cm/yr for hydrocal, a gypsum cement. Further study at the site (Sharp 1980) showed a substantial increase in abrasion rate due to the increased sand supply upwind of the abrasion test site caused by flooding of the Whitewater River.

The five cases described above have a range of rates from 10^{-5} to 6×10^{-1} cm/yr, and represent widely varying environments from cold and wet to hot and dry. These rates are comparable to those inferred by McCauley *et al.* (1981) and provide a basis for comparing rates on Earth with those on Mars.

Rate of wind abrasion on Mars
Rates of wind abrasion at the VL-1 site on Mars can be estimated from saltation fluxes and wind frequencies (Table 16.4) and susceptibilities to abrasion for potential Martian surface materials (Table 16.5). Basalt and hydrocal were considered as targets; basalt is the most likely material present on the Martian surface and hydrocal was used as an analog to indurated clay which might also be on Mars. Several impactors were considered: quartz, basalt, basaltic ash, and aggregates composed of finer grains. All particles were $\sim 100\,\mu$m in diameter, the size most easily moved on Mars (Fig. 16.2). For the calculation,

Table 16.4 Factors involved in estimating rates of eolian abrasion.

Factor	Value	Notes
wind velocity u measured at VL-1	25–30 m/s	threshold strength (and strongest) winds measured at height of VL-1 anemometer (Hess *et al.* 1977, J. A. Ryan, pers. comm. 1981)
threshold velocity u_{*t}	1.25 m/s	for $D_p = 100\,\mu m$, atmospheric surface pressure of 10 mb, 215°K (Greeley *et al.* 1980)
shear velocity u_*	1.5 m/s	20% above threshold
freestream velocity u_∞	70 m/s	based on $u = (0.4)(u_\infty)$ given by Leovy and Zurek (1979)
wind frequency f	10^{-4}	frequency of winds 25–30 m/s (J. A. Ryan, pers. comm. 1981)
flux Q	1.3×10^{-1} g/cm·sec	calculated from White (1979) for $u_* = 1.5$ and $u_{*t} = 1.25$
flux q at 10–12 cm height	6.6×10^{-3} g/cm²·sec	at 10–12 cm above surface, $q = (0.05)\,Q$, shown in Figure 16.15
particle velocity at 10–12 cm height	20 m/s	average particle velocity at height of 10–12 cm, $v_p = 0.3u_\infty$ (from Fig. 16.10)
particle size D_p	100 μm	particle size most easily moved by lowest strength wind (from Fig. 16.2)
impact angle	90°	for simplicity in the calculation, only 90° impacts were considered.

Table 16.5 Estimated rates of eolian abrasion at VL-1.

Target	Particle	S_a[†]	Abrasion Rate gm/cm²s	Abrasion Rate cm/yr[‡]
basalt	quartz	2×10^{-4}	1×10^{-10}	2×10^{-3}
basalt	basalt	2×10^{-4}	1×10^{-10}	2×10^{-3}
basalt	ash	4×10^{-5}	3×10^{-11}	3×10^{-4}
hydrocal	quartz	1×10^{-4}	6×10^{-11}	2×10^{-3}
hydrocal	basalt	5×10^{-4}	3×10^{-10}	8×10^{-3}
hydrocal	ash	3×10^{-4}	2×10^{-10}	5×10^{-3}

Rate = Susceptibility S_a × flux q × wind frequency f.
[†] from MRED experiments.
[‡] Direct conversion from mass loss; does not take into account differences in abrasion resulting from varying impact angles.

abrasion of rocks was considered for a height of 10–12 cm above the surface, corresponding to the size of many rocks at the VL-1 site. Impacts were considered to occur normal to vertically oriented rock surfaces due to the flat trajectories of saltating particles on Mars.

In order to use the appropriate values for susceptibility to abrasion, particle velocities for measured winds at the VL-1 site must be determined. The winds measured at the VL-1 site are about 40% of the geostrophic wind speeds (Leovy & Zurek, 1979). The 25–30 m/s observed wind speeds thus imply a $u_\infty = 62$–75 m/s. This range is consistent with the upper dust cloud speed of 50 m/s observed over VL-1 when the near-surface winds of 25–30 m/s were recorded (James & Evans 1981). For a height of 10–12 cm above the surface and a $u_\infty = 70$ m/s, we would expect the typical saltating particle to be moving at ~30% of u_∞, or ~20 m/s (Fig. 16.10).

The most difficult parameter to assess is the flux of saltating particles. The movement of particles at VL-1 would not be greatly inhibited by the presence of rocks. Even though rocks are common at VL-1, Moore *et al.* (1977) found that fine-grained material covers most of the surface and the rocks occupy less than 25% of the total area. On Earth a comparable rock concentration would affect eolian activity. Depending on the size and spacing of the rocks, particle movement could be enhanced in areas of high turbulence and retarded in wind shadows. Because of the low Martian atmospheric density, however, only surfaces with a roughness, z_o, greater than 1 cm would greatly impede motion. The value of z_o at VL-1 is only 0.1–1.0 cm (Sutton *et al.* 1978); hence from the standpoint of particle movement the VL-1 site should be the same as the 'sand only' case of Pollack *et al.* (1976). Therefore the saltation flux equation of White (1979) can be used to calculate Q for VL-1 (Table 16.4). Since wind speeds measured at VL-1 rarely exceed threshold and then only slightly, a value of u_* 20% larger than u_{*t} was used for the calculation. At 10–12 cm above the surface, q is about 5% of Q (derived from data of Fig. 16.15).

The remaining parameter required to calculate an eolian abrasion rate is f, the frequency of saltation-strength winds. Although only limited meteorological data are available from VL-1, an upper value is ~10^{-4} (J. A. Ryan, pers. comm.).

The rate of eolian abrasion can be calculated for VL-1 from Equation 1 and Table 16.4. Typical rates at VL-1 for basalt targets are 3×10^{-10} gm/cm^2 sec, which is equivalent to 4×10^{-3} cm/yr. These rates are enormous and, if allowed to operate for any reasonable length of time, result in drastic modification of the rocks at VL-1, much more than is observed. Furthermore the ejecta and rims of the small craters in the vicinity of VL-1 would also have been heavily eroded unless they were very young. However, there is no model of surface rock production or recent impact cratering flux that would allow the VL-1 and orbiter observations and the high calculated abrasion rate all to be correct.

Conflict between observed and calculated abrasion rates on Mars

We consider several possible solutions to the discrepancy between observations of the surface and the calculated abrasion rate. In order of increasing likelihood, these are: (1) climatic effects; (2) surface history; and (3) factors dealing with the targets, particles, and particle mobility on Mars.

Climatic effects

The estimated rates of eolian abrasion given here assume present-day meteorological conditions. Yet the frequency of eolian activity on Mars may have varied considerably over its history due to possible changes in atmospheric pressure. Atmospheric pressure at the surface, P_a, affects the frequency of eolian activity due to the dependence of both the wind speed distribution function and the threshold wind speed, u_{*t}, on P_a. Pollack (1979) argues that the former is expected to change little for large changes in P_a about the current value. However, measurements of Greeley *et al.* (1980) indicate that $u_{*t} \propto P_a^{-1/2}$ (Fig. 16.2). And the frequency of eolian activity is very sensitive to changes of u_{*t} (Eq. 16.33); currently it is only the relatively infrequent, highest wind speeds that set particles into motion. If P_a is less than about one quarter of the current pressure, no eolian activity can occur; if it is an order of magnitude larger, eolian activity will occur almost all the time.

Solar and planetary perturbations cause the eccentricity of Mars's orbit and the obliquity of its axis of rotation to undergo quasi-periodic oscillations on a time scale of 10^5–10^6 years and the orientation of its axis to precess with a period of 175 000 years, with the variation of eccentricity between extremes of 0 to 0.14 and the obliquity between 15° and 35° (Ward 1974). It has been suggested that a large amount of CO_2 is currently contained in the subpolar regolith; approximately an order of magnitude more CO_2 than is present in the atmosphere (Fanale & Cannon 1979). At times of high obliquity the annually averaged temperature at high latitudes increases and hence a larger amount of CO_2 is placed into the atmosphere and a smaller amount is taken up by the regolith. The opposite occurs at times of low obliquity.

Measurements of CO_2 absorption onto fine-grained dust suggest that the atmospheric pressure could rise to values of about 20 mb (3 times the current value) at times of the highest obliquity and could decrease to values of less than 1 mb at times of the lowest obliquity, due to the formation of permanent CO_2 polar caps (Fanale & Cannon 1979, Toon *et al.* 1980). Under these circumstances no eolian activity would be expected in the latter situation and a substantially elevated rate of eolian activity, perhaps as much as an order of magnitude, would be expected in the former situation. However, it should be remembered that an enhanced level of eolian activity may result mostly from the lowering of the threshold wind speed and thus involves chiefly lower wind speeds, for which abrasion occurs at a diminished rate. Nevertheless, global dust storms will be more frequent and may involve wind speeds comparable to today's values due to dust-wind feedback relationships.

The precession of the axis of rotation causes a change in the relationship between seasons in a given hemisphere and the position of Mars in its orbit. At present, summer solstice in the southern hemisphere occurs at a time when Mars is close to its perihelion position. This relationship may be responsible for the apparently greater eolian activity in the southern hemisphere than in the northern for the present epoch. If so, the northern hemisphere may have had greater eolian activity half a precession cycle earlier ($\sim$100 000 years) and more generally this difference will vary in an oscillatory fashion.

Over the lifetime of Mars, a few bars of CO_2 (that is, about 300 times the amount now in the atmosphere) may have been outgassed from the planet's interior (Pollack & Black 1979). Conceivably, much of the outgassed CO_2 remained in the atmosphere during the early history of Mars. However, due to the possible formation of carbonate rocks and the development of the regolith, the atmospheric pressure may have more or less monotonically declined with time over the last few billion years (Pollack & Yung 1980). In this case the level of eolian activity may have been substantially higher in its past than at the present time. We can place an upper boundary on the amount of enhancement by supposing the global dust storms occurred all the time, but allowing enough time between successive storms for a given one to decay enough to permit the next to occur.

From these considerations of the climatic history of Mars, eolian activity and abrasion could have been as much as an order of magnitude larger in the past than they are today. Thus not only do climatic considerations fail to account for the discrepancy between the preservation of the surface and the present eolian abrasion if holocrystalline materials were involved, these considerations drive the result in the opposite direction.

Surface history

The assumption is made that the surface of Chryse Planitia observed from orbit is very old because of the presence of abundant impact craters that were formed millions to hundreds of millions of years ago. These craters and the surface on which they occur could be preserved if they had been protected from the eolian regime by a mantling deposit and then recently exhumed, as proposed by Sagan *et al.* (1977) and Binder *et al.* (1977). Much of Mars appears to be mantled by deposits of unknown origin, as first proposed by Soderblom *et al.* (1973) based on Mariner 9 results. The characteristics of the mantling deposits suggests to most investigations (reviewed by Carr 1981) that they are of eolian origin and probably resulted from deposition of fine-grained materials from dust storms.

Viking Orbiter images show many parts of Mars that have been mantled and exhumed. The impact crater shown in Figure 16.31 has been partly exhumed; note that rather fine detail has been preserved in the ejecta field. Were it not for the fact that part of the crater remains buried, it is unlikely that the crater and surrounding surface would have given any evidence of having been mantled.

Parts of Chryse Planitia, site of the Viking Lander 1, also show evidence of

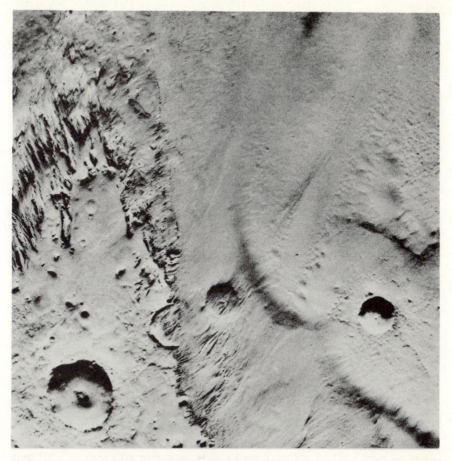

Figure 16.31 Martian landscape showing eolian mantle being stripped away, exhuming craters (Viking frame 438S01). The width of the frame is approximately 75 km.

having been mantled and partly exhumed (Greeley, Theilig *et al.* 1977). Furthermore, the deposits of fine-grained debris observed from the Viking 1 lander (Fig. 16.4) have been described as drifts of eolian deposits that have been eroded (Mutch *et al.* 1976). These deposits could be the last remnant of a much more extensive mantle (Sharp & Malin 1984).

On Earth, exhumed surfaces formerly blanketed with eolian deposits can reveal features of pristine form. Figure 16.32 shows a basaltic plain in Iceland that was buried by loess to a depth of several meters. The area is currently undergoing deflation and the basaltic flows are being exhumed. Fine structure on the scale of centimeters, such as pahoehoe ropes, has been preserved by the loess and despite the extensive eolian activity required for deflation, remains in excellent condition.

Thus mantling of the cratered basaltic plains of Chryse Planitia and sub-

Figure 16.32 Basalt lava flows in Iceland, northwest of Askja volcano, were buried by windblown silt to a depth exceeding 2 m and are currently being exhumed by deflation of the silt; primary flow features as small as a centimeter remain on the exhumed surfaces, evidenced by the pahoehoe ropes seen in the foreground. Mushroom-shaped features to the left of figure on horizon is remnant of eolian silt not yet eroded. (From Greeley *et al.* 1982. Used by permission of the American Geophysical Union).

sequent exhumation millions of years later could explain both the high current calculated rate of abrasion and the preservation of the cratered surface. However, high resolution Viking Orbiter images show that many areas of Mars preserve small, ancient impact craters, and it seems unlikely that all such areas could be explained by exhumation because there are no known 'sinks' that could accommodate the volume of deflated sediments.

Material properties
Rates of eolian abrasion are determined partly by the material properties of both the target and the windblown particles. Despite evidence to the contrary, perhaps the bedrock material on Mars is not basalt, but some other extremely resistant rock. However, we determined the S_a values for a wide variety of common rocks and even the most resistant yield a very high rate of abrasion for Mars. Thus the bedrock would have to be something completely foreign to terrestrial experience; from petrological considerations this seems highly unlikely.

The calculated eolian abrasion rate is based on the assumption of an unlimited and unrestricted sand supply. If there is little sand in the system, abrasion will occur only at a greatly reduced rate. Although sand sized material is present, indicated by thermal inertia properties (Christensen 1983), and inferred from the presence of dunes, the sand supply at the Viking sites could be limited. It is also possible that the free movement of sand is inhibited by cohesion in the uppermost layer (Binder *et al.* 1977) or by the protection from the wind by a dust fallout deposit. The dust would be difficult to move because of its small grain size and the corresponding large u_{*_t} needed to initiate movement (Christensen 1982).

Other important factors are the compositions of the abrasion agents and textures. Figure 16.30 shows S_a values based on abrasion by quartz sand, basalt grains, and basaltic ash—all of about 150 μm in diameter and all striking the same target type at the same velocity. Quartz and basalt grains are roughly comparable in their ability to abrade. Although basaltic ash appears to be less effective than quartz or basalt by a factor of approximately 2, rates of abrasion based on basaltic ash are still much higher than would be reasonable, based on the age of the cratered surface.

Some of the discrepancy may be accounted for by the texture of the abrasion agent. As we have stated when considering the susceptibility of rocks to eolian abrasion, aggregates of fine material are very inefficient as agents of abrasion. At particle speeds lower than about 15 m/s, aggregates form a veneer on the target surface (Greeley *et al.* 1982). Abrasion does not occur until velocities greater than 15 m/s are achieved. Average particle speeds in this range on Mars require freestream wind speeds in excess of 75 m/s. The maximum inferred geostrophic winds measured at the Viking Lander site are only about 62–75 m/s. Although winds speeds higher than those measured at the landing sites are predicted for other areas on Mars based on the global circulation model of Pollack *et al.* (1981), the frequency of geostrophic winds exceeding 75 m/s is extremely low.

Conclusions

Calculations of present rates of eolian abrasion yield unreasonably high values for the Viking lander sites when based on impacting grains composed of holocrystalline grains or basaltic ash. However, the surface displays an impact crater distribution that indicates its age to be several hundred million years old. The presence of craters and blocks that are relatively unmodified on a surface of great age argues against a continuous high abrasion rate. Although there are many uncertainties and assumptions in the analysis, we consider the best explanation for this discrepancy to be a combination of factors dealing with the particles and the surface history, at least for the region surrounding the Viking Lander 1 site. A model has been proposed (Greeley 1979, Greeley & Leach 1978) which suggests that fine particles ($\sim$ few μm) derived from the frequent dust storms and the rapid attrition of larger grains would aggregate to form sand-size grains. Such aggregates are found to be very inefficient agents of abrasion at the low wind speeds typical for Mars. However, some sand of holocrystalline grains are undoubtedly present, as evidenced by the presence of dunes on Mars. Thus, over the long period since the craters were formed, we consider that there would have been more erosion than is apparent, and therefore, given the photogeological evidence for mantling, we also consider the surface was covered with a blanket of sediments and effectively shielded from erosion.

Acknowledgments

This work began in 1974 when it became apparent that knowledge of rates of wind erosion was essential to understanding the surface history of Mars. Because of the complexity of the problem, it was decided to use a team approach. Experiments to determine the susceptibility to abrasion for different materials began with fabrication of a device to simulate eolian abrasion under terrestrial and Martian conditions by R. Leach and the instrument shops at NASA-Ames Research Center. Several different machines were tested, each contributing solutions to different parts of the problem. During the early phase W. McKinnon, then a student at the California Institute of Technology, carried out tests partly to determine the procedure for using laboratory devices for abrasion studies. He was followed by S. Williams (then a NASA Planetary Geology summer intern) who initiated systematic studies of rock abrasion. Concurrently, R. Leach conducted experiments on the erosion of the eolian particles. In 1975 D. Krinsley began SEM studies of both the particles and abraded rocks to compare their surfaces with geological materials that had been abraded under natural conditions in order to verify the use of laboratory devices. The final machine used in most of the abrasion studies is based on a design used at the Argonne National Laboratory for material testing, and was the result of efforts by T. McKee and D. Hamilton, both of Arizona State University. S. Williams and G. Stewart completed the series of abrasion experiments as part of their graduate studies.

Parallel with the investigations of eolian abrasion, wind tunnel experiments were conducted to determine various particle parameters under Earth and Martian environments. Using a design by R. Leach, in 1977 J. Burt from the California Institute of Technology fabricated a particle collector and carried out a series of tests on particle flux. Subsequent analyses of the results led to redesign of the collectors; the final collectors were tested using a smoke wind tunnel at the University of California-Davis to determine the most effective shape for collecting the particles with least disruption to air flow. Data were then systematically collected in the wind tunnel, first by M. Niemiller and M. Plummer, University of Santa Clara, then by K. Malone, a NASA Planetary Geology summer intern. Particle velocity measurements were made in wind tunnel tests using a device modified by R. Leach and the NASA-Ames instrument shops from a similar device used by the U.S. Forest Service to measure windblown snow particles. Experiments were conducted by R. Leach, K. Malone, and R. Leonard. Particle trajectories and velocities were determined by B. White and J. Schulz from high speed motion pictures obtained by the Photographic Technology Branch at NASA-Ames. Computational models of particle motion were developed by B. White. J. Marshall determined the manner in which geological materials abrade on the microscopic scale.

We thank Jack McCauley for his review of this manuscript and helpful discussion regarding wind erosion. All phases of the work were supported by the Planetary Geology Program, National Aeronautics and Space Administration, Washington, D.C., through Ames Research Center.

References

Adler, W. F., 1974. Erosion of fused silica by glass beads. In *Erosion, Wear, and Interface with Corrosion.* Am. Soc. Test. Mater., Spec. Tech. Publn.

Adler, W. F. 1976a. Analysis of particulate erosion. *Wear* 37, 345–52.

Adler, W. F., 1976b. Analytical modeling of multiple particle impacts on brittle materials. *Wear* 37, 353–64.

Andrews, J. P., 1931. Percussion figures. *Proc. Phys. Soc. London* 43, 18–25.

Arvidson, R., E. Guinness, and S. Lee 1979. Differential eolian redistribution rates on Mars. *Nature* 278, 533–35.

Bagnold, R. A., 1941. *The physics of blown sand and desert dunes.* London: Methuen.

Binder, A. B., R. E. Arvidson, E. A. Guinness, K. L. Jones, E. C. Morris, T. A. Mutch, D. C. Pieri, and C. Sagan 1977. The geology of the Viking Lander 1 site. *J. Geophys. Res.* 82, 4439–51.

Blake, W. P., 1855. On the grooving and polishing of hard rocks and minerals by dry sand. *Am. J. Sci.* 20, 2nd series, 178–81.

Briggs, G. A., W. A. Baum, and J. Barnes 1979. Viking Orbiter imaging observations of dust in the Martian atmosphere. *J. Geophys. Res.* 84, 2795–820.

Carr, M. H. 1981. *The surface of Mars.* New Haven: Yale University Press.

Christensen, P. R. 1982. Martian dust mantling and surface composition: interpretation of thermophysical properties. *J. Geophys. Res.* 87, no. B12, 9985–98.

Christensen, P. R., 1983. Aeolian intercrater deposits on Mars: physical properties and global distribution. *Icarus* 56, 496–518.

Cutts, J. A., and R. S. U. Smith 1973. Eolian deposits and dunes on Mars. *J. Geophys. Res.* 78, 4139–54.

Fanale, F. P., and W. A. Cannon 1979. Mars: CO_2 absorption and capillary condensation of clays—significance for volatile storage and atmospheric history. *J. Geophys. Res.* 84, 8404–14.

Finnie, I., 1960. Erosion of surfaces by solid particles. *Wear* 3, 87–103.

Goudie, A. S., R. U. Cooke, and J. C. Doornkamp 1979. The formation of silt from quartz dune sand by salt-weathering processes in deserts. *J. Arid Environments* 2, 105–12.

Greeley, R., 1979. Silt-clay aggregates on Mars. *J. Geophys. Res.* 84, 6248–54.

Greeley, R., J. D. Iversen, J. B. Pollack, N. Udovich, and B. White 1974. Wind tunnel simulations of light and dark streaks on Mars. *Science* 183, 847–49.

Greeley, R., B. R. White, J. B. Pollack, J. D. Iversen, and R. N. Leach 1977. *Dust storms of Mars: considerations and simulations.* NASA Tech. Memo. TM 78423.

Greeley, R., E. Theilig, J. E. Guest, M. H. Carr, H. Masursky, and J. A. Cutts 1977. Geology of Chryse Planitia. *J. Geophys. Res.* 82, 4093–109.

Greeley, R., and R. Leach 1978. A preliminary assessment of the affects of electrostatics on aeolian processes. *Rept. Plan. Geol. Prog.* 1977–1978, NASA TM-79729, 236–37.

Greeley, R., R. N. Leach, B. R. White, J. D. Iversen, and J. B. Pollack 1980. Threshold windspeeds for sand on Mars: wind tunnel simulations. *Geophys. Res. Lett.* 7, 121–24.

Greeley, R., B. R. White, J. B. Pollack, J. D. Iversen, and R. N. Leach 1981. Dust storms of Mars: considerations and simulations. *Geol. Soc. Am. Spec. Publ.* 186, 101–21.

Greeley, R., and P. D. Spudis 1981. Volcanism on Mars. *Rev. Geophys. Space Phys.* 19, 13–41.

Greeley, R., R. N. Leach, S. H. Williams, B. R. White, J. B. Pollack, D. H. Krinsley, and J. R. Marshall 1982. Rate of wind abrasion on Mars, *J. Geophys. Res.* **87**, no. B12, 10,009–24.

Greeley, R., S. H. Williams, and J. R. Marshall 1983. Velocities of windblown particles in saltation: preliminary laboratory and field measurements. In *Eolian sediments and processes*, M. E. Brookfield and T. S. Ahlbrandt (eds), 133–48. Amsterdam: Elsevier.

Guiness, E. A., R. E. Arvidson, D. C. Gehret, and L. K. Bolef 1979. Color changes at the Viking landing sites over a course of a Mars year. *J. Geophys. Res.* **84**, no. B14, 8355–64.

Henry, R. M., 1975. *Saltation on Mars and expected lifetime of Viking 75 wind sensors*. NASA Tech. Note, D–8035.

Hess, S. L., R. M. Henry, C. B. Leovy, J. L. Mitchell, J. A. Ryan, and J. E. Tillman 1976. Early meteorological results from the Viking 2 lander. *Science* **184**, 1352.

Hess, S. L., R. M. Henry, C. B. Leovy, J. A. Ryan, and J. E. Tillman 1977. Meteorological results from the surface of Mars: Viking 1 and 2. *J. Geophys. Res.* **82**, 4559–74.

Hickox, C. F., 1959. Formation of ventifacts in a moist, temperate climate. *Geol. Soc. Am. Bull.* **70**, 1489–90.

Huffman, G. G., and W. A. Price 1949. Clay dune formation near Corpus Christi, Texas, *J. Sed. Petrol.* **19**, 118–27.

Hunter, W. A., 1979. *Ventifacts on Garnet Hill*. Unpublished senior thesis, Dept. of Earth Sciences, Cal. State Polytechnic Univ., Pomona.

Iversen, J. D., J. B. Pollack, R. Greeley, and B. R. White 1976. Saltation threshold on Mars: the effect of interparticle force, surface roughness, and low atmospheric density. *Icarus* **29**, 381–93.

James, P. B., and N. Evans 1981. A local dust storm in the Chryse region of Mars: Viking orbiter observations. *Geophys. Res. Lett.* **8**, 903–06.

Johnson, K. L., J. J. O'Connor, and A. C. Woodward 1973. The effect of the indenter elasticity on the Hertzian fracture of brittle materials. *Proc. Royal Soc. London, A* **334**, 95–117.

Kawamura, R., 1951. Study on sand movement by wind. In *Report 5*, 95–112. Tokyo: Inst. of Science and Technology.

King, I. C., 1936. Wind-faceted stones from Marlborough, New Zealand. *J. Geol.* **44**, 201–13.

Knight, C. G., M. V. Swain, and M. N. Chaudhri 1977. Impact of small steel spheres on glass surfaces. *J. Material Sci.* **12**, 1573–86.

Krinsley, D. H., J. H. Fink, and R. Greeley 1980. Explosive volcanism: possible source of aggregate formation on Mars. *EOS, Trans. Am. Geophys. Union* (abst.) **61**, no. 46, 1138.

Kuenen, Ph. H., 1960. Experimental abrasion 4: eolian action. *J. Geology* **68**, 427–49.

Lawn, B., and R. Wilshaw 1975. Review—indentation fracture: principles and applications. *J. Material Sci.* **10**, 1049–81.

Leovy, C. B., and R. W. Zurek 1979. Thermal tides and martian dust storms: direct evidence for coupling. *J. Geophys. Res.* **84**, 2956–68.

Lobeck, A. K., 1939. *Geomorphology*. New York: McGraw-Hill.

Malin, M. C., 1974. Salt weathering on Mars. *J. Geophys. Res.* **79**, 3888–94.

Marshall, J. R., 1979. *The simulation of sand grain surface textures using a pneumatic gun and quartz plates*. Ph.D. thesis, Univ. College, London.

McCauley, J. F., 1973. Mariner 9 evidence for wind erosion in the equatorial and mid-latitude regions of Mars. *J. Geophys. Res.* **78**, 4123–37.

McCauley, J. F., C. S. Breed, F. El-Baz, M. I. Whitney, M. J. Grolier, and A. W. Ward 1979. Pitted and fluted rocks in the Western Desert of Egypt: Viking comparisons. *J. Geophys. Res.* **84**, 8222–31.

McCauley, J. F., C. S. Breed, and M. C. Grolier 1981. The interplay of fluvial, mass-wasting and eolian processes in the eastern Gilf Kebir region. *Annals Geol. Surv. Egypt* **11**, 207–39.

McLaughlin, D., 1954. Volcanism and aeolian deposition on Mars. *Bull. Geol. Soc. Am.* **65**, 715–17.

Mills, A. A., 1977. Dust clouds and frictional generation of glow discharges on Mars. *Nature* **268**, 614.

Moore, H. J., R. Hutton, C. Scott, C. Spitzer, and R. Shorthill 1977. Surface materials of the

Viking landing sites. *J. Geophys. Res.* **82**, 4497–523.

Morsi, S. A., and A. J. Alexander 1972. An investigation of particle trajectories in two-phase flow systems. *J. Fluid Mech.* part 2, **55**, 193–208.

Mutch, T. A., R. A. Arvidson, A. B. Binder, F. O. Huck, E. C. Levinthal, S. Liebes, E. C. Morris, D. Nummendal, J. B. Pollack, and C. Sagan 1976. Fine particles on Mars: observations with the Viking 1 lander cameras. *Science* **194**, 87–91.

Neilson, J. H., and A. Gilchrist 1968. Erosion by a stream of solid particles. *Wear* **11**, 111–22.

Oh, H. L., K. P. L. Oh, S. Vaidyanathan, and I. Finnie 1972. On the shaping of brittle solids by erosion and ultrasonic cutting. Proceedings of the Symposium on the Science of Ceramic Machining and Surface Finishing. *U.S. Natl. Bur. Stand. Spec. Publ.* **348**, 119–32.

Peterfreund, A. R., 1981. Visual and infrared observations of wind streaks on Mars. *Icarus* **45**, 447–67.

Peterfreund, A. R., and H. H. Kieffer 1979. Thermal infrared properties of the Martian atmosphere 3, local dust clouds. *J. Geophys. Res.* **84**, 2853–64.

Pettijohn, F. J., P. E. Potter, and R. Siever 1972. *Sand and sandstone.* New York: Springer-Verlag.

Pollack, J. B., 1979. Climatic change on the terrestrial planets. *Icarus* **37**, 479–553.

Pollack, J. B., R. Haberle, R. Greeley, and J. Iversen 1976. Estimates of the wind speeds required for particle motion on Mars. *Icarus* **29**, 395–417.

Pollack, J. B., and D. C. Black, 1979. Implications of the gas compositional measurements of Pioneer Venus for the origin of planetary atmospheres. *Science* **205**, 56–69.

Pollack, J. B., D. S. Colburn, F. M. Flasar, R. Kahn, C. E. Carlson, and D. Pidek 1979. Properties and effects of dust particles suspended in the Martian atmosphere. *J. Geophys. Res.* **84**, 2929–45.

Pollack, J. B., and Y. L. Yung 1980. Origin and evolution of planetary atmospheres. *Ann. Rev. Earth Planet. Sci.* **8**, 425–87.

Pollack, J. B., C. B. Leovy, P. W. Greiman, and Y. Mintz 1981. A Martian general circulation experiment with large topography. *J. Atmos. Sci.* **38**, 3–29.

Ryan, J. A., R. M. Henry, S. L. Hess, C. B. Leovy, J. E. Tillman, and C. Walcek 1978. Mars meteorology: three seasons at the surface. *Geophys. Res. Lett.* **5**, 715–18.

Sagan, C., 1973. Sandstorms and eolian erosion on Mars. *J. Geophys. Res.* **78**, 4155–61.

Sagan, C., J. Veverka, P. Fox, R. Dubisch, J. Lederberg, E. Levinthal, L. Quam, R. Tucker, J. B. Pollack, and B. A. Smith 1972. Variable features on Mars: preliminary Mariner 9 television results. *Icarus* **17**, 346–72.

Sagan, C., D. Pieri, P. Fox, R. Arvidson, and E. Guiness 1977. Particle motion on Mars inferred from the Viking lander cameras. *J. Geophys. Res.* **82**, 4430–38.

Scott, D. H., and M. H. Carr 1978. Geological map of Mars. *U.S. Geol. Survey, Misc. Geol. Inv. Map* I-1083.

Sharp, R. P., 1949. Pleistocene ventifacts east of the Big Horn Mountains, Wyoming. *J. Geol.* **57**, 175–95.

Sharp, R. P., 1964. Wind-driven sand in Coachella Valley, California. *Geol. Soc. Am. Bull.* **75**, 785–804.

Sharp, R. P. 1980. Wind-driven sand in Coachella Valley, California: Further data, *Geol. Soc. Am. Bull.* **91**, 724–30.

Sharp, R. P. and M. C. Malin 1984. Geology from the Viking landers on Mars: a second look. *Bull. Geol. Soc. Am.* **95**, 1398–412.

Sheldon, G. L., 1970. Similarities and differences in the erosion behavior of materials. *Trans. Am. Soc. Mech. Eng., J. Basic Eng.* **92**, 619–26.

Sheldon, G. L., and I. Finnie 1966a. On the ductile behavior of nominally brittle materials during erosive cutting. *Trans. Am. Soc. Mech. Eng., J. Eng. Ind.* Nov., 387–92.

Sheldon, G. L., and I. Finnie 1966b. The mechanism of material removal in the erosive cutting of brittle materials. *Trans. Am. Soc. Mech. Eng., J. Eng. Ind.* Nov., 393–400.

Sheldon, G. L., and A. Kanhere 1972. An investigation of impingement erosion using single particles. *Wear* **21**, 195–200.

Singer, R. B., T. B. McCord, R. N. Clark, J. B. Adams, and R. L. Huguenin 1979. Mars surface composition from reflectance spectroscopy: a summary, *J. Geophys. Res.* **84**, 8415–87.

Smalley, I. J., and D. H. Krinsley 1979. Aeolian sedimentation on earth and Mars: some comparisons. *Icarus* **40**, 276–88.

Smeltzer, C. E., M. E. Goulden, and W. A. Compton 1970. Mechanism of metal removal by impacting dust particles. *Trans. Am. Soc. Mech. Eng., J. Basic Eng.*, **92**, 639–54.

Soderblom, L. A., M. C. Malin, J. A. Cutts, and B. C. Murray 1973. Mariner 9 observations of the surface of Mars in the north polar regions. *J. Geophys. Res.* **78**, 4197–210.

Sutton, J. L., C. B. Leovy, and J. E. Tillman 1978. Diurnal variations of the martian surface layer meteorological parameters during the first 45 sols at two Viking lander sites. *J. Atmos. Sci.* **35**, 2346–55.

Suzuki, T., and K. Takahashi 1981. An experimental study of wind abrasion. *J. Geol.* **89**, 509–22.

Thomas, P., and J. Veverka 1979. Seasonal and secular variations of wind streaks on Mars: an analysis of Mariner 9 and Viking data. *J. Geophys. Res.* **84**, 8131–46.

Tillett, J. P. A., 1956. Fracture of glass by spherical indenters. *Proc. Phys. Soc.* **B69**, 47–54.

Tilly, G. P. 1969. Erosion caused by airborne particles. *Wear* **14**, 63–79.

Tilly, G. P., and W. Sage 1970. The interaction of particle and material behavior in erosion processes. *Wear* **16**, 447–65.

Toon, O. B., J. B. Pollack, W. Ward, J. Burns, and K. Bilski 1980. The astronomical theory of climatic change on Mars. *Icarus* **44**, 552–607.

Toulmin, P., III, A. K. Baird, B. C. Clark, K. Heil, H. J. Rose, R. P. Christian, P. H. Evans, and W. C. Kelliher 1977. Geochemical and mineralogic interpretations of the Viking inorganic chemical results. *J. Geophys. Res.* **82**, 4625–34.

Tsoar, H., R. Greeley, and A. R. Peterfreund 1979. Mars: the north polar sand sea and related wind patterns. *J. Geophys. Res.* **84**, 8167–80.

Tsuchiya, Y., 1969. Mechanics of the successive saltation of a sand particle on a granular bed in a turbulent stream. *Disas. Prev. Res. Inst. Kyoto Univ. Bull.* **19**, part 1(152), 31–44.

Veverka, J., and C. Sagan 1974. McLaughlin and Mars. *Am. Scientist* **62**, 44–53.

Ward, W., 1974. Climate variations on Mars 1. Astronomical theory of insolation. *J. Geophys. Res.* **79**, 3375–85.

White, B. R., 1979. Soil transport by wind on Mars. *J. Geophys. Res.* **84**, 4653–51.

White, B. R., R. Greeley, J. D. Iversen, and J. B. Pollack 1976. Estimated grain saltation in a Martian atmosphere. *J. Geophys. Res.* **81**, 5643–50.

White, B. R., and J. C. Schulz 1977. Magnus effect in saltation. *J. Fluid Mech.* **81**, 497–512.

Whitney, M. I., and R. V. Dietrich 1973. Ventifact sculpture by windblown dust. *Geol. Soc. Am. Bull.* **84**, 2561–82.

Whitney, M. I., 1979. Electron micrography of mineral surfaces subject to wind-blast erosion. *Geol. Soc. Am. Bull.* **90**, 917–34.

Williams, S. H., 1981. Calculated and inferred aeolian abrasion rates: Earth and Mars. M.S. thesis, Ariz. State Univ., Tempe.

Index